LOGIC AND SET T
WITH APPLICATIONS

SIXTH EDITION

MW00736606

MAI Publishing

LOGIC AND SET THEORY
WITH APPLICATIONS

SIXTH EDITION

Philip Cheifetz
Nassau Community College

Frank Avenoso
College of Charleston

Laurie Delitsky
Nassau Community College

Kenneth Lemp
Nassau Community College

Jay Martin
Nassau Community College

Ellen Schmierer
Nassau Community College

Michael Steuer
Nassau Community College

Theresa Vecchiarelli
Nassau Community College

Copyright © 2012 by MAI, Inc. All rights reserved.

No part of this publication may be reproduced, stored in a retrieval system, or transmitted in any form or by any means, electronic, mechanical, photocopying, recording, scanning, or otherwise, except as permitted under Section 107 or 108 of the 1976 United States Copyright Act, without the prior written permission of the publisher.

Cover Design: Eva Reck

ISBN: 978-0-916060-08-4

Printed in the United States of America
10 9 8 7 6 5 4 3 2 1

Preface

In the coming years, a solid foundation in logic and set theory will be essential to succeeding in an increasingly technological world. This textbook serves as an introduction to the basic ideas of these topics. Logic and Set Theory with Applications, 6th edition, is targeted toward non-science students and prospective elementary school teachers who seek to improve their skills in logical thinking and organization of information. Minimal mathematical background is necessary.

With the sixth edition of this text, we present a refinement of its predecessor. Chapter One introduces the student to the basic concepts of logic. It has been reformatted so that the conditional equivalence, the converse and the inverse are now grouped in one section, whereas all the negation equivalences are in another. Chapter Two introduces arguments and the testing of their validity by TF methods. Chapter Three covers valid argument forms and formal proofs. Chapter Four includes some of the more innovative material presented in this new edition. Sections on universal and existential quantifiers and electronic circuits are provided to demonstrate some of the applications of logic. Chapter Five introduces set theory and includes survey problems. Chapter Six presents several applications of set theory including a new section on functions and relations. Some of the more creative material found in the fifth edition such as the logic puzzles, "Just for Fun" problems and "Did you know?" which provides biographical sketches of mathematicians associated with logic, set theory or probability, are retained in the sixth edition.

Feedback from users of the fifth edition has prompted other revisions as well. Some changes and additions to the sixth edition include:

- the renaming of the section quizzes as "Check your Understanding" and the inclusion of their answers in the text,
- an updating of many examples,
- illustrative stories to assist students in the learning of the basic truth tables,
- a summary of equivalences with explanations,
- new exercises and problems.

Several variations are possible for those instructors not wishing to cover Chapters One through Six consecutively. For an earlier treatment of set theory, Chapter Five may be covered before Chapters One through Four. Instructors teaching a shorter course may omit any of the sections in Chapters Four or Six without loss of continuity. Chapter Four may be covered at any point after Chapter One.

Each section concludes with two groups of exercises: one designed to be used as illustrative examples or in-class practice problems, the other containing homework problems. Also, at the end of each chapter there is a cumulative set of exercises designed to reinforce the concepts covered, as well as a sample exam covering all topics in that chapter. Answers to all in-class exercises, chapter reviews and sample exams appear at the end of the text, as do the answers to odd-numbered homework problems.

We would like to thank our colleagues at Nassau Community College for their ideas and suggestions. We are particularly grateful to Emad Alfar, Carmine DeSanto, Jerry Honig, Jessica Bosworth, Abe Mantell, Lilia Orlova, James Peluso, David Sher and Gregory Spengler.

Philip Cheifetz
Frank Avenoso
Laurie Delitsky
Kenneth Lemp
Jay Martin
Ellen Schmierer
Michael Steuer
Theresa Vecchiarelli

Table of Contents

CHAPTER ONE

STATEMENTS AND TRUTH TABLES

A unifying concept in mathematics is the proof of arguments. To determine if an argument is valid we must examine its component parts, that is, the sentences or statements that comprise the argument. Such sentences are called *TF* statements. In this chapter we will examine individual *TF* statements and the relationship between combinations of *TF* statements.

Objectives

After completing Chapter One, the student should be able to:

- Identify True/False statements.

- Translate an English sentence into a symbolic logic statement.

- Translate a symbolic logic statement into an English sentence.

- Determine the truth value of a compound statement.

- Determine the truth value of a variable given the truth value of a compound statement.

- Construct truth tables for compound statements.

- Recognize logically equivalent statements.

- Write a statement that is logically equivalent to a given statement.

- Construct the converse, inverse, and contrapositive of a conditional statement.

1.1 Introduction to Logic

TF Statements

A *TF statement* is a declarative sentence to which we can assign a truth value of either true or false, but not both at the same time.

Example 1

The sentence "Veterans Day is in July" is a *TF* statement since it is a declaration, and is either true or false. In this case the *TF* statement is false.

■

Example 2

The sentence "Obama was the first African-American President of the United States" is a *TF* statement since, as in the previous example, it is a declarative sentence. In this case the *TF* statement is true.

■

Example 3

"Is that a Cadillac?" is not a *TF* statement since it is a question, not a declarative sentence.

■

Example 4

"Speak clearly" is not a *TF* statement. It is a command, not a declarative sentence.

■

Representing Statements using Variables

TF statements are often represented by *variables*. In this text, the variables used will be lowercase letters. As an illustration of our notation,

p: Precision is required in logic

would be read "p represents (or stands for) the statement 'Precision is required in logic.' " The choice of the letter p to symbolize this sentence is convenient, but any letter could have been used.

Negations

To negate a *TF* statement is to change its truth value. A *TF* statement may be negated by using the word *not*. It may also be negated by placing the words *it is not true that* in front of the statement. The negation is denoted by the symbol "~." Thus, the statement ~ p is read "not p." The "~" symbol is also used to mean *it is not true that*.

Example 5

Let o: The okapi is a close relative of the giraffe. Then ~o: It is not true that the okapi is a close relative of the giraffe, or simply "The okapi is not a close relative of the giraffe."

■

Example 6

Let c: The cavy is a rodent. Then $\sim c$: It is not true that the cavy is a rodent, or simply "The cavy is not a rodent." ∎

Connectives

A *connective* is a word or group of words used to join two *TF* statements. In our study of logic, the four basic connectives we will use are the conjunction, the disjunction, the conditional, and the biconditional. These connectives are discussed in detail in the following pages.

Simple and Compound Statements

If a *TF* statement does not contain a connective, it is called a *simple statement*. If two or more simple statements are joined by one or more connectives, the resulting statement is called a *compound statement*. We will investigate the truth values of compound statements for the remainder of this chapter.

Conjunctions

If we join two *TF* statements by the word *and*, we call the resulting compound statement the *conjunction* of the two *TF* statements. The symbol for the conjunction is $\wedge$. The statement $p \wedge q$ is read "*p and q*" or "*p* conjunction *q*."

Example 7

Let i: Interest rates rise. Let u: Unemployment rates fall. The conjunction "Interest rates rise and unemployment rates fall" is written symbolically as $i \wedge u$. ∎

Example 8

Let l: Logic is helpful. Let s: Statistics are useful. The conjunction "Logic is helpful and statistics are useful" can be written symbolically as $l \wedge s$. ∎

Example 9

Let d: Jared likes coffee from Dunkin' Donuts. Let s: Jared likes coffee from Starbucks. Then $d \wedge \sim s$ represents the compound statement "Jared likes coffee from Dunkin' Donuts and Jared does not like coffee from Starbucks," or simply "Jared likes coffee from Dunkin' Donuts but not from Starbucks." Notice that the word "but" really means "and." ∎

Disjunctions

If we join two *TF* statements by the word *or*, the resulting compound statement is called the *disjunction* of the two statements. The symbol for the disjunction is $\vee$. The statement $p \vee q$ is read "*p or q*" or "*p disjunction q*." Sometimes the word *either* is implied as the correlative of the word *or*.

Example 10

Let *p*: I will pass physics. Let *s*: I will go to summer school. Then the disjunction $p \vee s$ represents the compound statement "Either I will pass physics or I will go to summer school" or simply "I will pass physics or go to summer school."

■

Conditionals

Given two statements *p* and *q*, a compound statement of the form "*if p then q*" is called a *conditional* statement or an *implication*. The symbol for a conditional statement is $\rightarrow$. The statement $p \rightarrow q$ is read "*p implies q*" or "*if p then q*." The left hand side (LHS) of the conditional statement, *p*, is called its *antecedent*. The right hand side (RHS) of the conditional statement, *q*, is called its *consequent*.

Example 11

Let *j*: I have a part-time job. Let *p*: I pay for my insurance. The implication "If I have a part-time job then I pay for my insurance" is symbolically written as $j \rightarrow p$, while $p \rightarrow j$ is read "If I pay for my insurance then I have a part-time job." As we shall see, these two statements are not equivalent.

■

Example 12

Let *r*: I order spare ribs. Let *c*: I go to a Chinese restaurant. Consider the statement "I order spare ribs if I go to a Chinese restaurant." This statement may be rephrased as "If I go to a Chinese restaurant then I order spare ribs." Notice that the antecedent of the statement is *c*, and the consequent is *r*. Thus the statement is represented symbolically as $c \rightarrow r$. Informally, we say that wherever the "if" appears in the English sentence, it must be represented as the antecedent of the conditional statement.

■

Example 13

Using the underlined letters, the symbolization of "The food will spoil if left unrefrigerated" is $u \rightarrow s$.

■

Biconditionals

Given two statements p and q, a compound statement of the form $(p \rightarrow q) \wedge (q \rightarrow p)$ is called a *biconditional* statement. Because of this cumbersome notation, we denote the biconditional of p and q as $p \leftrightarrow q$. The statement $p \leftrightarrow q$ is read *"p if and only if q."*

Example 14

Let r: Roses are red, and let v: Violets are blue. Then $r \leftrightarrow v$ is read "Roses are red if and only if violets are blue."

∎

Example 15

Let s: I study, and g: I get a good grade. Then $s \leftrightarrow g$ is read "I study if and only if I get a good grade." The biconditional may also be read as "I get a good grade if and only if I study."

∎

The Role of Parentheses in Compound Statements

Often, compound statements have more than one connective. These connectives can sometimes be grouped in more than one way, and different groupings may convey different meanings. Consider the statement "Emeril will buy apples or bananas and cherries." This statement has three simple ideas: a: Emeril will buy apples, b: [Emeril will buy] bananas and c: [Emeril will buy] cherries. The statement contains an "or" as well as an "and" connective. Therefore, a reasonable question to ask is, "Does it matter whether we think of this statement as the disjunction $a \vee (b \wedge c)$, or as the conjunction $(a \vee b) \wedge c$?"

If the statement is written as the disjunction $a \vee (b \wedge c)$, Emeril may buy apples, or buy both bananas and cherries, or buy all three fruits. If the statement is written as the conjunction $(a \vee b) \wedge c$, Emeril will buy either apples or bananas (or both) and he will definitely buy cherries. The following table summarizes the possible ways Emeril could fill a shopping cart, depending on where the parentheses are placed.

Disjunction $a \vee (b \wedge c)$ What might be in the cart?	Conjunction $(a \vee b) \wedge c$ What might be in the cart?
apples	apples and cherries
bananas and cherries	bananas and cherries
apples and bananas and cherries	apples and bananas and cherries

The results are not the same. This is because there is ambiguity in the English statement as it was written. However, a mathematical statement must not have any ambiguity. It must have one and only one meaning. In an English sentence, the ambiguity is often remedied by using a comma to separate ideas. In order to construct compound statements that are non-ambiguous in logic, parentheses are often required to convey the intended meaning. Thus, if we wrote "Emeril will buy apples, or bananas and cherries," we mean $a \vee (b \wedge c)$, while if we wrote "Emeril will buy apples or bananas, and cherries," we mean $(a \vee b) \wedge c$.

Example 16

Let c: I like carrots. Let p: I like peas. Explain the difference between $\sim (c \wedge p)$ and $\sim c \wedge p$.

Solution

The statement $\sim (c \wedge p)$ is translated as "It is not true that I like both carrots and peas." The statement $\sim c \wedge p$ means "I don't like carrots but I *do* like peas."

■

Example 17

Is the following statement a conjunction or a disjunction?
"I will order the soup and the salad, or the club steak."

Solution

It is a disjunction. A symbolization could be $(s \wedge d) \vee c$.

■

Example 18

Is the statement a conditional or a conjunction?
"I will go to the night club, and if you are not there I will leave."

Solution

It is a conjunction. A symbolization might be $n \wedge (\sim y \to l)$.

■

Example 19

Is the following statement a conditional or a conjunction?
"If the team comes in last place, then they will fire their manager and trade for a better second baseman."

Solution

This is a conditional statement. A symbolization is $l \to (m \wedge s)$.

■

In-Class Exercises and Problems for Section 1.1

In-Class Exercises

In Exercises 1–15, use the underlined letter to write the sentences in symbolic form:

1. Aubrey likes to play <u>s</u>occer and <u>b</u>aseball.

2. If I am a <u>v</u>egetarian then I like <u>t</u>ofu.

3. Either I will <u>r</u>edo my English homework or study <u>l</u>ogic.

4. I will buy you <u>c</u>andy if and only if it's <u>V</u>alentine's Day.

5. It is not the case that <u>o</u>kapis are found in Canada.

6. It is noon in <u>S</u>an Francisco if it is 3 pm in <u>N</u>ew York.

7. It's not true that <u>d</u>octors and <u>l</u>awyers are wealthy.

8. It is recommended that you take <u>c</u>alcium supplements and lift <u>w</u>eights if you have been diagnosed with <u>o</u>steoporosis.

9. If they go to the <u>S</u>uper Bowl and <u>c</u>elebrate, then they <u>w</u>on.

10. Solving a <u>s</u>udoku is time-consuming but <u>e</u>njoyable.

11. If she has a <u>b</u>ronze medal she did not <u>w</u>in but if she has a <u>s</u>ilver medal, she came in secon<u>d</u>.

12. If the <u>S</u>enate and <u>H</u>ouse pass the bill, then it will become a <u>l</u>aw and the police will <u>e</u>nforce it.

13. The roads will become <u>i</u>cy if the rain changes to <u>s</u>leet.

14. If you <u>g</u>o to the <u>b</u>each and do not <u>a</u>pply sunscreen you will get a <u>s</u>unburn.

15. The rich dark Turkish coffee was poured into the large green ceramic mug. (Choose your own symbols)

In Exercises 16–21, let p: She works part-time. Let q: She owns a car. Let r: She goes to college. Let s: She lives alone. Express each of the following compound statements as an English sentence.

16. $q \land \sim r$

17. $\sim (s \rightarrow q)$

18. $\sim p \lor r$

19. $q \to (p \wedge s)$

20. $p \leftrightarrow (q \wedge r)$

21. $s \vee (p \to q)$

Problems

In Problems 22–37, classify each sentence as a simple *TF* statement, a conjunction, a disjunction, a conditional, or a biconditional. Then use the underlined letters to express the statement symbolically.

22. Brown <u>b</u>ears hibernate.

23. The polar ice <u>c</u>aps are melting quickly.

24. You get <u>v</u>accinated or <u>r</u>isk becoming infected.

25. The vegetables are <u>h</u>ard if and only if they are not <u>s</u>teamed.

26. <u>N</u>ine times one equals two, and <u>s</u>ix divided by two equals three.

27. If you <u>s</u>ing, then you like <u>m</u>usic.

28. The dog <u>b</u>arks all night if she doesn't get her <u>e</u>xercise, but will beg for foo<u>d</u> nonetheless.

29. Her <u>e</u>yes are green but her hair is <u>b</u>lond.

30. <u>W</u>eeds grow quickly if they are not <u>a</u>ttended.

31. Many <u>c</u>ommuters do not take the Long Island Rail Road into Manhattan.

32. If you want to go to <u>l</u>aw school, you must take the LSAT <u>e</u>xam.

33. We will go to the <u>c</u>oncert at Carnegie Hall if we get tic<u>k</u>ets, or we will go to <u>d</u>inner.

34. Melissa will buy a <u>d</u>og if and only if it doesn't <u>s</u>hed.

35. If the salary is a negative <u>n</u>umber, then the computer prints an <u>e</u>rror message.

36. Either you will be able to write a <u>c</u>omputer program if you do well in <u>l</u>ogic or you will become a computer <u>o</u>perator.

37. You will become <u>s</u>lim and <u>t</u>rim if and only if you take <u>k</u>ickboxing classes and go on a <u>c</u>arbohydrate-free diet.

In Problems 38–47, each sentence is an example of a compound statement. Symbolize each sentence, using appropriate letters as symbols for the simple statements.

38. Economic equilibrium is achieved if and only if supply is equal to demand.

39. If today is Saturday, we will either go on a picnic or go to the game.

40. You eat at Beefy Delight and not at Charlie's Crab House if you like steak but not lobster.

41. It is not true that if a woman swims fast then she can run fast.

42. If a woman cannot swim fast then she can run fast.

43. I will go out with him if and only if he has a good job and he likes to dine out.

44. Averi will not grow tall if she eats her dessert but not her spinach.

45. If she gets a new car she will withdraw her savings, and if she buys a new boat she will get a loan or not quit her weekend job.

46. If you are married and filing jointly then you will use Table X, but if you are married and not filing jointly then you will use Table Y.

47. The computer will perform Routine-A if the salary is greater than one hundred dollars and not greater than one thousand dollars, or if the employee has worked overtime.

1.2 Connectives and their Truth Tables

Negations

The following table is called a *truth table*. A truth table indicates the instances when a *TF* statement is true and when it is false. Truth values are displayed within the table using the capital letters *T* and *F*.

The following truth table shows us that the negation of a true statement is false, while the negation of a false statement is true.

p	$\sim p$
T	F
F	T

This truth table is used to summarize our previous results about negations. A common misconception is that the negation of a *TF* statement must be false. Many *TF* sentences that use the word *not* are true statements. Consider the false statement p: There has been a female president of the U.S. Then $\sim p$: There has not been a female president of the U.S. Surely this is a true statement.

Conjunctions

Let us examine the truth values of a conjunction. By the nature of the word "and", if a conjunctive statement is to be true, both of its component sentences must be true. If either of the component sentences is false, the conjunction is false. *Thus, the only time a conjunction is true is when p and q are both true.*

The following truth table summarizes the results about conjunctions. Notice that there are exactly four ways that the truth values of the two statements can combine.

p	q	$p \wedge q$
T	T	T
T	F	F
F	T	F
F	F	F

Example 1

A prestigious organization at Rafael's college is known as the Gold Club. Requirements for membership are listed in the club's charter, and read: "The student must have a GPA of at least 3.6 and must be an officer of a campus club." Rafael wishes to apply for membership in this club.

Let's let p: Rafael's GPA is at least 3.6. Let q: Rafael is an officer of a campus club. Determine the truth values of p and q such that Rafael will be eligible for membership in the Gold Club.

Solution

Let's examine the first row of our truth table for the conjunction. In this case, Rafael does have a GPA of at least 3.6 (since p is true) and Rafael is an officer of a campus club (since q is true). Without question, Rafael will be eligible as a member of the Gold Club. Thus we see that when p is true and q is true, the conjunction is true.

Examining the second row of our truth table, Rafael does have a GPA of at least 3.6 (since p is true) but he is not an officer of a campus club (since q is false). He is not eligible for membership in the Gold Club since he must satisfy both of two requirements. Therefore when p is true and q is false, the conjunction is false.

In the next scenario, corresponding to the third row of our truth table, Rafael does not have a GPA of at least 3.6 (since p is false), although he is an officer of a campus club (since q is true). His application for membership in the Gold Club will again be denied since he satisfied only the second of these. It is clear that when p is false and q is true, the conjunction is false.

In the last row, both p and q are false. This corresponds to the case when Rafael neither has the required GPA nor is an officer of a campus club. Rafael has not even satisfied one of these requirements. He will certainly be denied membership in the Gold Club. Thus, when p is false and q is false, the conjunction is false.

∎

Disjunctions

In the English language, the word "or" is often used in the exclusive sense. When we say "Either we will go to the movies or go out to dinner," we often mean that we will do *one* of these two activities, but not *both*. In logic, the "or" is inclusive. In the above example, we mean that we will go to the movies, or go to dinner, or do both. The following truth table summarizes our results about disjunctions. *Notice that a disjunction is false only when p and q are both false.*

p	q	$p \vee q$
T	T	T
T	F	T
F	T	T
F	F	F

Example 2

Another organization at the college in Example 1 is known as the Silver Club. Requirements for membership are listed in the Silver Club's charter, and read: "The student must have a GPA of at least 3.6 or must be a member of a sports team on campus." Suzie wishes to apply for membership in this club. Let p: Suzie's GPA is at least 3.6. Let q: Suzie is a member of a sports team on campus. Determine the truth values of p and q such that Suzie will be accepted for membership in the Silver Club.

Solution

Let's examine the first row of our truth table for the disjunction. In this case, Suzie does indeed have a GPA of at least 3.6 (since p is true) and Suzie is a member of a sports team on campus (since q is true). Surely Suzie will be accepted as a member of the Silver Club. Thus when p is true and q is true, the disjunction $p \vee q$ is true.

Now examine the second row of our truth table for the disjunction. In this scenario, Suzie does have a GPA of at least 3.6 (since p is true) but she is not a member of a sports team (since q is false). Suzie is still eligible for membership in the Silver club since she satisfied one of the two requirements. Thus we see that when p is true and q is false, the disjunction $p \vee q$ is true.

In the next scenario, corresponding to the third row of our truth table for the disjunction, Suzie does not have a GPA of at least 3.6 (since p is false), although she is a member of a sports team on campus (since q is true). Once again, Suzie is eligible for membership in the Silver Club since she satisfied one of the two requirements. So, when p is false and q is true, the disjunction $p \vee q$ is true.

In the last row of our truth table for the disjunction, both p and q are false. This corresponds to the case in which Suzie neither has the required GPA of 3.6 (since p is false) nor is a member of a sports team on campus (since q is false). Suzie has not satisfied either condition, so she will be denied membership in the Silver Club. Thus, when p is false and q is false, the disjunction $p \vee q$ is false. ■

Conditionals

The truth table for a conditional statement is less intuitive. Mathematicians feel that *the only time a conditional statement is false is when the LHS is true and the RHS is false.* The following truth table shows the truth values for a conditional statement.

p	q	$p \rightarrow q$
T	T	T
T	F	F
F	T	T
F	F	T

Example 3

Charnelle has not been feeling well lately. She asks the advice of her doctor, who prescribes a certain medication. The doctor promises Charnelle that if she takes the medication, she will feel better by the end of the week.

Let p: Charnelle takes the medication and let q: Charnelle feels better by the end of the week. Then the doctor's promise is represented symbolically as $p \rightarrow q$. Examine all four sets of truth values for p and q, and in each case, decide if the doctor told Charnelle the truth.

Solution

Examining the first row of our truth table for the conditional, both p and q are true. In this scenario, Charnelle did take the medication (since p is true) and she did indeed feel better by the end of the week (since q is true). The doctor's advice was correct. His conditional statement was true. We observe that when the antecedent and the consequent of a conditional statement are both true, the conditional statement is true.

Next, let's look at the second row of the truth table, where p is true and q is false. In this second scenario, Charnelle did take the medication prescribed by her doctor, but she did not feel better by the end of the week. Apparently, the doctor's statement was not true. Thus, when its antecedent is true and its consequent is false, a conditional statement is false.

Let's examine the last two rows of our truth table collectively. In each of these scenarios, p is false, indicating that Charnelle did not take the medication prescribed by her doctor. Regardless of whether she feels better by the end of the week (where q would be true) or does not feel better by the end of the week (where q would be false), Charnelle would be in error if she were to claim that her doctor lied to her. After all, her doctor never made any promises about what might happen if she did not take the medication. Since, in each case the doctor's statement is not a lie, it must be true. Therefore, any conditional statement whose antecedent is false is a true statement. ■

Biconditionals

The truth table for the biconditional is shown below.

p	q	$p \rightarrow q$	$q \rightarrow p$	$(p \rightarrow q) \wedge (q \rightarrow p)$
T	T	T	T	T
T	F	F	T	F
F	T	T	F	F
F	F	T	T	T

Noting that the heading of the last column is the definition of $p \leftrightarrow q$, our results may be summarized in the truth table below.

p	q	$p \leftrightarrow q$
T	T	T
T	F	F
F	T	F
F	F	T

Notice that for a biconditional to be true the LHS and the RHS must have the same truth value. That is, *for a biconditional to be true both sides of the biconditional must be true, or both sides must be false*. Notice that unlike a conditional statement, the order in the biconditional is unimportant. This can be seen from the truth table for the biconditional.

Example 4

A meteorologist alerts the tri-state area of a cold front headed our way. He warns that there will be a significant snowfall if and only if the cold front brings subzero temperatures. Let c: The cold front brings subzero temperatures. Let s: There is a significant snowfall.
When is the the meteorologist's forecast correct?

Solution

The meteorologist's forecast would be considered correct if the front brought in subzero temperatures and there was a significant snowfall or, if the temperature did not drop below zero and there was not a significant snowfall. This corresponds to lines one and four of the biconditional truth table. But, if the front brought in subzero temperatures and there was not a significant snowfall or the temperatures were above zero and there was a significant snowfall, the forecast would be considered incorrect. This corresponds to lines two and three of the truth table.

■

Truth Table Summary

The truth value of a compound statement depends on the truth value of its individual simple sentences, and the connective used. The following is a summary of the truth tables for the connectives we have discussed.

p	q	$p \wedge q$	$p \vee q$	$p \rightarrow q$	$p \leftrightarrow q$
T	T	T	T	T	T
T	F	F	T	F	F
F	T	F	T	T	F
F	F	F	F	T	T

There are four sentences that stress the main points found in the truth tables for conjunction, disjunction, conditional and biconditional statements. They are:

$p \wedge q$	is true only when both parts are true.
$p \vee q$	is false only when both parts are false.
$p \rightarrow q$	is false only if the LHS is true and the RHS is false.
$p \leftrightarrow q$	is true only when the LHS and the RHS have the same truth value.

Example 5

Suppose p is a statement that we know is true and q is a statement that we know is false. Then

$$p \vee q \quad \text{is true}$$
$$p \wedge q \quad \text{is false}$$
$$p \rightarrow q \quad \text{is false}$$
$$p \leftrightarrow q \quad \text{is false}$$

■

In-Class Exercises and Problems for Section 1.2

In-Class Exercises

In Problems 1–21, fill in each blank with either true, false, or can't be determined due to insufficient information.

1. If an implication is false, then its LHS must be _____ and its RHS must be _____.

2. If a conjunction is true then both its parts must be _____.

3. If an implication is false, its RHS must be _____.

4. If a disjunction is false, both its parts must be _____.

5. If a biconditional is true and its LHS is false then its RHS is

   _____.

6. If p is true and q is false then $p \vee q$ must be _____.

7. If p is true and q is false then $p \rightarrow q$ must be _____.

8. If p is true and q is false then $p \leftrightarrow q$ must be _____.

9. If p is false and q is false then $p \leftrightarrow q$ must be _____.

10. If p is false and q is false then $p \rightarrow q$ must be _____.

11. If p is false then $p \rightarrow q$ must be _____.

12. If p is true then $p \vee q$ must be _____.

13. If q is true then $p \rightarrow q$ must be _____.

14. If q is false then $p \wedge q$ must be _____.

15. If $p \leftrightarrow q$ is false and p is true then q must be _____.

16 If $p \leftrightarrow q$ is true and p is true then q must be _____.

17. If q is false and $p \rightarrow q$ is true then p must be _____.

18. Suppose the truth value of r is unknown, but s is true. The truth value of $r \rightarrow s$ must be_____.

19. Suppose the truth value of r is unknown, but s is true. The truth value of $r \vee s$ must be_____.

20. Suppose the truth value of r is unknown, but s is false. The truth value of $r \wedge s$ must be_____.

21. Suppose the truth value of r is unknown, but s is false. The truth value of $r \leftrightarrow s$ must be _____.

Problems

22. Fill in the following truth table.

p	q	$p \wedge q$	$p \vee q$	$p \rightarrow q$	$p \leftrightarrow q$
T	T				
T	F				
F	T				
F	F				

In Exercises 23–38, enter true, false, or can't be determined because of insufficient information.

23. If $\sim p$ is true and $\sim q$ is true then $p \vee q$ must be_____.

24. If $\sim p$ is true and q is true then $p \wedge q$ must be _____.

25. If p is false and q is false then $p \rightarrow q$ must be _____.

26. If p is false and q is true then $p \leftrightarrow q$ must be _____.

27. If $\sim p$ is true then $p \vee q$ must be _____.

28. If $p \rightarrow q$ is false, then q must be _____.

29. If $p \leftrightarrow q$ is false and $\sim p$ is false, then q must be _____.

30. If $p \vee q$ is false then $p \rightarrow q$ must be _____.

31. If $p \rightarrow (q \rightarrow r)$ is false then $p \rightarrow q$ must be _____.

32. If $p \leftrightarrow (\sim q \vee r)$ is false and p is true, $r \rightarrow s$ must be_____.

33. If $\sim p$ is true, the truth value of $q \rightarrow p$ must be_____.

34. If q is true, the truth value of $q \vee p$ must be _____.

35. If $q \rightarrow \sim p$ is false, then p must be _____.

36. If $q \rightarrow \sim p$ is true then q must be _____.

37. If $\sim p \leftrightarrow q$ and $\sim q \wedge r$ are both true, $w \rightarrow p$ must be _____.

38. If $\sim (p \vee q)$ is true then the value of $q \rightarrow p$ must be _____.

Check Your Understanding

For questions 1–4, answer true or false.

1. If $\sim r \rightarrow s$ is false then r must be_____.

2. If $a \leftrightarrow \sim c$ is true and $c \vee d$ is false, then a must be_____.

3. If $r \wedge (s \vee \sim w)$ is true and s is false, then w must be _____.

4. If $g \wedge \sim d$ is true then $k \rightarrow \sim d$ must be_____.

For questions 5–8, symbolize each sentence using the underlined letters as symbols for the *TF* statements.

5. If Laurie <u>r</u>uns daily and lifts <u>w</u>eights every other day, she will <u>e</u>nter in the 5K race._____

6. I will <u>b</u>uy a smartphone if and only if my current cell phone stops <u>w</u>orking and I don't <u>s</u>pend all my savings._____

7. Either I will use <u>C</u>ablevision as my internet provider, or if I can save <u>m</u>oney then I will switch to <u>V</u>erizon._____

8. Either Fara will accept the offer from Company <u>A</u> if they pay for her <u>m</u>edical benefits, or she will accept the offer from Company <u>B</u> if they increase her <u>s</u>alary by $10,000._____

For questions 9–10, let r: I am registered to vote, p: I take a political science class, and v: I will vote in the next presidential election. Create sentences for each of the symbolizations.

9. $v \leftrightarrow (r \wedge p)$

10. $\sim p \vee (r \rightarrow v)$

JUST FOR FUN

Two US coins total fifty-five cents. One of the coins is not a fifty cent piece. What are the two coins?

1.3 Evaluating *TF* Statements

We have seen that the truth value of a compound statement depends on the truth value of its individual simple sentences and the particular connective used. Sometimes we wish to know the truth value when we combine more than two simple statements into a compound statement. The strategy we use to obtain the truth value is to consider the truth values two at a time.

Example 1

Suppose we are told that p and q are true but r is false. Determine the truth value of the compound statement $(p \rightarrow q) \wedge r$.

Solution

We begin by noticing that the compound statement is a conjunction. For a conjunction to be true, both of its components must be true. Since r is known to be false, the compound statement $(p \rightarrow q) \wedge r$ is false for the given truth values. ∎

Example 2

Let a and b be false and c be true. What is the truth value of $(a \rightarrow c) \vee b$?

Solution

The statement $(a \rightarrow c) \vee b$ is a disjunction. Disjunctions are true if either side is true. Since b is not true, the truth value of the disjunction will depend on $(a \rightarrow c)$. Since the LHS of this implication is false and the RHS is true, $(a \rightarrow c)$ is true. Therefore, $(a \rightarrow c) \vee b$ is true for the given truth values. ∎

The simplest way to resolve the truth value of the compound statement is to resolve the truth value of its components. First, replace each simple sentence with its given truth value. Then, resolve each component, working from the inside out. The next three examples show how this is done.

Example 3

Suppose r: The computer has 4 gigabytes of RAM, d: The computer has a 15 inch monitor, and c: The computer has a wireless keyboard. Assume r, d, and c are all true. Find the truth value of $\sim [(r \wedge \sim d) \vee c]$.

Solution

First, replace each statement with its truth values. Then resolve each component, working outward from the innermost statements.

$$\sim[(r \wedge \sim d) \vee c]$$
$$\sim[(T \wedge \sim T) \vee T]$$
$$\sim[(T \wedge F) \vee T]$$
$$\sim[F \vee T]$$
$$\sim[T]$$
$$F$$

Therefore, the statement $\sim[(r \wedge \sim d) \vee c]$ is false for the given truth values.

∎

Example 4

If p is true while q and r are false, find the truth value of $(p \rightarrow q) \rightarrow \sim r$.

Solution

$$(p \rightarrow q) \rightarrow \sim r$$
$$(T \rightarrow F) \rightarrow \sim F$$
$$F \rightarrow T$$
$$T$$

Therefore, $(p \rightarrow q) \rightarrow \sim r$ is a true statement for the given truth values.

∎

Example 5

Let c: She buys a new coat, w: She withdraws money from her savings, b: She buys a used car, and q: She quits her part-time job. Suppose c and w are true while b and q are false. What is the truth value of the compound statement "If she buys a new coat she will withdraw money from her savings, but if she buys a used car she will not quit her part-time job"?

Solution

First, symbolize the compound statement as $(c \rightarrow w) \wedge (b \rightarrow \sim q)$. Then replace each statement with its truth value and then resolve each component, working outward from the innermost statements.

$$(c \rightarrow w) \wedge (b \rightarrow \sim q)$$
$$(T \rightarrow T) \wedge (F \rightarrow \sim F)$$
$$T \wedge (F \rightarrow T)$$
$$T \wedge T$$
$$T$$

Therefore, the compound statement is true for the given truth values. ■

Example 6

Suppose p and q are both true. What are the truth vaues of $p \rightarrow (q \vee r)$ and $p \rightarrow (q \wedge r)$?

Solution

The truth value of r is not given, yet we are able to determine the truth value of $p \rightarrow (q \vee r)$. The RHS is a disjunction, and we know that the disjunction is true since one of its parts is true. The only way an implication is false is if the LHS is true and the RHS is false. Therefore, since both the LHS and RHS of the implication are true, $p \rightarrow (q \vee r)$ is true. In the implication $p \rightarrow (q \wedge r)$, the truth value of the RHS is unknown, since it is a conjunction. Therefore, we can not determine the truth value of $p \rightarrow (q \wedge r)$. ■

In-Class Exercises and Problems for Section 1.3

In-Class Exercises

In Exercises 1–15 determine whether the compound statements are true, false or can't be determined because of insufficient information. Assume that p and q are true and r and s are false.

1. $\sim s \rightarrow q$

2. $\sim s \rightarrow r$

3. $p \leftrightarrow q$

4. $\sim q \rightarrow \sim r$

5. $q \vee z$

6. $p \wedge r$

7. $p \wedge y$

8. $\sim p \leftrightarrow s$

9. $(q \rightarrow s) \rightarrow (a \vee p)$

10. $r \vee (\sim s \vee n)$

11. $q \wedge (\sim p \vee b)$

12. $s \rightarrow (p \wedge e)$

13. $\sim [p \rightarrow (q \wedge s)] \wedge r$

14. $(p \wedge s) \leftrightarrow \sim (q \vee \sim r)$

15. $[(w \vee p) \wedge (\sim q \vee s)] \rightarrow [(p \leftrightarrow r) \vee \sim q]$

For Exercises 16–25, answer true, false or can't be determined because of insufficient information.

16. Suppose $(p \wedge s) \rightarrow r$ is false. What is the truth value of r?

17. What is the truth value of r if $(s \vee \sim r) \vee (p \rightarrow q)$ is false?

18. If $(s \vee \sim r) \wedge (p \rightarrow q)$ is true, what is the truth value of s?

19. Suppose $(p \leftrightarrow \sim r) \rightarrow (s \wedge \sim p)$ and p are both true. What is the truth value of r?

20. If $(a \wedge \sim b) \leftrightarrow (\sim c \rightarrow d)$ and b are both true, what is the truth value of c?

21. What is the truth value of m if $(m \rightarrow n) \vee (\sim p \wedge r)$ is true and r is false?

22. Suppose r is false. Then what is the truth value of the statement $\sim (p \vee \sim r) \rightarrow (r \leftrightarrow w)$?

23. What is the truth value of a if $\sim [(c \rightarrow \sim a) \vee (\sim p \leftrightarrow s)]$ is true?

24. If $\sim [(p \vee \sim r) \wedge (\sim s \rightarrow b)]$ and p are both false, what is the truth value of r?

25. If $[(\sim r \rightarrow s) \vee \sim d] \leftrightarrow (w \wedge p)$ is true and p is false, what is the truth value of d?

Problems

For Problems 26–45, determine whether the compound statements are true, false or can't be determined because of insufficient information. Assume that p and q are true and r and s are false.

26. $q \vee \sim s$

27. $(p \wedge q) \rightarrow s$

28. $s \rightarrow (r \vee q)$

29. $(r \vee s) \leftrightarrow q$

30. $q \rightarrow \sim (p \wedge s)$

31. $\sim (q \rightarrow r)$

32. $\sim q \rightarrow r$

33. $(z \wedge s) \rightarrow p$

34. $p \rightarrow (m \rightarrow q)$

35. $(p \rightarrow q) \leftrightarrow (\sim p \vee q)$

36. $[p \rightarrow (q \wedge r)] \leftrightarrow [(p \rightarrow q) \wedge (p \rightarrow r)]$

37. $\sim [(\sim s \vee p) \wedge (q \vee \sim r)]$

38. $[(s \vee \sim r) \wedge (p \rightarrow \sim q)] \rightarrow (\sim p)$

39. $(p \rightarrow q) \wedge z$

40. $z \rightarrow (\sim r \rightarrow q)$

41. $(\sim r \vee p) \rightarrow (s \wedge q)$

42. $\sim [(\sim r \rightarrow s) \wedge (p \wedge q)]$

43. $(p \rightarrow r) \rightarrow \sim s$

44. $[(m \leftrightarrow n) \wedge c] \rightarrow p$

45. $[(a \vee p) \wedge (r \rightarrow k)] \leftrightarrow (c \vee s)$

For Problems 46–55, answer true, false or can't be determined because of insufficient information.

46. If $(c \rightarrow \sim d) \wedge b$ is true, and c is true, what is the truth value for d?

47. What is the truth value of $m \rightarrow (a \vee p)$ if m is false?

48. If $(n \vee \sim p) \leftrightarrow (c \wedge g)$ is true and p is false, what is the truth value of c?

49. Let $(a \vee \sim b) \rightarrow (\sim b \vee c)$ be true. What is the truth value of a?

50. If $(r \rightarrow \sim s) \rightarrow (a \wedge \sim w)$ is false and $\sim w$ is true, then what is the truth value of a?

51. Suppose p is false. Find the truth value of $\sim p \leftrightarrow [(p \rightarrow r) \wedge p]$.

52. What is the truth value of $(\sim m \rightarrow \sim n) \vee (a \wedge \sim b)$ if we know that n is false?

53. If $(a \wedge \sim b) \vee (p \wedge \sim r)$ is true and $\sim a$ is true, what is the truth value of r?

54. What is the truth value of $\sim [(\sim c \rightarrow d) \wedge (a \rightarrow \sim b)]$ given that a is true and b is false?

55. If both $(\sim p \wedge \sim c) \leftrightarrow (\sim d \wedge p)$ and p are false, what is the truth value of c?

Check Your Understanding

For questions 1–5, determine whether the following compound statements are true, false or can't be determined because of insufficient information. Assume p and q are true and r and s are false.

1. $(p \rightarrow q) \leftrightarrow (\sim p \vee q)$ _____

2. $(p \vee \sim q) \wedge (\sim r \rightarrow s)$ _____

3. $q \rightarrow (p \rightarrow k)$ _____

4. $(q \vee w) \rightarrow \sim (a \rightarrow p)$ _____

5. $(r \leftrightarrow \sim p) \rightarrow [(a \vee q) \wedge (a \wedge p)]$ _____

For questions 6–10, answer true, false or can't be determined because of insufficient information.

6.   What is the truth value of $(\sim w \wedge q) \vee (p \rightarrow \sim r)$ given that r is false? _____

7.   If $(\sim b \wedge c) \rightarrow (a \vee \sim w)$ is false then what is the truth value of $a \wedge c$? _____

8.   Suppose $\sim m \vee r$ and $r \leftrightarrow q$ are true statements. If q is true, then what is the truth value of $\sim m$? _____

9.   If $(w \vee \sim k) \leftrightarrow (\sim c \wedge h)$ is true and k is false, what is the truth value of c? _____

10.   If $\sim (a \wedge \sim b)$ is false and $a \rightarrow \sim d$ is true, what is the truth value of d? _____

JUST FOR FUN

What is next letter in the following sequence?

O, T, T, F, F, S, S, E, _____

1.4 Truth Tables with Two Simple Statements

Introduction

More often than not, we do not know the truth values of the components of a compound statement. We can, however, construct a truth table that lists all the possible combinations of truth values for the compound statement. We have already seen this technique when we constructed truth tables for negations, conjunctions, disjunctions, implications, and biconditionals.

When the truth table for negation was constructed for a single simple statement, the truth table only had two rows, one row for true and one for false. When we constructed truth tables for compound statements that contained two simple statements, the truth tables had four rows, one for each possible combination of truth values.

Example 1

Construct a truth table for $(p \vee q) \rightarrow p$.

Solution

We begin by noticing that there are only two simple statements involved, p and q. Therefore, the truth table will have four rows. We will have a column for each of the simple statements, one column for $p \vee q$ and finally one column for our entire statement $(p \vee q) \rightarrow p$.

p	q	$p \vee q$	$(p \vee q) \rightarrow p$
T	T		
T	F		
F	T		
F	F		

Now we fill in the table by columns.

p	q	$p \vee q$	$(p \vee q) \rightarrow p$
T	T	T	T
T	F	T	T
F	T	T	F
F	F	F	T

In this example, the compound statement $(p \vee q) \rightarrow p$ is true except when p is false and q is true. ∎

Example 2

Construct a truth table for the statement $(r \lor \sim s) \to (\sim r \lor s)$.

Solution

We begin by noticing that there are only two simple statements involved, r and s. Therefore, the truth table will have four rows. We will have a column for each of the simple statements, one column for each of the negations, one for $(r \lor \sim s)$, one for $(\sim r \lor s)$, and finally, one for the entire compound statement $(r \lor \sim s) \to (\sim r \lor s)$.

r	s	$\sim r$	$\sim s$	$r \lor \sim s$	$\sim r \lor s$	$(r \lor \sim s) \to (\sim r \lor s)$
T	T					
T	F					
F	T					
F	F					

Now we fill in the table by columns, using the truth values for each component as we proceed from left to right. We obtain:

r	s	$\sim r$	$\sim s$	$r \lor \sim s$	$\sim r \lor s$	$(r \lor \sim s) \to (\sim r \lor s)$
T	T	F	F	T	T	T
T	F	F	T	T	F	F
F	T	T	F	F	T	T
F	F	T	T	T	T	T

Using the truth table, it is now a simple matter to determine the conditions under which the compound statement is true. In this case, $(r \lor \sim s) \to (\sim r \lor s)$ is always true, except for the case when r is true and s is false. ■

Many times the column for the negation of simple *TF* statements is omitted from the truth table, as is shown in the next example.

Example 3

Construct a truth table for $(p \to \sim q) \to (q \to \sim p)$.

Solution

Once again the truth table will have four rows, but this time we will not include a column for the negations of p and q. Thus, the truth table will have only five columns.

p	q	$p \to\, \sim q$	$q \to\, \sim p$	$(p \to\, \sim q) \to (q \to\, \sim p)$
T	T	F	F	T
T	F	T	T	T
F	T	T	T	T
F	F	T	T	T

In this case, no matter what the truth values of the simple statements, the compound statement $(p \to\, \sim q) \to (q \to\, \sim p)$ is always true. ∎

Compact Truth Tables

Rather than constructing a column in a truth table for every *TF* statement in a complicated compound statement, and then a column for each sub-statement, we often construct a more compact truth table that yields the same result.

Example 4

Determine when $(p \to q) \leftrightarrow (q \vee \sim p)$ is true and when it is false.

Solution

We begin by noting that the truth table will have four rows since there are two simple sentences. Previously, the truth table would have looked like this:

p	q	$p \to q$	$q \vee \sim p$	$(p \to q) \leftrightarrow (q \vee \sim p)$
T	T			
T	F			
F	T			
F	F			

Instead of the above truth table, we will use the compact form.

p	q	$p \to q$	$\leftrightarrow$	$q \vee \sim p$
T	T			
T	F			
F	T			
F	F			

Notice how the table is slightly different than in previous examples.

We have already entered the truth values for the statements p and q. Now, we will resolve the third and fifth columns.

p	q	$p \rightarrow q$	$\leftrightarrow$	$q \vee \sim p$
T	T	T		T
T	F	F		F
F	T	T		T
F	F	T		T

Finally, we resolve the fourth column.

p	q	$p \rightarrow q$	$\leftrightarrow$	$q \vee \sim p$
T	T	T	T	T
T	F	F	T	F
F	T	T	T	T
F	F	T	T	T

The fourth column is the resolution of the compound statement and it is always true. Therefore, $(p \rightarrow q) \leftrightarrow (q \vee \sim p)$ is always true. ■

Tautologies

One of the main uses of truth tables is to determine if a compound statement is *always* true. Statements that are always true, regardless of the truth values of the simple statements involved are called *tautologies*. The compound statement we examined in Example 3, $(p \rightarrow \sim q) \rightarrow (q \rightarrow \sim p)$, is a tautology.

> Statements that are always true, regardless of the truth values of the simple statements involved are called tautologies.

Example 5

Determine whether or not the statement $\sim (z \vee x) \leftrightarrow (\sim z \wedge \sim x)$ is a tautology.

Solution

z	x	$\sim z$	$\sim x$	$(z \vee x)$	$\sim(z \vee x)$	$\sim z \wedge \sim x$	$\sim(z \vee x) \leftrightarrow (\sim z \wedge \sim x)$
T	T	F	F	T	F	F	T
T	F	F	T	T	F	F	T
F	T	T	F	T	F	F	T
F	F	T	T	F	T	T	T

The statement $\sim(z \vee x) \leftrightarrow (\sim z \wedge \sim x)$ is a tautology since it is always true.   ■

Example 6

Use a compact truth table to determine whether or not the biconditional statement in Example 5 is a tautology.

Solution

We construct a compact truth table for $\sim(z \vee x) \leftrightarrow (\sim z \wedge \sim x)$ using the steps below.

z	x	$\sim(z \vee x)$		$\leftrightarrow$	$(\sim z \wedge \sim x)$
T	T	F	T	T	F
T	F	F	T	T	F
F	T	F	T	T	F
F	F	T	F	T	T

Again, we see that $\sim(z \vee x) \leftrightarrow (\sim z \wedge \sim x)$ is a tautology.   ■

In-Class Exercises and Problems for Section 1.4

In-Class Exercises

In Exercises 1–10, construct a truth table to determine if the given compound statement is a tautology.

1. $p \wedge \sim q$

2. $p \rightarrow (p \wedge q)$

3. $(\sim p \wedge q) \vee \sim(\sim q \rightarrow \sim p)$

4. $(\sim p \rightarrow q) \rightarrow (p \vee q)$

5. $\sim(p \to q) \to (p \wedge \sim q)$

6. $[p \wedge (p \vee q)] \leftrightarrow p$

7. $[p \wedge (\sim p \vee q)] \to q$

8. $[p \vee (\sim p \wedge q)] \leftrightarrow [(p \vee \sim p) \wedge (p \vee q)]$

9. $[\sim q \wedge (\sim p \to q)] \vee [(q \leftrightarrow \sim p) \wedge p]$

10. $[(p \to q) \wedge (q \to p)] \to (\sim p \leftrightarrow \sim q)$

Problems

In Problems 11–30, use a truth table to determine if the given compound statement is a tautology.

11. $(p \wedge q) \to p$

12. $(p \wedge q) \to (q \vee p)$

13. $q \to (p \vee q)$

14. $(q \to p) \to (p \to q)$

15. $\sim(p \to q) \to (\sim p \to \sim q)$

16. $(p \to q) \to (\sim p \vee q)$

17. $[p \wedge (p \to q)] \leftrightarrow q$

18. $[p \wedge (p \to q)] \to q$

19. $[(\sim p \vee \sim q) \wedge q] \to p$

20. $(p \to \sim q) \to \sim(p \wedge q)$

21. $\sim(\sim p \to q) \leftrightarrow (p \to \sim q)$

22. $\sim(p \vee q) \leftrightarrow (\sim p \vee \sim q)$

23. $(p \leftrightarrow q) \leftrightarrow [(p \to q) \wedge (q \to p)]$

24. $[(p \to q) \wedge q] \to p$

25. $[(p \to q) \wedge \sim p] \to \sim q$

26. $\sim(p \to q) \leftrightarrow (\sim p \vee q)$

27. $[(q \to p) \to p] \leftrightarrow (\sim q \vee p)$

28. $\sim(p \leftrightarrow q) \leftrightarrow [(p \wedge \sim q) \vee (q \wedge \sim p)]$

29. $\sim(p \wedge q) \leftrightarrow (p \to \sim q)$

30. $(p \wedge q) \leftrightarrow \sim(p \to \sim q)$

1.5 Truth Tables with Three Simple Statements

As we discussed earlier, when only two variables are involved in the construction of a compound sentence, four different sets of truth values must be considered. This results in a truth table that has four rows. If a compound statement involves three simple statements, there will be exactly eight ways that the truth values of these variables can combine. This will result in a truth table that has eight rows, as shown below.

p	q	r
T	T	T
T	T	F
T	F	T
T	F	F
F	T	T
F	T	F
F	F	T
F	F	F

Observe the *TF* entries in each column. The first column has four entries of T followed by four entries of F. The second column has the truth values alternating in groups of two. The last column has the truth values alternating, beginning with T. If we fill in a truth table for a compound statement that contains three simple statements in this way, we are assured of accounting for all possible truth values.

Example 1

Construct a truth table for $p \rightarrow (q \wedge r)$.

Solution

We begin by listing a column for each simple statement, as well as for each compound component. We then assign truth values to the simple sentences, entering our truth values by columns. The first column will have four entries of T followed by four entries of F. The second column will have the truth values alternating in groups of two, beginning with T. The last column has the truth values alternating, beginning with T.

p	q	r	$q \wedge r$	$p \to (q \wedge r)$
T	T	T		
T	T	F		
T	F	T		
T	F	F		
F	T	T		
F	T	F		
F	F	T		
F	F	F		

We now fill in the truth values for $q \wedge r$, by considering the truth values of q and r in each row.

p	q	r	$q \wedge r$	$p \to (q \wedge r)$
T	T	T	T	
T	T	F	F	
T	F	T	F	
T	F	F	F	
F	T	T	T	
F	T	F	F	
F	F	T	F	
F	F	F	F	

Finally, we resolve the truth values of the entire compound statement by combining the truth values of the first column with the truth values in the fourth column using the conditional connective. For example, in the first row of the truth table, $p \to (q \wedge r)$ corresponds to $T \to T$, and is therefore true. In the second row, $p \to (q \wedge r)$ corresponds to $T \to F$, which is false. We continue to resolve the the compound statement row by row.

p	q	r	$q \wedge r$	$p \rightarrow (q \wedge r)$
T	T	T	T	T
T	T	F	F	F
T	F	T	F	F
T	F	F	F	F
F	T	T	T	T
F	T	F	F	T
F	F	T	F	T
F	F	F	F	T

We see that the statement $p \rightarrow (q \wedge r)$ is true five out of eight times. ∎

Example 2

For what values of p, q, and r is the statement $(p \vee q) \vee (p \rightarrow r)$ true?

Solution

We construct a truth table and include all the components. We will need columns for the three simple statements p, q, and r, as well as the statements $p \vee q$, $p \rightarrow r$, and $(p \vee q) \vee (p \rightarrow r)$. Therefore, the truth table will have eight rows and six columns. The completed table is shown below.

p	q	r	$p \vee q$	$p \rightarrow r$	$(p \vee q) \vee (p \rightarrow r)$
T	T	T	T	T	T
T	T	F	T	F	T
T	F	T	T	T	T
T	F	F	T	F	T
F	T	T	T	T	T
F	T	F	T	T	T
F	F	T	F	T	T
F	F	F	F	T	T

The statement $(p \vee q) \vee (p \rightarrow r)$ is always true, regardless of the truth values of the variables. Therefore, it is a tautology. ∎

Example 3

Determine if the statement $(p \wedge q) \leftrightarrow r$ is a tautology.

Solution

We construct a truth table with eight rows and five columns.

p	q	r	$p \wedge q$	$(p \wedge q) \leftrightarrow r$
T	T	T	T	T
T	T	F	T	F
T	F	T	F	F
T	F	F	F	T
F	T	T	F	F
F	T	F	F	T
F	F	T	F	F
F	F	F	F	T

The statement $(p \wedge q) \leftrightarrow r$ is not a tautology. ∎

Example 4

Use a truth table to determine if $[p \wedge (q \vee r)] \leftrightarrow [(p \wedge q) \vee (p \wedge r)]$ is a tautology.

Solution

We construct a truth table as shown below. Notice that aside from the three columns needed for our variables p, q and r, only three more columns are required in this compact table: one for the LHS of the biconditional, one for its RHS, and one for the symbol $\leftrightarrow$ itself which appears between these two sides.

p	q	r	$p \wedge (q \vee r)$	$\leftrightarrow$	$(p \wedge q) \vee (p \wedge r)$
T	T	T			
T	T	F			
T	F	T			
T	F	F			
F	T	T			
F	T	F			
F	F	T			
F	F	F			

To determine the truth values of $p \wedge (q \vee r)$ we first resolve the truth values of $(q \vee r)$.

p	q	r	$p \wedge (q \vee r)$	$\leftrightarrow$	$(p \wedge q) \vee (p \wedge r)$
T	T	T	T		
T	T	F	T		
T	F	T	T		
T	F	F	F		
F	T	T	T		
F	T	F	T		
F	F	T	T		
F	F	F	F		

Next, resolve the truth values of $p \wedge (q \vee r)$.

p	q	r	$p \wedge (q \vee r)$		$\leftrightarrow$	$(p \wedge q) \vee (p \wedge r)$
T	T	T	T	T		
T	T	F	T	T		
T	F	T	T	T		
T	F	F	F	F		
F	T	T	F	T		
F	T	F	F	T		
F	F	T	F	T		
F	F	F	F	F		

The boxed truth values are those of the LHS of the biconditional. We now determine the truth values of $(p \wedge q)$ and of $(p \wedge r)$.

p	q	r	$p \wedge (q \vee r)$		$\leftrightarrow$	$(p \wedge q) \vee (p \wedge r)$	
T	T	T	T	T		T	T
T	T	F	T	T		T	F
T	F	T	T	T		F	T
T	F	F	F	F		F	F
F	T	T	F	T		F	F
F	T	F	F	T		F	F
F	F	T	F	T		F	F
F	F	F	F	F		F	F

Next, we use the truth values of $(p \wedge q)$ and of $(p \wedge r)$ to determine the truth values of the disjunction connecting these statements.

p	q	r	$p \wedge (q \vee r)$		$\leftrightarrow$	$(p \wedge q) \vee (p \wedge r)$		
T	T	T	T	T		T	T	T
T	T	F	T	T		T	T	F
T	F	T	T	T		F	T	T
T	F	F	F	F		F	F	F
F	T	T	F	T		F	F	F
F	T	F	F	T		F	F	F
F	F	T	F	T		F	F	F
F	F	F	F	F		F	F	F

The fourth and seventh columns of truth values are those of the LHS and the RHS, respectively, of the biconditional statement. Using these boxed truth values, we finally resolve the truth values of the biconditional itself, recalling that a biconditional is true if and only if its LHS and RHS have the same truth value.

p	q	r	$p \wedge (q \vee r)$		$\leftrightarrow$	$(p \wedge q) \vee (p \wedge r)$		
T	T	T	T	T	T	T	T	T
T	T	F	T	T	T	T	T	F
T	F	T	T	T	T	F	T	T
T	F	F	F	F	T	F	F	F
F	T	T	F	T	T	F	F	F
F	T	F	F	T	T	F	F	F
F	F	T	F	T	T	F	F	F
F	F	F	F	F	T	F	F	F

The truth values in the sixth column reveal that the statement $[p \wedge (q \vee r)] \leftrightarrow [(p \wedge q) \vee (p \vee r)]$ is always true. Hence, it is a tautology. ∎

Example 5

Determine if $[(p \wedge q) \to r] \to [(p \to r) \wedge (q \to r)]$ is a tautology.

Solution

We will use a compact truth table. The numbers at the bottom of the columns refer to the order in which we resolved the truth values of the component statements.

p	q	r	$(p \wedge q) \to r$		$\to$	$[(p \to r)$	$\wedge$	$(q \to r)]$
T	T	T	T	T	T	T	T	T
T	T	F	T	F	T	F	F	F
T	F	T	F	T	T	T	T	T
T	F	F	F	T	F	F	F	T
F	T	T	F	T	T	T	T	T
F	T	F	F	T	F	T	F	F
F	F	T	F	T	T	T	T	T
F	F	F	F	T	T	T	T	T
1	2	3	4	5	9	6	8	7

The columns numbered 5 and 8 contain the truth values of the LHS and RHS respectively. The column numbered 9 contains the truth values of the entire conditional statement. Since this conditional statement is not always true, it is not a tautology.

■

In-Class Exercises and Problems for Section 1.5

In-Class Exercises

For Exercises 1–10, construct a truth table to determine if the given statement is a tautology.

1. $[p \vee (q \wedge r)] \leftrightarrow [(p \vee q) \wedge (p \vee r)]$
2. $[p \wedge (q \vee r)] \leftrightarrow [(p \wedge q) \vee r]$
3. $[\sim p \rightarrow \sim (q \wedge \sim r)] \leftrightarrow [\sim q \vee (r \vee p)]$
4. $[p \rightarrow (q \vee r)] \leftrightarrow [(p \rightarrow q) \vee (p \rightarrow r)]$
5. $[\sim p \vee (\sim q \vee \sim r)] \vee [\sim p \rightarrow \sim (q \wedge \sim r)]$
6. $[p \vee (q \rightarrow r)] \rightarrow [(p \vee q) \rightarrow (p \vee r)]$
7. $[\sim (p \vee q) \wedge r] \rightarrow [(\sim q \rightarrow r) \vee p]$
8. $[p \rightarrow (q \rightarrow r)] \leftrightarrow [(p \rightarrow q) \rightarrow r)]$
9. $\{p \vee [q \wedge \sim (r \vee p)]\} \leftrightarrow \sim (\sim q \vee r)$
10. $\{[q \wedge \sim (\sim p \wedge r)] \vee [(\sim q \wedge p) \vee (\sim q \wedge \sim r)]\} \leftrightarrow (p \vee \sim r)$

Problems

For Problems 11–20, construct a truth table to determine if the given statement is a tautology.

11. $[p \wedge (q \vee r)] \leftrightarrow [(p \wedge q) \vee (p \wedge r)]$
12. $(\sim p \vee q) \vee (\sim q \vee r)$
13. $[p \rightarrow (q \wedge r)] \leftrightarrow [(p \rightarrow q) \wedge (p \rightarrow r)]$
14. $[(r \leftrightarrow \sim p) \wedge q] \vee \sim (q \wedge \sim r)$
15. $[p \wedge (q \rightarrow r)] \rightarrow [(p \wedge q) \rightarrow (p \wedge r)]$
16. $[(p \rightarrow q) \rightarrow r] \vee (r \rightarrow p)$
17. $[p \rightarrow (q \rightarrow r)] \rightarrow [(p \rightarrow q) \rightarrow r)]$
18. $\{r \wedge \sim [p \vee (r \rightarrow \sim q)]\} \wedge (\sim p \rightarrow \sim r)$
19. $[p \rightarrow (\sim q \vee r)] \leftrightarrow [\sim p \rightarrow (q \rightarrow r)]$
20. $\sim [p \rightarrow (q \rightarrow r)] \leftrightarrow [\sim p \rightarrow (\sim q \rightarrow \sim r)]$

For Problems 21–29, fill in the blanks.

21. We have seen that if a compound statement involves two simple statements, its truth table will have four rows. If a compound statement involves three simple statements, its truth table will have _____ rows.

22. Notice that $2^2 = 4$, and $2^3 = 8$. Using similar reasoning, if a compound statement involves 4 simple statements, its truth table will have _____ rows.

23. In constructing a truth table for a compound statement that involves two variables, we can ensure that all possible arrangements of truth values will be accounted for by filling in the first column with two entries of T followed by _____ entries of F. The second column will have alternating values of T and _____.

24. When constructing a truth table that involves three simple statements we can ensure that all possible arrangements of truth values will be accounted for by filling in the first column of truth values with four entries of T followed by _____ entries of F. The second column will contain entries of T and F alternating _____ at a time. The third column will contain entries of T and F alternating _____.

25. When a truth table has four simple statements we can ensure that all possible arrangements of truth values will be accounted for by filling in the first column of truth values with eight entries of T followed by _____.
The second column will have entries of T and F alternating four at a time. The third column will have entries of T and F alternating _____ at a time. The last column will have entries of T and F alternating _____ at a time.

26. Compound statements that are always true, regardless of the truth values of the variables involved, are called _____.

27. The compound statement $[p \wedge (q \vee r)] \leftrightarrow [(p \wedge q) \vee (p \wedge r)]$ is a tautology. Hence, $[a \wedge (b \vee c)] \leftrightarrow$ _____ is a tautology.

28. The statement $[p \to (q \wedge r)] \leftrightarrow [(\sim p \vee (q \wedge r)]$ is a tautology. Therefore, the statement $[\sim a \to (b \wedge \sim c)] \leftrightarrow$ _____ is a tautology.

29. The statement $[\sim p \vee (\sim q \wedge r)] \leftrightarrow [(\sim p \vee \sim q) \wedge (\sim p \vee r)]$ is a tautology. Hence, $[m \vee (l \wedge w)] \leftrightarrow$ _____ is a tautology.

Check Your Understanding

1. Construct a truth table to determine if the given statement is a tautology.

$$[(p \rightarrow \sim q) \vee p] \rightarrow (p \rightarrow q)$$

p	q	

2. Construct a truth table to determine if the given statement is a tautology.

$$[(\sim p \vee q) \wedge (\sim q \wedge r)] \rightarrow (\sim p \leftrightarrow r)$$

p	q	r	

1.6 Equivalences–Part I

Introduction

Two compound statements that always have the same truth value, regardless of the truth values of the variables involved, are said to be logically equivalent. When two such statements are joined by the symbol $\Leftrightarrow$, the resulting expression is referred to as an *equivalence*.

Equivalences allow us to express the same thought in two different, but logically equivalent ways. Equivalences are used extensively in the remainder of this chapter and in many of the chapters that follow. While there are infinitely many equivalences, a few are used with such great frequency as to warrant special attention. A list of the frequently used equivalences is summarized on page 73.

Double Negation Equivalence

The *double negation equivalence* is expressed symbolically as $\sim(\sim p) \Leftrightarrow p$. This equivalence states that the negation of a negation of a statement is equivalent to the original statement.

> The double negation equivalence:
> $$\sim(\sim p) \Leftrightarrow p$$

Example 1

Let d: I like to dance. Then $\sim d$: I don't like to dance. The statement "It is not true that I don't like to dance," symbolized as $\sim(\sim d)$, is logically equivalent to the statement "I like to dance." The truth table verifies this relationship.

d	$\sim d$	$\sim(\sim d)$
T	F	T
F	T	F

Since the columns for d and $\sim(\sim d)$, are identical, these statements are equivalent. That is, $\sim(\sim d) \Leftrightarrow d$.

Commutative Equivalence

The *commutative equivalences,* which are expressed symbolically as $(p \wedge q) \Leftrightarrow (q \wedge p)$ or $(p \vee q) \Leftrightarrow (q \vee p)$, pertain only to the *conjunction* and *disjunction* connectives.

The commutative equivalences:

$$(p \wedge q) \Leftrightarrow (q \wedge p)$$

$$(p \vee q) \Leftrightarrow (q \vee p)$$

This equivalence tells us that the order in which the statements occur in a conjunction or in a disjunction is immaterial. You encountered commutative relationships when dealing with addition and multiplication of real numbers. The order in which you add or multiply two numbers has no effect on the resulting sum or product.

Example 2

Verify that $(p \wedge q) \Leftrightarrow (q \wedge p)$ is an equivalence.

Solution

p	q	$p \wedge q$	$q \wedge p$
T	T	T	T
T	F	F	F
F	T	F	F
F	F	F	F

Since $p \wedge q$ and $q \wedge p$ always have the same truth value, these two conjunctions are logically equivalent.

■

Example 3

Let e: The Earth revolves around the sun. Let m: The moon revolves around the Earth. Write a statement equivalent to "The Earth revolves around the sun and the moon revolves around the Earth," using the commutative equivalence.

Solution

The original conjunction can be expressed symbolically as $e \wedge m$. Using the commutative equivalence we obtain $m \wedge e$. Therefore, an equivalent statement is "The moon revolves around the Earth and the Earth revolves around the sun."

■

Associative Equivalence

The *associative equivalences* are expressed symbolically as $[(p \wedge q) \wedge r] \Leftrightarrow [p \wedge (q \wedge r)]$ or $[(p \vee q) \vee r] \Leftrightarrow [p \vee (q \vee r)]$. Like the commutative equivalences, the associative equivalences pertain only to the *conjunction* and *disjunction* connectives. Notice that the order in which the statements appear is unchanged—only the order in which the operations are performed is altered.

The associative equivalences:

$$[(p \wedge q) \wedge r] \Leftrightarrow [p \wedge (q \wedge r)]$$
$$[(p \vee q) \vee r] \Leftrightarrow [p \vee (q \vee r)]$$

You encountered associative relationships when dealing with addition and multiplication. Recall that the different grouping of three numbers has no effect on the sum or product of the numbers, i.e., $(1 + 2) + 3 = 1 + (2 + 3)$ and $2 \cdot (3 \cdot 4) = (2 \cdot 3) \cdot 4$.

Example 4

Verify that $[(p \vee q) \vee r] \Leftrightarrow [p \vee (q \vee r)]$ is an equivalence.

Solution

We will construct a compact truth table. The numbers at the bottom of each column indicate the order in which these columns were filled in. Notice that column 5 is the resolution of the LHS of the equivalence and column 7 is the resolution of the RHS of the equivalence.

p	q	r	[(p	$\vee$	q)	$\vee$	r]		[p	$\vee$	(q	$\vee$	r)]
T	T	T		T		T				T		T	
T	T	F		T		T				T		T	
T	F	T		T		T				T		T	
T	F	F		T		T				T		F	
F	T	T		T		T				T		T	
F	T	F		T		T				T		T	
F	F	T		F		T				T		T	
F	F	F		F		F				F		F	
1	2	3		4		5				7		6	

Observing the entries in columns 5 and 7, we notice that they are identical. Therefore, $[(p \vee q) \vee r] \Leftrightarrow [p \vee (q \vee r)]$ is an equivalence. ∎

Example 5

Write a statement equivalent to "In the summer, either I play golf or baseball, or I go sailing" using the associative equivalence.

Solution

The statement can be symbolized as $(g \vee b) \vee s$. Using the associative equivalence, this is equivalent to $g \vee (b \vee s)$. In words, we would say "In the summer, I play golf, or I either play baseball or go sailing." ∎

Distributive Equivalence

The *distributive equivalences* involve two different connectives, and they take two forms. One form tells us that a conjunction distributes over a disjunction. This form can be expressed symbolically as $[p \wedge (q \vee r)] \Leftrightarrow [(p \wedge q) \vee (p \wedge r)]$.

The other form of the distributive equivalence tells us that a disjunction distributes over a conjunction. It is expressed as $[p \vee (q \wedge r)] \Leftrightarrow [(p \vee q) \wedge (p \vee r)]$.

The distributive equivalences:

$$[p \wedge (q \vee r)] \Leftrightarrow [(p \wedge q) \vee (p \wedge r)]$$

$$[p \vee (q \wedge r)] \Leftrightarrow [(p \vee q) \wedge (p \vee r)]$$

Example 6

Show that $p \wedge (q \vee r)$ is logically equivalent $(p \wedge q) \vee (p \wedge r)$.

Solution

We construct a compact truth table. The order in which the columns were completed is again indicated by the numbers you see at the bottom of each column.

p	q	r	[(p	∧	(q ∨ r)]	[(p∧q)	∨	(p∧r)]
T	T	T		T	T	T	T	T
T	T	F		T	T	T	T	F
T	F	T		T	T	F	T	T
T	F	F		F	F	F	F	F
F	T	T		F	T	F	F	F
F	T	F		F	T	F	F	F
F	F	T		F	T	F	F	F
F	F	F		F	F	F	F	F
1	2	3		5	4	6	8	7

Observing the entries in columns 5 and 8, we notice that they always have the same truth value. Therefore, the statements $p \wedge (q \vee r)$ and $(p \wedge q) \vee (p \wedge r)$ are logically equivalent. ∎

Example 7

Write a statement equivalent to "She will sleep late and eat either lunch or brunch."

Solution

The statement can be symbolized as $s \wedge (l \vee b)$. We may rewrite $s \wedge (l \vee b)$ as $(s \wedge l) \vee (s \wedge b)$ by using the distributive equivalence. In English, we would then say "She will sleep late and eat lunch, or she will sleep late and eat brunch." ∎

Example 8

Write an equivalent statement for "They will practice and win the gold medal, or they will practice and win the silver medal."

Solution

The sentence can be symbolized as $(p \wedge g) \vee (p \wedge s)$. Using the distributive equivalence we can rewrite this as $p \wedge (g \vee s)$. In English, we would say, "They will practice and either win the gold medal or win the silver medal." ∎

In-Class Exercises and Problems for Section 1.6

In-Class Exercises

For Exercises 1–10, complete each statement by applying the indicated equivalence.

1. $\sim(\sim p) \Leftrightarrow$ (Double Negation)

2. $(w \wedge \sim r) \Leftrightarrow$ (Commutative)

3. $[p \wedge (q \vee r)] \Leftrightarrow$ (Distributive)

4. $\sim[\sim(\sim p \rightarrow s)] \Leftrightarrow$ (Double Negation)

5. $[g \vee (s \vee \sim p)] \Leftrightarrow$ (Associative)

6. $[(\sim c \vee d) \wedge (\sim c \vee \sim b)] \Leftrightarrow$ (Distributive)

7. $[(s \rightarrow r) \wedge p] \Leftrightarrow$ (Commutative)

8. $[(\sim a \wedge b) \wedge \sim c] \Leftrightarrow$ (Associative)

9. $[\sim b \vee (\sim p \wedge q)] \Leftrightarrow$ (Distributive)

10. $\sim w \Leftrightarrow$ (Double Negation)

For Exercises 11–13, write a statement equivalent to the given statement using the indicated equivalence.

11. I will parallel park or pull into a lot. (Commutative)

12. We can paint the town red or we can paint it blue and green. (Distributive)

13. The swan's long neck glistened in the noonday sun. (Double Negation)

Problems

For Problems 14–21, state the name of the equivalence that is illustrated in each statement.

14. $(a \vee \sim b) \Leftrightarrow (\sim b \vee a)$

15. $[\sim m \vee (p \wedge \sim s)] \Leftrightarrow [(\sim m \vee p) \wedge (\sim m \vee \sim s)]$

16. $(p \rightarrow q) \Leftrightarrow \sim[\sim(p \rightarrow q)]$

17. $[(r \vee s) \vee p] \Leftrightarrow [r \vee (s \vee p)]$

18. $[(r \wedge s) \vee p] \Leftrightarrow [(r \vee p) \wedge (s \vee p)]$

19. $[(p \rightarrow w) \vee r] \Leftrightarrow [r \vee (p \rightarrow w)]$

20. $\{[(a \rightarrow b) \wedge c] \wedge d\} \Leftrightarrow [(a \rightarrow b) \wedge (c \wedge d)]$

21. $\sim[\sim(\sim s \leftrightarrow w)] \Leftrightarrow (\sim s \leftrightarrow w)$

For Problems 22–27, match each numbered statement with its lettered equivalent.

22. $\sim(w \wedge r)$ a. $(\sim r \vee s) \wedge (w \vee s)$

23. $\sim(\sim w \wedge \sim r)$ b. $(\sim r \wedge w) \vee (\sim r \wedge s)$

24. $\sim[\sim(r \wedge w)]$ c. $(\sim r \wedge w) \wedge s$

25. $\sim r \wedge (w \vee s)$ d. $\sim(r \wedge w)$

26. $\sim r \wedge (w \wedge s)$ e. $\sim(\sim r \wedge \sim w)$

27. $(\sim r \wedge w) \vee s$ f. $r \wedge w$

For Problems 28–34, write a statement equivalent to the given statement using the indicated equivalence.

28. I will either buy you flowers or take you out for dinner. (Commutative)

29. The euro is the currency of the European Union. (Double Negation)

30. In the winter, either I go skiing or ice skating, or I go tobogganing. (Associative)

31. Before you enter the restaurant you must wear shoes and, either a jacket or a tie. (Distributive)

32. It is not true that pearls are not found in clams. (Double Negation)

33. The raspberry is plump and the banana is ripe. (Commutative)

34. I will eat pasta and chicken or pasta and shrimp. (Distributive)

DID YOU KNOW?

There are 293 different ways to make change for one US dollar!
How many can you list?

Check Your Understanding

For questions 1–3, state the name of the equivalence that is illustrated in each of the compound statements below.

1. $[s \wedge (\sim w \vee r)] \Leftrightarrow [(s \wedge \sim w) \vee (s \wedge r)]$ _____

2. $[(s \wedge \sim w) \wedge r] \Leftrightarrow [s \wedge (\sim w \wedge r)]$ _____

3. $[(s \wedge \sim w) \wedge r] \Leftrightarrow [(\sim w \wedge s) \wedge r]$ _____

For questions 4–7, complete each statement by applying the indicated equivalence.

4. $[(p \rightarrow w) \vee r] \Leftrightarrow$ _____   Commutative Equivalence

5. $[(\sim a \wedge b) \wedge \sim c] \Leftrightarrow$ _____   Associative Equivalence

6. $[\sim b \wedge (\sim p \vee q)] \Leftrightarrow$ _____   Distributive Equivalence

7. $\sim [\sim (\sim b \rightarrow c)] \Leftrightarrow$ _____   Double Negation Equivalence

For questions 8–10, write a statement that is equivalent to the given statement using the indicated equivalence.

8. In 2008, an African-American won the US presidential election. (Double Negation)

_____

_____

9. I will pay my credit card bill on time or, I will incur a late fee and a finance fee. (Distributive)

_____

_____

10. Roger will either exercise at the gym or jog five miles. (Commutative)

_____

_____

1.7 Equivalences–Part II: The Conditional Statement

Introduction

A conditional statement has two equivalent statements that are usually associated with it. There are also two related statements that may appear to be equivalent to a conditional statement but are not. In this section we will consider these four cases.

Conditional Equivalence

The *conditional equivalence* is expressed symbolically as $(p \rightarrow q) \Leftrightarrow (\sim p \vee q)$, that is, $p \rightarrow q$ and $\sim p \vee q$ are logically equivalent. The conditional equivalence allows us to express a conditional statement as a disjunction to which it is logically equivalent. Notice that the left side of the disjunction is the negation of the LHS of the original conditional statement.

$$\boxed{\begin{array}{c} \text{The conditional equivalence:} \\ (p \rightarrow q) \Leftrightarrow (\sim p \vee q) \end{array}}$$

Although the equivalence of $p \rightarrow q$ and $\sim p \vee q$ is not immediately obvious, the following truth table will convince us that they are.

p	q	$p \rightarrow q$	$\sim p \vee q$
T	T	T	T
T	F	F	F
F	T	T	T
F	F	T	T
1	2	3	4

Since columns 3 and 4 always have the same truth values, $p \rightarrow q$ and $\sim p \vee q$ are logically equivalent statements.

Example 1

Let p: I get pasta, and m: I want a meatball. Write a disjunctive statement equivalent to "If I get pasta then I want a meatball."

Solution

The statement "If I get pasta then I want a meatball" is expressed symbolically as $p \rightarrow m$. We may write a conditional statement as an equivalent disjunction provided we negate the LHS of the conditional statement. So, $p \rightarrow m$ is equivalent to $\sim p \vee m$, which can be translated into "I don't get pasta or I want a meatball." ∎

Contrapositive Equivalence

The *contrapositive equivalence* is expressed symbolically as $(p \rightarrow q) \Leftrightarrow (\sim q \rightarrow \sim p)$. It allows us to express one conditional statement as another conditional statement. The contrapositive of a conditional statement is formed by interchanging its LHS with its RHS, and negating both sides of the new implication. Thus, the contrapositive of $p \rightarrow q$ is $\sim q \rightarrow \sim p$. The truth table that proves this equivalence on page 53. It is critical to realize that one can only form the contrapositive of a *conditional* statement.

The contrapositive equivalence:
$$(p \rightarrow q) \Leftrightarrow (\sim q \rightarrow \sim p)$$

Example 2

Write a statement logically equivalent to "If I work overtime I will arrive to the party late," using the contrapositive equivalence.

Solution

The implication can be symbolized as $w \rightarrow l$. Its contrapositive is symbolized as $\sim l \rightarrow \sim w$. The English translation would be "If I don't arrive to the party late, then I didn't work overtime." ∎

Example 3

Write the contrapositive of the statement "If I don't exercise, then I gain weight."

Solution

The implication can be symbolized as $\sim e \rightarrow w$. Its contrapositive is symbolized as $\sim w \rightarrow e$. The English translation would be "If I don't gain weight, then I exercise," which is logically equivalent to the original statement. ∎

Given the conditional statement $p \rightarrow q$, two other associated conditional statements may be formed. Neither of these, as we shall see, is logically equivalent to the original conditional $p \rightarrow q$.

The Converse

We define the *converse* of the conditional statement $p \rightarrow q$ as the conditional statement $q \rightarrow p$. In words, we would say that to form the converse of a conditional statement, we simply interchange its LHS with its RHS.

> The converse of the conditional statement $p \rightarrow q$ is the conditional statement $q \rightarrow p$.

Example 4

Write the converse of $r \rightarrow \sim s$.

Solution

To write the converse, we interchange the LHS with the RHS of the conditional statement. Therefore, the converse of $r \rightarrow \sim s$ is $\sim s \rightarrow r$. ∎

Example 5

Write an English sentence that is the converse of the statement "If he called me then he liked me."

Solution

We first symbolize the given sentence as $c \rightarrow l$. Now we interchange the LHS with the RHS to produce $l \rightarrow c$. In English, this translates into "If he liked me then he called me." ∎

Example 6

Is a conditional statement equivalent to its converse?

Solution

To answer this question, we will use a truth table to determine if $p \rightarrow q$ and $q \rightarrow p$ are equivalent.

p	q	$p \rightarrow q$	$q \rightarrow p$
T	T	T	T
T	F	F	T
F	T	T	F
F	F	T	T
1	2	3	4

Since columns 3 and 4 are not the same for *all* values of p and q, $p \rightarrow q$ and $q \rightarrow p$ are not logically equivalent. That is, a conditional statement and its converse are *not* equivalent. ∎

The Inverse

We define the *inverse* of the conditional statement $p \rightarrow q$ as the conditional statement $\sim p \rightarrow \sim q$. Notice that to form the inverse of a conditional statement, we simply negate both the LHS and the RHS of the given conditional statement.

> The inverse of the conditional statement $p \rightarrow q$ is the conditional statement $\sim p \rightarrow \sim q$.

Example 7

Write, in English, the inverse of "If I don't pay the toll I will get a summons."

Solution

The implication is symbolized as $\sim p \rightarrow s$. If we negate the LHS and RHS of this conditional, we obtain $\sim (\sim p) \rightarrow \sim s$. Using the double negation equivalence, this is equivalent to $p \rightarrow \sim s$. Therefore, the English sentence would be "If I pay the toll then I will not get a summons."

∎

Example 8

Is a conditional statement equivalent to its inverse?

Solution

We will again use a truth table to see if $p \rightarrow q$ and $\sim p \rightarrow \sim q$ are logically equivalent.

p	q	$p \rightarrow q$	$\sim p \rightarrow \sim q$
T	T	T	T
T	F	F	T
F	T	T	F
F	F	T	T
1	2	3	4

Since $p \rightarrow q$ and $\sim p \rightarrow \sim q$ are not the same for all values of p and q, a conditional statement and its inverse are *not* equivalent.

∎

Students often erroneously assume that the negation of a conditional $p \rightarrow q$ is logically equivalent to its inverse $\sim p \rightarrow \sim q$. While this may seem reasonable, a truth table will show that it is *not* true.

Summary

The following truth table summarizes the four concepts discussed in this section.

p	q	original $p \rightarrow q$	cond. equiv. $\sim p \vee q$	contrapositive $\sim q \rightarrow \sim p$	converse $q \rightarrow p$	inverse $\sim p \rightarrow \sim q$
T	T	T	T	T	T	T
T	F	F	F	F	T	T
F	T	T	T	T	F	F
F	F	T	T	T	T	T
		1	2	3	4	5

Looking at the table row for row, columns 1, 2, and 3 all have the same truth values. Therefore, they are equivalent. Column 4 (the converse) and 5 (the inverse) do not have the same truth values as columns 1, 2, and 3. However, notice that the inverse and converse of the original statement found in columns 4 and 5 are equivalent since they are contrapositives of each other. This can be seen by noticing that columns 4 and 5 have the same truth values.

In-Class Exercises and Problems for Section 1.7

In-Class Exercises

In Exercises 1–10, use the conditional equivalence to complete each statement.

1. $(c \rightarrow k) \Leftrightarrow$

2. $(p \vee \sim r) \Leftrightarrow$

3. $(\sim p \rightarrow \sim s) \Leftrightarrow$

4. $(\sim w \vee \sim z) \Leftrightarrow$

5. $(\sim b \rightarrow a) \Leftrightarrow$

6. $[\sim m \rightarrow (a \wedge b)] \Leftrightarrow$

7. $[p \vee (c \wedge d)] \Leftrightarrow$

8. $[\sim (p \wedge r) \rightarrow k] \Leftrightarrow$

9. $[\sim a \rightarrow (r \wedge s)] \Leftrightarrow$

10. $[\sim (w \wedge h) \vee (a \wedge \sim c)] \Leftrightarrow$

In Exercises 11–15, use the contrapositive equivalence to complete each statement.

11. $(a \rightarrow \sim h) \Leftrightarrow$

12. $(\sim c \rightarrow s) \Leftrightarrow$

13. $(\sim p \rightarrow \sim q) \Leftrightarrow$

14. $[p \rightarrow (w \wedge a)] \Leftrightarrow$

15. $[\sim (m \vee n) \rightarrow b] \Leftrightarrow$

In Exercises 16–20, match each numbered statement to its lettered equivalent.

16. $d \rightarrow \sim k$	a. $\sim d \rightarrow \sim k$
17. $k \rightarrow d$	b. $\sim d \rightarrow k$
18. $\sim k \rightarrow \sim d$	c. $\sim k \rightarrow d$
19. $\sim k \rightarrow d$	d. $k \vee \sim d$
20. $\sim d \rightarrow k$	e. $\sim d \vee \sim k$

For Exercises 21–30, write the associated converse or inverse as indicated.

21. $r \rightarrow \sim s$	(Converse)
22. $\sim q \rightarrow \sim p$	(Converse)
23. $\sim p \rightarrow q$	(Inverse)
24. $m \rightarrow \sim r$	(Inverse)
25. $(m \vee p) \rightarrow \sim s$	(Converse)
26. $d \rightarrow \sim (b \wedge p)$	(Inverse)
27. $\sim (p \leftrightarrow q) \rightarrow (c \vee d)$	(Converse)
28. $(p \vee \sim q) \rightarrow \sim (p \wedge a)$	(Converse)
29. $\sim (p \leftrightarrow q) \rightarrow (c \vee d)$	(Inverse)
30. $(r \wedge \sim s) \rightarrow \sim (w \vee c)$	(Inverse)

In Exercises 31–35, write a statement that is logically equivalent to the given statement using the indicated equivalence.

31. I do homework or I don't pass. (Conditional Equivalence)

32. If I win the lottery, I'll quit my job. (Contrapositive Equivalence)

33. You'll get a fortune cookie if you go to the Chinese restaurant. (Conditional Equivalence)

34. Either the bird gets up early or it doesn't catch the worm. (Conditional Equivalence)

35. I'll fit into my jeans if I go on a diet. (Contrapositive Equivalence)

In Exercises 36–40, write the inverse and the converse of each given conditional statement.

36. If you live in a glass house then you shouldn't throw stones.

37. You should bring a present if you go to the birthday party.

38. If you can't stand the heat, then you stay out of the kitchen.

39. You don't take your umbrella if it's not raining.

40. The stars can't be seen if the sun is shining.

Problems

In Problems 41–47, state the equivalence illustrated in each statement.

41. $(\sim q \vee \sim r) \Leftrightarrow (q \to \sim r)$

42 $(p \to q) \Leftrightarrow (\sim q \to \sim p)$

43. $(\sim p \to q) \Leftrightarrow (\sim q \to p)$

44. $(p \vee s) \Leftrightarrow (\sim p \to s)$

45. $[(q \wedge \sim p) \to \sim s] \Leftrightarrow [\sim (q \wedge \sim p) \vee \sim s]$

46. $[q \to (\sim p \vee s)] \Leftrightarrow [\sim q \vee (\sim p \vee s)]$

47. $[q \to (\sim p \vee s)] \Leftrightarrow [\sim (\sim p \vee s) \to \sim q]$

In Problems 48–54, match each numbered statement to its lettered equivalent.

48. $\sim b \to \sim a$ a. $\sim a \to b$

49. $a \vee \sim b$ b. $\sim a \to \sim b$

50. $a \to \sim b$ c. $\sim b \to a$

51. $\sim a \vee \sim b$ d $b \to a$

52. $a \vee b$ e. $\sim a \vee \sim b$

53. $\sim a \rightarrow \sim b$ f. $a \rightarrow b$

54. $\sim a \rightarrow b$ g. $a \rightarrow \sim b$

In Problems 55–59, rewrite each sentence first using the contrapositive equivalence and then using the conditional equivalence. Finally, write the inverse and converse of the original statement.

55. If revenue exceeds cost I will make a profit.

56. If I buy a digital camera I will post your picture on the web.

57. A cord of wood will come in handy if it keeps snowing.

58. Weeds grow quickly if they are unattended.

59. The polar ice caps will melt if global warming is not controlled.

JUST FOR FUN

There are three people in a room, all facing in the same direction. Let's call them person 1, 2, and 3. They are shown six hats. Four of the hats are black, the other two are white. The people are then blindfolded. One hat is placed on each person, and the remaining three hats are taken from the room. The blindfolds are then removed. Each person is asked to determine the color of the hat they are wearing without looking at it.

Person 1 said: "I see the color of the hat on person 2 and I see the color of the hat on person 3, but I don't know the color of my hat."

Person 2 said: "I heard what person 1 said. I see the color of the hat on person number 3, but I don't know the color of my hat."

Person 3 said: "I heard what persons 1 and 2 said and I know the color of my hat. It is _____."

What color hat is person 3 wearing? Why?

Everyone is facing to your right

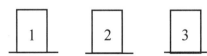

Check Your Understanding

For questions 1–4, state the equivalence illustrated in each statement.

1. $[\sim (a \wedge w) \to \sim c] \Leftrightarrow [(a \wedge w) \vee \sim c]$ _____

2. $[\sim c \to (a \wedge w)] \Leftrightarrow [\sim (a \wedge w) \to c]$ _____

3. $[c \vee \sim (a \wedge w)] \Leftrightarrow [\sim c \to \sim (a \wedge w)]$ _____

4. $[(a \wedge w) \to c] \Leftrightarrow [\sim c \to \sim (a \wedge w)]$ _____

For questions 5–8, write a statement that is equivalent to the given statement using the indicated equivalence.

5. $[k \to \sim (p \vee q)] \Leftrightarrow$ _____ Contrapositive Equivalence

6. $[\sim (p \vee q) \to k] \Leftrightarrow$ _____ Conditional Equivalence

7. $[\sim k \vee (p \vee q)] \Leftrightarrow$ _____ Conditional Equivalence

8. $[(p \vee q) \to \sim k] \Leftrightarrow$ _____ Contrapositive Equivalence

For questions 9–11, circle the correct answer.

9. $\sim (g \vee h) \to k$ is the (converse/inverse/contrapositive) of $k \to \sim (g \vee h)$.

10. $\sim k \to (g \vee h)$ is the (converse/inverse/contrapositive) of $k \to \sim (g \vee h)$.

11. $(g \vee h) \to \sim k$ is the (converse/inverse/contrapositive) of $k \to \sim (g \vee h)$.

For questions 12 and 13, write the indicated form associated with each statement.

12. $\sim r \to q$ Converse

13. $\sim l \to \sim d$ Inverse

For questions 14 and 15, refer to the statement "Children grow confident if they are praised."

14. Write the converse of the statement.

15. Write the contrapositive of the statement.

1.8 Negations of Compound Statements

DeMorgan's Equivalence

DeMorgan's equivalences provide us with rules for negating disjunctions and conjunctions. DeMorgan's equivalences are expressed as $\sim(p \vee q) \Leftrightarrow (\sim p \wedge \sim q)$ and $\sim(p \wedge q) \Leftrightarrow (\sim p \vee \sim q)$. This means that the negation of a disjunction is a *conjunction* with each of its components negated, and that the negation of a conjunction is a *disjunction* with each of its components negated.

DeMorgan's equivalences:

$$\sim(p \vee q) \Leftrightarrow (\sim p \wedge \sim q)$$

$$\sim(p \wedge q) \Leftrightarrow (\sim p \vee \sim q)$$

Example 1

Express the negation of "I will drive or I will take the bus" as a conjunction.

Solution

The disjunction can be symbolized as $d \vee b$ so its negation would be symbolized as $\sim(d \vee b)$. DeMorgan's equivalence allows us to rewrite this as $\sim d \wedge \sim b$. In English, we would say "I will not drive *and* I will not take the bus."

■

Example 2

Express the negation of "I work part-time and pay for auto insurance" as a disjunction.

Solution

Using DeMorgan's equivalence, $\sim(w \wedge p)$ is equivalent to $\sim w \vee \sim p$. In English, this becomes "I don't work part-time *or* I don't pay for auto insurance."

■

Example 3

Write a sentence equivalent to "I don't drink and I don't smoke."

Solution

DeMorgan's equivalence allows us to write $\sim d \wedge \sim s$ as $\sim(d \vee s)$. The corresponding English sentence is "It is not true that I drink or smoke."

■

Example 4

Write the negation of "Milton is short and bald."

Solution

We can symbolize the original statement as $s \wedge b$. Its negation is $\sim(s \wedge b)$. DeMorgan's equivalence allows us to rewrite this as $\sim s \vee \sim b$ which translates into English as "Milton is not short or he is not bald." This statement can appear in other forms. Sometimes it sounds better to begin a disjunction with the word "either." Then the negation becomes "Either Milton is not short or he is not bald." We also can leave out some words, provided that the meaning is the same and no ambiguity results. For example, our answer could be rewritten as "Milton is not short or not bald."

■

Example 5

Write the negation of "It is cold or raining."

Solution

We symbolize the original statement as $c \vee r$. Its negation is $\sim(c \vee r)$. Using DeMorgan's equivalence, we get $\sim c \wedge \sim r$ which is rewritten in English as "It is not cold and it is not raining." This can be shortened to "It is not cold and not raining." Sometimes, this is written in the form "It is neither cold nor raining." Notice that the phrase "neither a nor b" really means "not a and not b".

■

Example 6

Write the negation of "The chicken is fried but not roasted."

Solution

Remember that the word "but" has the same logical meaning as "and," so the original statement is $f \wedge \sim r$. The negation is $\sim(f \wedge \sim r)$. Using DeMorgan's equivalence, this becomes $\sim f \vee r$. Translating back into English, we must remember that f represents the statement "The chicken is fried," while r represents the statement "The chicken is roasted." This means that the negation is "The chicken is not fried or the chicken *is* roasted." In this case, care must be taken not to shorten the sentence by eliminating words. Notice that if we eliminate the word "is" and write "The chicken is not fried or roasted," its meaning is *incorrect*.

■

The Negation of a Conditional Statement

The *conditional negation equivalence* is expressed symbolically as $\sim(p \rightarrow q) \Leftrightarrow (p \wedge \sim q)$. In order to see why this is so, we will apply both the conditional equivalence and DeMorgan's equivalence.

As we have learned, the conditional equivalence states that $p \rightarrow q$ is equivalent to $\sim p \vee q$. This means that negating $p \rightarrow q$ is equivalent to negating $\sim p \vee q$. Using DeMorgan's equivalence, $\sim (\sim p \vee q)$ is logically equivalent to $p \wedge \sim q$. Summarizing our results,

$$(p \rightarrow q) \quad \Leftrightarrow \quad (\sim p \vee q)$$
$$\sim (p \rightarrow q) \quad \Leftrightarrow \quad \sim (\sim p \vee q)$$
$$\sim (p \rightarrow q) \quad \Leftrightarrow \quad (p \wedge \sim q)$$

Therefore, the negation of an implication is the *conjunction* of its LHS with the negation of its RHS.

> The conditional negation equivalence:
> $$\sim (p \rightarrow q) \Leftrightarrow (p \wedge \sim q)$$

Example 7

Use a compact truth table to show that the negation of $p \rightarrow q$ is logically equivalent to $p \wedge \sim q$.

Solution

p	q	$[\sim$	$(p \rightarrow q)]$		$p \wedge \sim q$
T	T	F	T		F
T	F	T	F		T
F	T	F	T		F
F	F	F	T		F
1	2	4	3		5

Since columns 4 and 5 are identical, $\sim (p \rightarrow q)$ is equivalent to $p \wedge \sim q$. ∎

Example 8

Construct a truth table for a conditional statement, its negation and its inverse to determine if the negation of a conditional statement is logically equivalent to its inverse.

Solution

p	q	$(p \rightarrow q)$	$\sim (p \rightarrow q)$	$\sim p \rightarrow \sim q$
T	T	T	F	T
T	F	F	T	T
F	T	T	F	F
F	F	T	F	T
1	2	3	4	5

Observe that the statements $\sim (p \rightarrow q)$ and $\sim p \rightarrow \sim q$ do not always have the same truth value, so they are not equivalent. Therefore, the negation of a conditional statement *is not* logically equivalent to its inverse.

■

Example 9

In words, what is the negation of the sentence "If you use Microsoft Word you can format documents easily"?

Solution

First, we symbolize the sentence as $w \rightarrow d$. The negation of $w \rightarrow d$ is symbolized as $\sim (w \rightarrow d)$ and is equivalent to $w \wedge \sim d$. In English, this is "You use Microsoft Word *but* you can't format documents easily."

■

Sometimes, expressing the negation of a given statement involves the use of more than one equivalence.

Example 10

Write the negation of "There will be rain or hail if the barometer falls."

Solution

The original statement is symbolized as $b \rightarrow (r \vee h)$. Negating this statement, we have $\sim [b \rightarrow (r \vee h)]$. Since we are negating a conditional, we first apply the conditional negation equivalence to obtain $b \wedge \sim (r \vee h)$. This sentence contains the negation of a disjunction. In order to simplify the statement further, we now apply DeMorgan's equivalence to obtain $b \wedge (\sim r \wedge \sim h)$. After translating into English, we have "The barometer falls, and there is no rain and there is no hail" or "The barometer fell but there was no rain and no hail."

■

Example 11

Write the negation of "The college will be open if it doesn't snow."

Solution

Recall that since the antecedent must come first in the symbolization, we must symbolize the original statement as $\sim s \rightarrow o$. Its negation is $\sim (\sim s \rightarrow o)$. When simplified, using the conditional negation equivalence, we get $\sim s \wedge \sim o$. This translates into "It doesn't snow and the college is not open."

■

Sometimes it might sound better if we change the tense of a sentence to past or future, while still preserving the logical meaning. Thus, the negation in Example 11 could have been written as "It didn't snow but the college wasn't open" rather than "It doesn't snow and the college is not open."

The Negation of a Biconditional Statement

The *biconditional negation equivalence* can be expressed as $\sim (p \leftrightarrow q) \Leftrightarrow [(p \wedge \sim q) \vee (q \wedge \sim p)]$. Recall that in Section 1.1, we defined $p \leftrightarrow q$ as $(p \rightarrow q) \wedge (q \rightarrow p)$. Note that a biconditional statement is a conjunction of an implication and its converse. To negate the statement $p \leftrightarrow q$, we actually negate the conjunctive statement $(p \rightarrow q) \wedge (q \rightarrow p)$. We therefore apply DeMorgan's equivalence, followed by the negation of the conditional equivalence. Symbolically, we write

$$\sim (p \leftrightarrow q) \Leftrightarrow \sim [(p \rightarrow q) \wedge (q \rightarrow p)]$$
$$\Leftrightarrow \sim (p \rightarrow q) \vee \sim (q \rightarrow p) \quad \text{(DeMorgan's)}$$
$$\Leftrightarrow (p \wedge \sim q) \vee (q \wedge \sim p) \quad \text{(Conditional Negation)}$$

The biconditional negation equivalence:

$$\sim (p \leftrightarrow q) \Leftrightarrow [(p \wedge \sim q) \vee (q \wedge \sim p)]$$

Example 12

Write, in English, the negation of "I will board the plane if and only if they screen the luggage."

Solution

We can symbolize the sentence to be negated as $b \leftrightarrow s$. Its negation would be $\sim (b \leftrightarrow s)$, which is equivalent to $(b \wedge \sim s) \vee (s \wedge \sim b)$. Translating back to English, we have "I board the plane and they do not screen the luggage, or they screen the luggage and I don't board the plane." ∎

Example 13

Write the negation of "The NY Giants will make the playoffs if and only if they don't lose their last two games."

Solution

The original statement is the biconditional statement $p \leftrightarrow \sim l$. Its negation is $\sim (p \leftrightarrow \sim l)$. The biconditional negation equivalence makes this logically equivalent to $(p \wedge l) \vee (\sim l \wedge \sim p)$. Since this is a lengthy disjunction, we may want to begin with the word "either", although the use of this word is optional. Translation into English results in "Either the NY Giants make the playoffs and lose their last two games or they didn't make the playoffs and didn't lose their last two games." ■

Example 14

Write the negation of "Moriarty is the murderer if and only if the fingerprints match and the patio door was not locked."

Solution

In symbolic form, the original statement is symbolized as $m \leftrightarrow (f \wedge \sim l)$. Its negation is $\sim [m \leftrightarrow (f \wedge \sim l)]$. Since we are negating a biconditional statement, we use the biconditional negation equivalence to obtain the disjunctive statement $[m \wedge \sim (f \wedge \sim l)] \vee [(f \wedge \sim l) \wedge \sim m]$. We next apply DeMorgan's equivalence to obtain $[m \wedge (\sim f \vee l)] \vee [(f \wedge \sim l) \wedge \sim m]$. Finally, translating into English, we have "Moriarity was the murderer but either the fingerprints did not match or the patio door was locked, or, the fingerprints matched and the patio door was not locked, but Moriarity was not the murderer." Although this is a long statement, it very clearly presents a choice of exactly two possibilities for disproving (negating) the original allegation. ■

Three of the equivalences we have discussed involve the negation of a given statement. These are:
- DeMorgan's Equivalence, for negating a conjunction or disjunction.
- The Conditional Negation Equivalence, for negating a conditional.
- The Biconditional Negation Equivalence, for negating a biconditional.

The following table summarizes how these equivalences are used to express the negation of a given statement.

A Summary of the Negation Rules

Given Statement	Negation of Given Statement	Equivalence Used	Negation after Equivalence is used
$p \wedge q$ p and q	$\sim (p \wedge q)$ It is not true that p and q	DeMorgan's $\sim (p \wedge q) \Leftrightarrow (\sim p \vee \sim q)$	$\sim p \vee \sim q$ 1. Not p or not q 2. Either not p or not q

$p \vee q$
 	
1. p or q
2. Either p or q
	$\sim (p \vee q)$
	
	1. It is not true that
	   p or q
	2. It is not true that
	   either p or q
		DeMorgan's
		
		$\sim (p \vee q) \Leftrightarrow (\sim p \wedge \sim q)$
			$\sim p \wedge \sim q$
			
			1. Not p and not q
			2. Neither p nor q

$p \rightarrow q$
 	
1. If p then q
2. p implies q
	$\sim (p \rightarrow q)$
	
	1. It is not true that
	   if p then q
	2. It is not true that
	   p implies q
		Conditional Negation
		
		$\sim (p \rightarrow q) \Leftrightarrow (p \wedge \sim q)$
			$p \wedge \sim q$
			
			1. p and not q
			2. p but not q

$p \leftrightarrow q$
 	
p if and only if q
	$\sim (p \leftrightarrow q)$
	
	It is not true that
	p if and only if q
		Biconditional Negation
		
		$\sim (p \leftrightarrow q) \Leftrightarrow [(p \wedge \sim q) \vee (q \wedge \sim p)]$
			$(p \wedge \sim q) \vee (q \wedge \sim p)$
			
			1. p and not q, or
			   q and not p
			2. p but not q, or
			   q but not p
			3. Either p and not q
			   or q and not p
			4. Either p but not q
			   or q but not

In-Class Exercises and Problems for Section 1.8

In-Class Exercises

For Exercises 1–14 write a statement that is logically equivalent to the given statement using DeMorgan's Equivalence.

1. $\sim (p \vee \sim q)$ 2. $\sim (w \wedge \sim s)$

3. $\sim(\sim p \vee \sim s)$

4. $b \wedge \sim r$

5. $\sim a \wedge \sim b$

6. $q \vee \sim s$

7. $\sim[m \vee (a \to b)]$

8. $\sim[(p \leftrightarrow q) \wedge \sim r]$

9. $[p \vee \sim (a \to c)]$

10. $\sim[\sim (w \wedge r) \wedge (p \to q)]$

11. It is not true that I will take p̲iano lessons and not practice the s̲cales.

12. Travelling abroad requires a p̲assport or b̲irth certificate.

13. We will not go out for d̲inner, but we will go to the c̲oncert.

14. It is not true that doctors advise us not to eat h̲ealthy foods or not to e̲xercise.

In Exercises 15–22, use the conditional negation equivalence to write an equivalent statement.

15 $\sim(m \to b) \Leftrightarrow$

16. $\sim(\sim s \to w) \Leftrightarrow$

17. $(a \wedge \sim d) \Leftrightarrow$

18. $(\sim k \wedge c) \Leftrightarrow$

19. $\sim[(r \vee s) \to w] \Leftrightarrow$

20. $[p \wedge (w \vee q)] \Leftrightarrow$

21. If the ozone layer is not p̲rotected, life on Earth is d̲oomed.

22. The e̲conomy will improve if disposable income r̲ises.

In Exercises 23–27, express the negation of each biconditional statement as the disjunction of two conjunctions.

23. $\sim(p \leftrightarrow \sim r)$

24. $\sim(\sim w \leftrightarrow s)$

25. $\sim(\sim m \leftrightarrow \sim n)$

26. You buy a new c̲omputer if and only if your current computer is more than two y̲ears old.

27. You d̲ial 911 if and only if there is no other a̲lternative.

Problems

For Problems 28-39, use the negation rules to simplify each statement.

28. $\sim(p \vee \sim r)$

29. $\sim(\sim k \vee w)$

30. $\sim(\sim p \wedge q)$

31. $\sim (c \wedge \sim p)$

32. $\sim (p \rightarrow \sim r)$

33. $\sim (\sim p \rightarrow \sim a)$

34. $\sim (c \leftrightarrow \sim d)$

35. $\sim (\sim d \leftrightarrow g)$

36. $\sim [a \wedge (b \leftrightarrow c)]$

37. $\sim [\sim q \vee (a \rightarrow p)]$

38. $\sim [a \rightarrow (b \vee \sim r)]$

39. $\sim [\sim a \rightarrow \sim (p \vee m)]$

In Problems 40–45, match each numbered statement to its lettered equivalent.

40. $\sim b \vee c$	a. $\sim (\sim b \wedge \sim c)$
41. $\sim b \wedge c$	b. $\sim (\sim b \rightarrow \sim c)$
42. $\sim (b \vee c)$	c. $b \wedge \sim c$
43. $\sim b \wedge \sim c$	d. $\sim (\sim b \rightarrow c)$
44. $\sim (b \rightarrow c)$	e. $\sim (b \wedge \sim c)$
45. $b \vee c$	f. $\sim b \wedge \sim c$

For Problems 46–55, use the negation rules to write, in English, the negation of each statement.

46. Barbara is fluent in English and French.

47. The sun was out, but it wasn't raining.

48. The economy is down, and people aren't happy.

49. If I don't pass mathematics, I will have to drop out of school.

50. Ramon will not vacation in Florida if he spends all his money.

51. The best music is jazz, but if you don't like it, you will like country.

52. You must invest your money wisely, or you will suffer the consequences if there is a recession.

53. The temperature outside is 110 degrees, and if you don't hydrate you will become dizzy.

54. There won't be any firings if and only if the Yankees win the World Series.

55. A peace treaty is to be signed if and only if there is a cease fire and the President attends the conference.

For Problems 56–60, analyze each statement.

56. The defendant's attorney stated, "My client has no prior arrests, but even if he did, there was no motive for committing the crime." The prosecutor objected saying, "That is not true." What did the prosecutor mean?

57. A claim against a politician running for office said, "If he is elected, the economy will suffer and if the latter happens, you will be out of a job." The politician disputed the claim saying "It's a lie." What did this mean?

58. A sportswriter claimed "The Yankees or Mets will make the playoffs but if the Mets do, the Phillies won't." The editor claimed this wasn't so. What was the editor saying?

59. Two physicians were arguing about a diagnosis. Dr. A said, "The patient has signs of vertigo and anxiety, which together imply a neurological condition." Dr. B stated, "That is completely wrong." What was Dr. B saying?

60. Darlene plans to watch television if and only if Bill doesn't call or there's an important test the next day. Darlene's friend Claire thinks Darlene's plan won't happen. What is Claire suggesting?

DID YOU KNOW?

Zeno of Elea (ca. 490–430 BCE) was a pre-Socratic Greek philosopher. Aristotle called him the inventor of the dialectic, and Bertrand Russell credited him with having laid the foundations of modern logic. He is best known for his paradoxes, which you may have encountered in a philosophy course.

In the paradox of Achilles and the tortoise, Achilles is in a race with a tortoise. Achilles allows the tortoise a head start of 100 feet. Suppose Achilles runs at some constant, very fast speed, and the tortoise runs at a constant speed, but very slowly. Then after some finite time, Achilles will have run 100 feet, bringing him to where the tortoise began. During this time, the tortoise would have run a much shorter distance, for example, 10 feet. It will then take Achilles some more time to run that distance, in which time the tortoise will have advanced farther; and then more time still to reach this third point, while the tortoise again moved ahead. Thus, whenever Achilles reaches somewhere the tortoise has been, he still has farther to go. Therefore, he can never overtake the tortoise.

Comment on the reasoning used in the above paradox. Do you believe Achilles will ever reach the tortoise?

Chapter 1 Review

For questions 1–10, match each numbered statement with its lettered equivalent.

1. $\sim(\sim p \to \sim q)$ a. $\sim p \vee \sim q$

2. $\sim(\sim p \vee q)$ b. $q \to p$

3. $p \vee q$ c. $\sim p \to q$

4. $\sim(p \wedge q)$ d. $p \wedge \sim q$

5. $\sim p \vee q$ e. $(p \wedge \sim q) \vee (p \wedge r)$

6. $\sim p \to \sim q$ f. $\sim p \wedge q$

7. $p \vee (\sim q \wedge r)$ g. $p \vee (\sim q \vee r)$

8. $(p \vee \sim q) \vee r$ h. $\sim(\sim p \to q)$

9. $\sim p \wedge \sim q$ i. $p \to q$

10. $p \wedge (\sim q \vee r)$ j. $(p \vee \sim q) \wedge (p \vee r)$

For questions 11–23, name the equivalence illustrated.

11. $(p \vee q) \Leftrightarrow (q \vee p)$

12. $[p \wedge (q \wedge r)] \Leftrightarrow [(p \wedge q) \wedge r]$

13. $[p \vee (q \wedge r)] \Leftrightarrow [(p \vee q) \wedge (p \vee r)]$

14. $(\sim p \leftrightarrow q) \Leftrightarrow [(\sim p \to q) \wedge (q \to \sim p)]$

15. $\sim(p \to q) \Leftrightarrow (p \wedge \sim q)$

16. $\sim(\sim p \wedge q) \Leftrightarrow (p \vee \sim q)$

17. $(\sim p \to q) \Leftrightarrow (p \vee q)$

18. $(\sim p \vee \sim q) \Leftrightarrow \sim(p \wedge q)$

19. $(p \vee \sim q) \Leftrightarrow (\sim p \to \sim q)$

20. $[(p \to q) \vee (r \wedge s)] \Leftrightarrow \{[(p \to q) \vee r] \wedge [(p \to q) \vee s]\}$

21. $\sim[\sim(p \to q)] \Leftrightarrow (p \to q)$

22. $(\sim p \to q) \Leftrightarrow (\sim q \to p)$

23. $[(\sim p \to q) \vee r] \Leftrightarrow [r \vee (\sim p \to q)]$

For questions 24–31, symbolize each sentence using the underlined letters as symbols for the simple statements.

24. If I do a <u>c</u>hemistry experiment and it smells like <u>r</u>otten eggs then I used <u>s</u>ulfur dioxide.

25. I'm going to stay in <u>N</u>ew York, or if I <u>l</u>eave I will <u>r</u>elocate to Los Angeles.

26. Native Americans hunted buffalo for their <u>s</u>kins and their <u>m</u>eat or they were <u>c</u>old and <u>h</u>ungry.

27. I will not be able to eat <u>d</u>essert if I have the <u>s</u>alad and <u>m</u>ain course.

28. A <u>l</u>awn will grow if and only if it is <u>s</u>eeded and <u>w</u>atered.

29. Computers are <u>f</u>ast and <u>a</u>ccurate, but they can't <u>r</u>eflect.

30. If a flag is <u>w</u>aved and a <u>p</u>arade is in progress then a <u>b</u>and is playing or a <u>f</u>loat passes by.

31. <u>W</u>orking after school is not fun, but I earn <u>m</u>oney.

For questions 32–38 express the negation of each given statement as a conjunction or as a disjunction if possible. Simplify each statement using equivalences where appropriate. Your answer should not involve any conditional or biconditional symbols.

32. Avi will smile if he is happy.

33. I will play poker if and only if I don't have a test.

34. The computer is cheap and the monitor is not expensive.

35. If the answer to part three is true then the answer to part four is also true.

36. It is a sentence if and only if it has a subject and a verb.

37. In her psychology class she will study Skinner or Freud.

38. If stay in school I will study chemistry and work in the lab.

For questions 39–48, write a statement that is logically equivalent to the given statement using the indicated equivalence.

39. She wants to be a doctor or a lawyer. (Conditional Equivalence)

40. It is not true that they will visit and stay for two weeks. (DeMorgan's Equivalence)

41. If she has a manicure then she will have a pedicure. (Contrapositive Equivalence)

42. The dog will not bite if you feed her. (Conditional Equivalence)

43. Money is both a necessity and an evil. (Commutative Equivalence)

44. Calculators have made statistics easier but you have to do the homework. (DeMorgan's Equivalence)

45. If a number is even and prime then the number is two. (Contrapositive Equivalence)

46. It is not true that the enemy will be defeated if they are not supplied with missiles. (Conditional Negation Equivalence)

47. You use the Internet, and either your research skills will be enhanced or you will learn a great deal. (Distributive Equivalence)

48. I will go to the play if and only if the tickets are not expensive. (Biconditional Equivalence)

For questions 49–53, express the sentence "I will wear my <u>h</u>at and <u>g</u>loves if it is <u>c</u>old outside" symbolically using the underlined letters. First symbolically, and then as an English sentence, write its:

49. Contrapositive

50. Conditional Equivalence

51. Negation

52. Inverse

53. Converse

Sample Exam: Chapter 1

1. A statement logically equivalent to "It is not true that I like apples and bananas" is
   a. I don't like apples and I don't like bananas.
   b. If I don't like apples then I don't like bananas.
   c. I don't like apples or I like bananas.
   d. Either I don't like apples or I don't like bananas.

2. A statement logically equivalent to "I will buy you flowers or take you out for dinner if I get a raise" is
   a. If I buy you flowers or take you out for dinner then I get a raise.
   b. If I don't buy you flowers and I don't take you out for dinner then I don't get a raise.
   c. If I don't get a raise then I don't buy you flowers and I don't take you out for dinner.
   d. Either I don't buy you flowers and I don't take you out for dinner, or I get a raise.

3. The negation of $\sim(p \vee q) \rightarrow r$ is
   a. $(p \vee q) \rightarrow \sim r$ b. $(p \vee q) \vee \sim r$
   c. $(\sim p \wedge \sim q) \wedge \sim r$ d. $\sim r \rightarrow (p \vee q)$

4. A statement logically equivalent to $\sim[\sim a \vee (b \rightarrow c)]$ is
   a. $a \vee \sim(b \rightarrow c)$ b. $a \wedge (b \wedge \sim c)$
   c. $a \wedge \sim(b \vee \sim c)$ d. $a \rightarrow \sim(b \rightarrow c)$

5. If $(p \wedge q) \leftrightarrow (\sim r \vee s)$ is true and r is false then the truth value of p is
   a. True   b. False   c. Indeterminable

6. If $p \vee r$ is true and $\sim p \rightarrow q$ is false, then which of the following is true?
   a. $r \rightarrow q$ b. $\sim p \wedge \sim r$ c. $p \leftrightarrow \sim q$ d. $\sim q \leftrightarrow r$

7. Is $[(p \vee q) \rightarrow r] \leftrightarrow [(\sim p \wedge \sim q) \rightarrow \sim r]$ a tautology?
   a. Yes   b. No   c. Cannot be determined

8. Given that $[(\sim p \rightarrow q) \wedge \sim p] \rightarrow q$ is a tautology, if p is false, then q is
   a. True   b.   False   c. Indeterminable

9. If the converse of a statement is "If you like tennis then you will go to the U.S. Open," then the inverse of the original statement is
   a. If you do not go to the U.S. Open, then you do not like tennis.
   b. If you don't like tennis, then you will not go to the U.S. Open.
   c. You go to the U.S. Open if you like tennis.
   d. If you go to the U.S. Open, then you like tennis.

10. A statement equivalent to "If you go to culinary school, then you like to cook" is
    a. If you like to cook then you go to culinary school.
    b. Either you do not go to culinary school or you like to cook.
    c. If you do not go to culinary school then you do not like to cook.
    d. If you do not go to culinary school then you like to cook.

11. Is $p \to (r \wedge s)$ equivalent to $(\sim r \vee \sim s) \to \sim p$?
    a. Yes   b. No   c. Cannot be determined

12. The negation of "We will win if we try" is
    a. We will not win if we don't try.
    b. If we don't win we did not try.
    c. We win and we don't try.
    d. We try but we don't win.

13. A symbolization of "The grass will look beautiful if it is watered and mowed" is
    a. $g \to (w \wedge m)$ b. $(g \to w) \wedge m$
    c. $(w \wedge m) \to g$ d. $w \wedge (m \to g)$

14. The statement $\sim [k \to (g \vee h)]$ is equivalent to
    a. $\sim k \to \sim (g \vee h)$ b. $\sim k \vee (g \vee h)$
    c. $k \wedge (g \vee h)$ d. $k \wedge \sim (g \vee h)$

For questions 15–20, assume that a and b are true and c and d are false. Determine the truth value of each statement.

15. $(a \wedge b) \to \sim c$
    a. True   b. False   c. Indeterminable

16. $\sim c \leftrightarrow (a \wedge d)$
    a. True   b. False   c. Indeterminable

17. $\sim [c \to \sim (\sim a \vee \sim s)]$
    a. True   b. False   c. Indeterminable

18. $\sim (a \vee b) \to s$
    a. True   b. False   c. Indeterminable

19. $s \to [(a \vee c) \wedge \sim d]$
    a. True   b. False   c. Indeterminable

20. $r \to [(d \to w) \leftrightarrow (a \wedge c)]$
    a. True   b. False   c. Indeterminable

Frequently Used Equivalences

1. **Double Negation Equivalence**
   $\sim(\sim p) \Leftrightarrow p$
   The negation of the negation of a statement is logically equivalent to the original statement.

2. **Commutative Equivalences**
   With conjunctions and disjunctions, changing the order of the variables does not change the truth value of the statement.
   $(p \wedge q) \Leftrightarrow (q \wedge p)$
   $(p \vee q) \Leftrightarrow (q \vee p)$

3. **Associative Equivalences**
   With two conjunctions or two disjunctions, changing the grouping does not change the truth value of the statement.
   $[(p \wedge q) \wedge r] \Leftrightarrow [p \wedge (q \wedge r)]$
   $[(p \vee q) \vee r] \Leftrightarrow [p \vee (q \vee r)]$

4. **Distributive Equivalences**
   Conjunctions distribute over disjunctions and disjunctions distribute over conjunctions.
   $[p \wedge (q \vee r)] \Leftrightarrow [(p \wedge q) \vee (p \wedge r)]$
   $[p \vee (q \wedge r)] \Leftrightarrow [(p \vee q) \wedge (p \vee r)]$

5. **Conditional Equivalence**
   A conditional statement is equivalent to a disjunction with its LHS negated, and vice versa.
   $(p \rightarrow q) \Leftrightarrow (\sim p \vee q)$

6. **Contrapositive Equivalence**
   A conditional statement is equivalent to another conditional statement with the LHS and RHS interchanged and both negated.
   $(p \rightarrow q) \Leftrightarrow (\sim q \rightarrow \sim p)$

7. **Biconditional Equivalence**
   A biconditional statement is equivalent to a conjunction of a conditional statement and its converse.
   $(p \leftrightarrow q) \Leftrightarrow [(p \rightarrow q) \wedge (q \rightarrow p)]$

8. **DeMorgan's Equivalences**
   The negation of a disjunction is equivalent to a conjunction with both parts negated, while the negation of a conjunction is equivalent to a disjunction with both parts negated
   $\sim(p \vee q) \Leftrightarrow (\sim p \wedge \sim q)$
   $\sim(p \wedge q) \Leftrightarrow (\sim p \vee \sim q)$

9. **Conditional Negation Equivalence**
   The negation of a conditional statement is equivalent to a conjunction with its RHS negated.
   $\sim(p \rightarrow q) \Leftrightarrow (p \wedge \sim q)$

10. **Biconditional Negation Equivalence**
    The negation of a biconditional is equivalent to the disjunction of two statements: The conjunction of the LHS with the negation of the RHS, united with the conjunction of the RHS with the negation of the LHS.
    $\sim(p \leftrightarrow q) \Leftrightarrow [(p \wedge \sim q) \vee (q \wedge \sim p)]$

ARGUMENTS

Arguments form the foundation for all reasoning in mathematics and indeed for all the sciences. In this chapter, we will examine the concept of valid and invalid arguments and the process by which we test them for validity.

Objectives

After completing Chapter Two, the student should be able to:

- Define an argument.

- Represent an argument symbolically.

- Construct a truth table to determine whether an argument is valid or invalid.

- Construct a counterexample for an invalid argument.

- Determine if an argument is valid or invalid using the direct or indirect *TF* method, along with the equivalences discussed in Chapter One.

2.1 Using Truth Tables for Validity and Counterexamples

Introduction

An *argument* is a collection of *TF* statements, called *premises*, which are always assumed to be true, followed by a final statement, called the *conclusion* of the argument. The conclusion is often preceded by the word "therefore."

To show that an argument is *valid*, we must prove that, *based on the truth of its premises*, its conclusion *must* be true. It is crucial to understand that for an argument to be valid it is not sufficient to find one, two, or even several cases for which true premises lead to a true conclusion. For the argument to be valid, the conclusion must *always* be true if the premises are true. If true premises lead to a conclusion that is false even in one instance, the argument is *invalid*.

Consider the argument below that consists of two premises, followed by a conclusion.

> We will go to the game or study.
> We will not study.
> Therefore, we will go to the game.

In order to determine if an argument is valid or invalid, it is usually helpful to express the argument symbolically. When we do so, it is customary to list all the premises above a horizontal line, and to place the conclusion below the line. This format is shown below.

$$g \lor s$$
$$\frac{\sim s}{g}$$

Constructing Truth Tables for Validity

Example 1

Determine whether the argument below is valid or invalid.

> We will go to the game or study.
> We will not study.
> Therefore, we will go to the game.

Solution

We write the argument symbolically.

$$g \lor s$$
$$\frac{\sim s}{g}$$

We next construct a truth table whose headings are the simple statements involved, the premises, and the conclusion, as shown below.

		Premise 1	Premise 2	Conclusion
g	s	$g \vee s$	$\sim s$	g
T	T	T	F	T
T	F	T	T	T
F	T	T	F	F
F	F	F	T	F

To test the validity of the argument, we need only concern ourselves with those cases for which all the premises are true. In the table above, this occurs only in row two. This row is boxed in the table below.

		Premise 1	Premise 2	Conclusion
g	s	$g \vee s$	$\sim s$	g
T	T	T	F	T
T	F	$\boxed{T}$	T	T
F	T	T	F	F
F	F	F	T	F

In this row, the conclusion is also true. Therefore, the argument is valid, since whenever all the premises are true, the conclusion is true. ∎

Example 2

Determine whether the argument below is valid or invalid.

> If I lift weights I will become stronger.
> I became stronger.
> Therefore, I lifted weights.

Solution

We first symbolize the argument.

$$w \rightarrow s$$
$$\underline{s}$$
$$w$$

Now we construct a truth table whose headings are the simple statements, premises, and conclusion.

		Premise 1	Premise 2	Conclusion
s	w	$w \to s$	s	w
T	T	T	T	T
T	F	T	T	F
F	T	F	F	T
F	F	T	F	F

We are only concerned with those cases in which all the premises are true. In the table above, this occurs in the first and second rows. However, in this example the conclusion is not always true when both premises are true, as can be seen in row two. *If there is any case in which true premises lead to a false conclusion, the argument is invalid.* Therefore, this argument is invalid.

		Premise 1	Premise 2	Conclusion
s	w	$w \to s$	s	w
T	T	T	T	T
T	F	T	T	F
F	T	F	F	T
F	F	T	F	F

Example 3

Determine whether the following argument is valid or invalid.

> If I don't go on a date then I will watch the game or take a nap.
> I don't go on a date or I wash my car.
> I don't wash my car.
> Therefore, I take a nap.

Solution

The argument is symbolized below.

$$\sim d \to (g \vee n)$$

$$\sim d \vee w$$

$$\underline{\sim w}$$

$$n$$

Construct a truth table whose headings are the simple statements, premises, and the conclusion. Since four variables are involved, this truth table will have $2^4 = 16$ rows. The premises are all true only in rows 10, 12, and 14. The conclusion is not always true in these three cases. Therefore, the argument is invalid.

d	g	n	w	Premise 1 $\sim d \rightarrow (g \vee n)$	Premise 2 $\sim d \vee w$	Premise 3 $\sim w$	Conclusion n
T	T	T	T	T	T	F	T
T	T	T	F	T	F	T	T
T	T	F	T	T	T	F	F
T	T	F	F	T	F	T	F
T	F	T	T	T	T	F	T
T	F	T	F	T	F	T	T
T	F	F	T	T	T	F	F
T	F	F	F	T	F	T	F
F	T	T	T	T	T	F	T
F	T	T	F	T	T	T	T
F	T	F	T	T	T	F	F
F	T	F	F	T	T	T	F
F	F	T	T	T	T	F	T
F	F	T	F	T	T	T	T
F	F	F	T	F	T	F	F
F	F	F	F	F	T	T	F

■

Example 4

Is the following argument valid or invalid?

$$p \rightarrow q$$
$$\sim q \vee r$$
$$\underline{\sim r}$$
$$\sim p$$

Solution

Since the argument involves three variables, our truth table will have $2^3 = 8$ rows.

p	q	r	Premise 1 $p \rightarrow q$	Premise 2 $\sim q \vee r$	Premise 3 $\sim r$	Conclusion $\sim p$
T	T	T	T	T	F	F
T	T	F	T	F	T	F
T	F	T	F	T	F	F
T	F	F	F	T	T	F
F	T	T	T	T	F	T
F	T	F	T	F	T	T
F	F	T	T	T	F	T
F	F	F	T	T	T	T

There is only one case in which all the premises are true. This occurs in row eight. Since the conclusion is also true in this case, the argument is valid.

■

Counterexamples for Invalid Arguments

We know that if an argument is valid, its conclusion must be true based on the truth of the premises. If an argument is invalid, there must be at least one case in which the conclusion is false, even though the premises are true. A *counterexample* is an illustration that shows that the argument is not valid. To produce a counterexample, we must state the truth values of the variables such that all the premises are true, yet the conclusion is false.

Example 5

Is the given argument valid or invalid? If it is invalid, produce a counterexample.

> If I lift weights I will become stronger.
> I became stronger.
> Therefore, I lifted weights.

Solution

This is the invalid argument from Example 2. The symbolized argument is expressed as:

$$w \to s$$

$$s$$

$$\overline{\quad w \quad}$$

Reproducing the truth table associated with this argument, we notice that when s is true and w is false, the premises are both true but the conclusion is false.

		Premise 1	Premise 2	Conclusion
s	w	$w \to s$	s	w
T	T	T	T	T
T	F	T	T	F
F	T	F	F	T
F	F	T	F	F

■

Counterexample Charts

Row two in the truth table in Example 5 provides us with our counterexample. We display this result in the counterexample chart.

s	w
T	F

Example 6

Consider the following argument. Is the argument valid or invalid? If it is invalid, produce a counterexample chart.

> If I lift weights I will become stronger.
> I didn't become stronger.
> Therefore, I didn't lift weights.

Solution

The argument is expressed symbolically as

$$w \to s$$

$$\sim s$$

$$\overline{\quad \sim w \quad}$$

Constructing a truth table whose headings are the simple statements, premises, and conclusion, we obtain:

		Premise 1	Premise 2	Conclusion
s	w	$\sim s$	$w \rightarrow s$	$\sim w$
T	T	F	T	F
T	F	F	T	T
F	T	T	F	F
F	F	T	T	T

No counterexample can be produced since the conclusion is always forced to be true whenever the premises are true. This is exhibited in the fourth row of the truth table. Hence, the argument is valid. ∎

Example 7

Is the following argument valid or invalid? If it is invalid, produce a counterexample chart.

$$w \rightarrow j$$
$$j \vee b$$
$$\underline{\sim b}$$
$$w$$

Solution

We construct the truth table whose headings are the simple statements, premises, and conclusion.

			Premise 1	Premise 2	Premise 3	Conclusion
b	j	w	$w \rightarrow j$	$j \vee b$	$\sim b$	w
T	T	T	T	T	F	T
T	T	F	T	T	F	F
T	F	T	F	T	F	T
T	F	F	T	T	F	F
F	T	T	T	T	T	T
F	T	F	T	T	T	F
F	F	T	F	F	T	T
F	F	F	T	F	T	F

Noting the entries in the sixth row, we observe that when b is false, j is true, and w is false, the argument has true premises with a false conclusion and is therefore invalid. The truth values of the variables that provide us with our counterexample are displayed in the counterexample chart below.

b	j	w
F	T	F

■

Example 8

Is the following argument valid or invalid? If it is invalid, produce a counterexample chart.

$$p \to q$$
$$(\sim p \vee q) \to r$$
$$\overline{ p }$$

Solution

First, the truth table is constructed.

			Premise 1	Premise 2	Conclusion
p	q	r	$p \to q$	$(\sim p \vee q) \to r$	p
T	T	T	T	T	T
T	T	F	T	F	T
T	F	T	F	T	T
T	F	F	F	T	T
F	T	T	T	T	F
F	T	F	T	F	F
F	F	T	T	T	F
F	F	F	T	F	F

Observing the entries in rows five and seven, we note that there are two instances where true premises produce a false conclusion. Therefore, we have two equally correct counterexamples.

p	q	r
F	T	T
F	F	T

Either of the two counterexamples is sufficient to show that the argument is invalid.

■

In-Class Exercises and Problems for Section 2.1

In-Class Exercises

For Exercises 1–10, use a truth table to determine whether the argument is valid or invalid. If the argument is invalid produce a counterexample chart.

1. $\sim(c \to g)$

$$\frac{\sim g}{c}$$

2. $p \to q$

$$\frac{q \vee \sim r}{\sim r}$$

3. $r \to (w \vee \sim s)$

$$\frac{\sim(w \vee \sim s)}{r}$$

4. $(p \wedge r) \vee s$

$$\frac{\sim s}{\sim r}$$

5. $(b \to k) \wedge r$

$$\frac{\sim r \vee b}{k}$$

6. $(q \vee r) \leftrightarrow \sim p$

$$\frac{\sim p \wedge r}{q}$$

7. $\sim a$

$\sim a \to b$

$$\frac{c \to b}{c}$$

8. $\sim b \to d$

$\sim b \vee c$

$$\frac{c \to d}{d}$$

9. $b \wedge \sim a$

$a \vee (b \to c)$

$$\frac{c \vee d}{d}$$

10. $p \vee (q \wedge s)$

r

$$\frac{r \to \sim p}{s}$$

Problems

For Problems 11–26, construct a truth table to determine whether the argument is valid or invalid. If the argument is invalid produce a counterexample chart.

11. $\sim r \to s$

$$\frac{s}{r}$$

12. $\sim(w \to r)$

$$\frac{w \leftrightarrow s}{s}$$

13. $\sim (c \wedge \sim k)$
$$\frac{k \rightarrow p}{p}$$

14. $(p \rightarrow q) \rightarrow r$
$$\frac{\sim q \rightarrow \sim p}{r}$$

15. $\sim a \rightarrow \sim b$
$$\frac{\sim b \rightarrow c}{\sim c \rightarrow a}$$

16. $\sim (n \leftrightarrow p) \vee c$
$$\frac{\sim c}{\sim p \rightarrow \sim n}$$

17. $(m \rightarrow \sim q) \vee d$
$$\frac{q \wedge \sim d}{\sim m}$$

18. $(q \rightarrow p) \wedge c$
$$\frac{\sim p}{q}$$

19. $\sim a$
$\sim m \vee a$
$$\frac{\sim m \rightarrow p}{p \vee a}$$

20. $p \wedge r$
$$\frac{\sim p \rightarrow q}{q}$$

21. r
$$\frac{}{r \vee (s \rightarrow g)}$$

22. $\sim (p \vee s)$
$$\frac{\sim p \rightarrow q}{q}$$

23. $p \rightarrow q$
$\sim s \vee q$
$$\frac{p}{s}$$

24. $p \rightarrow q$
$q \rightarrow r$
$$\frac{r}{p}$$

25. $p \rightarrow q$
$s \rightarrow r$
$$\frac{p \vee s}{q \vee r}$$

26. $\sim p \vee q$
$q \rightarrow r$
$$\frac{r \rightarrow s}{p \rightarrow s}$$

Check Your Understanding

For questions 1–3, choose the correct counterexample chart for each invalid argument.

1. $\sim q \to \sim p$
   $\sim (q \to r)$
   ─────────
   $p \vee r$

   a.
p	q	r
f	f	f

   b.
p	q	r
f	t	f

   c.
p	q	r
t	t	t

2. $\sim (m \wedge n)$
   $c \to \sim m$
   $c \vee \sim n$
   ─────────
   $\sim n$

   a.
c	m	n
t	f	t

   b.
c	m	n
f	t	f

   c.
c	m	n
f	f	t

3. $\sim b \vee s$
   b
   $(s \vee q) \to w$
   ─────────
   $\sim w$

   a.
b	q	s	w
t	f	f	t

   b.
b	q	s	w
t	f	t	f

   c.
b	q	s	w
t	f	t	t

4. Consider the following argument. Is it possible for the table of associated truth values shown to be a counterexample chart for the given argument? Explain your answer.

   $\sim (a \to \sim b)$
   $\sim a \vee c$
   $c \to \sim d$
   ─────────
   $d \to a$

a	b	c	d
t	t	t	f

For question 5, construct a truth table to determine whether the argument is valid or invalid. If the argument is invalid produce a counterexample chart.

5. m
   $\sim m \leftrightarrow w$
   $w \vee \sim s$
   ─────────
   $s \to w$

2.2 Testing for Validity Using the *TF* Method–The Direct Approach

You have seen that as the number of variables involved in an argument increases, the truth table used for testing validity becomes unwieldy. Another technique for testing the validity of an argument, called the *TF* method, makes the process easier. With this method, we assume all the premises are true and try to arrive at the truth value of the conclusion.

Example 1

Determine whether the following argument is valid or invalid.

> We will go to the game or study.
> We will not study.
> Therefore, we will go to the game.

Solution

By assumption, each premise must be true. This is illustrated below.

$$\begin{array}{cc} g \vee s & T \\ \dfrac{\sim s}{g} & T \end{array}$$

Notice that since the first premise is a disjunction and is assumed true, we can be sure that g and s are not both false. Unfortunately, we cannot be sure whether only g is true, only s is true, or they are both true. Examining our second premise however, we note that there is only one truth value of s such that $\sim s$ is true. In order for $\sim s$ to be true, s must be false. Now we consider the first premise again. Since we know that s is false, our assumption that $g \vee s$ is true requires that g must be true. Note that the conclusion of the argument is g. Hence, the conclusion is forced to be true when all the premises are true. Thus, the argument is valid. ∎

In general, if a disjunction is true, and the negation of one side of this disjunction is also true, the other side of the disjunction must also be true. This kind of pattern recognition will be expanded upon in Chapter Three.

Example 2

Determine whether the following argument is valid or invalid.

If I traveled at a speed of over 70 miles per hour I received a summons.
I received a summons.
Therefore, I traveled at a speed of over 70 miles per hour.

Solution

First we represent the argument symbolically.

$$s \to r$$
$$\frac{r}{s}$$

By assumption, each premise must be true.

$$s \to r \quad T$$
$$\frac{r}{s} \qquad T$$

The first premise is an implication. There are three different sets of truth values for r and s that make this conditional statement true. Thus we cannot, at this point, determine truth values for s and r. We know only that s cannot be true while r is false. The assumption that our second premise is true however, requires that r be true. But if r is true, the implication $s \to r$ is true *regardless* of the truth value of s. Therefore, the conclusion, which is also s, is not forced to be true. Consequently, this argument is invalid. ■

The fact that the argument in Example 2 is invalid implies that based on the truth of the premises the conclusion is not forced to be true. Notice that when r is true and s is false, each premise is true, yet the conclusion is not true. The counterexample chart is shown below.

r	s
T	F

Example 3

Determine if the argument is valid or invalid.

If six plus six is twelve then three times three is nine.
Three times three is nine.
Therefore, six plus six is twelve.

Solution

Using the letters *s* and *r* we symbolize the argument.

$$s \rightarrow r$$

$$\frac{r}{s}$$

Notice that the argument has exactly the same form as the argument in Example 2, so the argument must be invalid. Recall that for an argument to be valid, we must be able to show that, *based on the truth of its premises,* the conclusion must be true. We do not dispute the fact that six plus six is twelve, but rather that one cannot conclude this fact from the given premises. ∎

Example 4

Determine if the argument is valid or invalid.

If six plus six is twelve then three times three is nine.
Six plus six is twelve.
Therefore, three times three is nine.

Solution

We first represent the argument symbolically.

$$s \rightarrow r \quad T$$

$$\frac{s}{r} \qquad T$$

The first premise is an implication. Again, we know that *s* cannot be true while *r* is false. The assumption that our second premise is true however, requires that *s* must be true. Thus, *r* is forced to be true in order for $s \rightarrow r$ to be true. Since *r* is the conclusion, and based on our reasoning, it was forced to be true, this argument is valid. In this case, we are able to show, based on the truth of the premises only, that three times three is nine. ∎

In general, arguments may have many premises. The next example will show us how to deal with an argument that has three premises. This same procedure can be used with arguments involving any number of premises.

Example 5

Determine whether the argument below is valid or invalid.

$$\sim p \vee q$$
$$r \rightarrow p$$
$$\underline{\sim q}$$
$$\sim r$$

Solution

We begin our reasoning by noting that the third premise, $\sim q$, must be true. Note that the first premise involves q. Since $\sim p \vee q$ is assumed true and $\sim q$ is true, $\sim p$ must be true and thus p is false. Since $r \rightarrow p$ is a premise, it too must be true. But since p is false, r must be false. Hence, the conclusion, $\sim r$, is forced to be true. The argument is valid.

■

Example 6

Is the following argument valid or invalid? If it is invalid, produce a counterexample chart.

If my salary increases by five percent I will go to Rome.
Either I go to Rome or I buy a computer.
I didn't buy a computer.
Therefore, my salary didn't increase by five percent.

Solution

First, we represent the argument symbolically.

$$s \rightarrow r$$
$$r \vee c$$
$$\underline{\sim c}$$
$$\sim s$$

The premises $\sim c$ and $r \vee c$ are both true, forcing r to be true. Since $s \rightarrow r$ is true, and r is true, s can either be true or false; the implication is true regardless of the truth value of s. The conclusion, $\sim s$, is not forced to be true. Therefore, the argument is invalid.

A counterexample will provide us with truth values for c, r, and s such that every premise is true yet the conclusion is false. We already know that c must be false and r must be true in order for the last two premises to be true. If s is true, the conclusion of the argument will be false, and the first premise will still be true. The counterexample chart

c	r	s
F	T	T

provides us with a set of truth values for the variables such that every premise will be true, yet the conclusion will be false.   ∎

Example 7

Is the following argument valid or invalid? If it is invalid, produce a counterexample chart.

> If I don't go on a date then I will watch the game or take a nap.
> I don't go on a date or I wash my car.
> I don't wash my car.
> Therefore, I take a nap.

Solution

First, symbolize the argument.

$$\sim d \rightarrow (g \vee n)$$
$$\sim d \vee w$$
$$\underline{\sim w}$$
$$n$$

Both $\sim w$ and $\sim d \vee w$ are true, so $\sim d$ must be true. The first premise must be true, so $g \vee n$ must be true. However, there are three ways for a disjunction to be true. The statement $g \vee n$ is true when g and n are both true, when g is true and n is false and when g is false and n is true. Thus, the conclusion, n, is not forced be true. The argument is invalid. The counterexample chart for this argument is shown below.

d	g	n	w
F	T	F	F

∎

In-Class Exercises and Problems for Section 2.2

In-Class Exercises

For Exercises 1–11, answer true, false or can't be determined because of insufficient information.

1. If $m \wedge \sim n$ is a premise, what is the truth value of $\sim m \to s$?

2. If $\sim m \vee n$ is a premise, what is the truth value of $n \to p$?

3. If $\sim r \vee w$ and r are both premises, what is the truth value of $r \leftrightarrow w$?

4. If $\sim (a \vee \sim b)$ is a premise, what is the truth value of $\sim a \leftrightarrow \sim b$?

5. If $\sim (d \to \sim c)$ is a premise, what is the truth value of $d \vee p$?

6. If $w \wedge \sim s$ is a premise, what is the truth value of $\sim w \vee \sim s$?

7. If $a \to \sim b$ is true, what is the truth value of b?

8. If $\sim a \wedge b$ is true, what is the truth value of $b \vee c$?

9. If $\sim m \vee p$ is true, what is the truth value of p?

10. If $\sim (q \vee p)$ is true, what is the truth value of $p \to k$?

11. If $\sim (w \to \sim s)$ is true, what is the truth value of s?

For Exercises 12–21, use the *TF* method to determine if each argument is valid or invalid. If it is invalid, produce a counterexample chart.

12. $(p \wedge q) \vee r$

$\underline{\sim r }$

p

13. $(s \vee d) \to \sim k$

$\underline{k }$

$\sim s$

14. $a \to (b \vee c)$

$\underline{\sim a \wedge c }$

b

15. $(m \leftrightarrow n) \to r$

$\underline{\sim r \wedge m }$

$n \vee c$

16. $(\sim r \vee s) \to \sim q$

$\sim s \vee p$

$\underline{q }$

p

17. $w \vee \sim b$

$b \wedge z$

$\underline{(\sim w \vee z) \to a}$

$a \vee p$

18. $(p \wedge s) \vee e$

    $\sim e$

    $\underline{p \rightarrow w}$

      $w \rightarrow e$

19. $(c \wedge g) \rightarrow \sim m$

    $m \wedge g$

    $\underline{a \rightarrow c}$

      $\sim a$

20. $(m \rightarrow n) \leftrightarrow q$

    $\sim q$

    $\underline{(\sim n \vee w) \rightarrow s}$

      $a \rightarrow s$

21. $\sim (p \vee \sim s) \rightarrow r$

    $\sim r \wedge s$

    $p \rightarrow w$

    $\underline{c \rightarrow w}$

      c

For Exercises 22–26, express each argument symbolically. Then use the *TF* method to determine if it is valid or invalid. If it is invalid, produce a counterexample chart.

22. The streets are not wet. If it rains the streets get wet. Therefore, it did not rain.

23. If the Rangers win I will go to the playoffs. Either the Rangers win or the Islanders win. The Islanders do not win. Therefore, I will not go to the playoffs.

24. If I quit my job I cannot pay for insurance. I quit my job or I get a scholarship. I didn't get a scholarship. Therefore, I cannot pay for insurance.

25. We will play basketball if and only if we have five players. We don't have five players but one of our players is seven feet tall. Therefore, if one of our players is seven feet tall, we will played basketball.

26. A candidate for state office offered his views on the economy as follows: "I am in favor of decreasing income taxes but increasing sales taxes. If neither of these measures is enacted then no new jobs will be created. Therefore, if no new jobs are created, recession will follow." Is his argument valid?

Problems

For Problems 27–36, answer either true, false, or can't be determined because of insufficient information.

27. If $\sim p$ is true and $\sim p \rightarrow q$ is a premise, then q must be _____.

28. If $r \rightarrow s$ is a premise and s is false, then r must be _____.

29. If $p \leftrightarrow q$ and $\sim q$ are both premises, p must be _____.

30. If $p \rightarrow q$ is false, p is _____ and q is _____.

31. If $\sim(p \wedge q)$ is a premise, then p must be _____.

32. Suppose $\sim(p \leftrightarrow \sim q)$ and q are premises of an argument. Then $\sim p$ must be _____.

33. If $\sim p \wedge q$ is a premise, then $\sim p$ is _____.

34. If $s \wedge r$ is a premise, then $\sim(s \rightarrow r)$ is _____.

35. Let $p \wedge q$ be a premise. Then $r \rightarrow q$ is _____.

36. Suppose $\sim p \vee \sim q$ is a premise and p is a premise as well. Then q is _____.

For Problems 37–46, answer yes or no.

37. If $p \rightarrow q$ is true, must p be true? _____.

38. If $p \wedge q$ is true, must p be true? _____.

39. If q is true, must $p \rightarrow q$ be true? _____.

40. If p is true, must $p \vee q$ be true? _____.

41. If q is true, must $p \wedge q$ be true? _____.

42. If $p \rightarrow q$ is true, must p and q both be true? _____.

43. If $p \leftrightarrow q$ is false, must p and q both be false? _____.

44. If p is false, must $p \rightarrow q$ be true? _____.

45. If $p \vee q$ is true, must p be true? _____.

46. If $p \rightarrow q$ is true and p is false, must q be false? _____.

For Problems 47–54, use the *TF* method to determine whether each argument is valid or invalid. If it is invalid, produce a counterexample chart.

47. $r \vee (p \rightarrow \sim q)$
    $\dfrac{\sim(p \rightarrow r)}{\sim r \rightarrow \sim q}$

48. $s \vee \sim(d \wedge p)$
    $\dfrac{d \wedge p}{\sim s}$

49. $w \rightarrow s$

   $\sim w \rightarrow \sim p$

   $c \wedge \sim s$

   $\overline{\quad \sim p \quad}$

50. $a \rightarrow (\sim a \vee b)$

   a

   $b \rightarrow \sim c$

   $\overline{\quad \sim c \quad}$

51. $\sim (w \vee \sim p)$

   $p \rightarrow a$

   $a \vee b$

   $\overline{\quad b \quad}$

52. $l \rightarrow (m \vee p)$

   $\sim (m \vee p)$

   $\sim l \rightarrow \sim w$

   $\overline{\quad \sim w \quad}$

53. $s \vee r$

   $\sim r \vee g$

   $g \rightarrow q$

   $\sim s$

   $\overline{\quad q \quad}$

54. $\sim (a \rightarrow \sim w)$

   $c \leftrightarrow d$

   $w \rightarrow (b \wedge c)$

   $d \rightarrow (c \vee p)$

   $\overline{\quad p \quad}$

For Problems 55–58, express each argument symbolically. Then use the *TF* method to determine if the argument is valid or invalid. If it is invalid, produce a counterexample chart.

55. If I take the subway to work, I spend money. If I don't take the subway to work, then I get some exercise. I walk to work and do not spend money. Therefore, I get some exercise.

56. Either the pizza does not have anchovies or it has onions. If the pizza has onions I will have indigestion. I don't have indigestion. Therefore, the pizza has anchovies.

57. If it is not the case that the dish is hot and spicy then Nick will order it. The dish is not hot. Therefore, Nick will order it.

58. If I read it in the newspaper or see it on the evening news then the story is true. If I don't see it on the evening news, then it is not a big story. It is a big story or it is a developing story. It is not a developing story. Therefore, the story is true.

Check Your Understanding

For questions 1–4, answer yes or no.

1. If $\sim d \to q$ and $\sim q$ are premises, must d be true. _____

2. If $\sim a \lor c$ is a premise, then c must be true. _____

3. If $\sim(k \leftrightarrow \sim j)$ and $j \land s$ are premises then k must be true. _____

4. If $s \land \sim p$ and $p \lor \sim w$ are premises then w must be true. _____

For questions 5–8, answer true, false or can't be determined because of insufficient information.

5. If both $\sim(\sim p \to q)$ and $\sim q \to r$ are premises, then the truth value of r is forced to be _____.

6. If $\sim m \land \sim q$ is a premise, then the statement $m \lor k$ must be _____.

7. If $\sim s \land r$ is a premise, then what is the truth value of $\sim(\sim r \to s)$? _____

8. Suppose $(a \lor \sim b) \to (\sim c \leftrightarrow d)$ is a premise and $\sim(\sim c \leftrightarrow d)$ is true, then what is the truth value of b? _____

For questions 9–10, use the *TF* method to determine whether the given argument is valid or invalid. If it is valid, circle the word valid. If it is invalid, produce a counterexample chart.

9. $(\sim a \lor b) \to c$ valid   or

    $\sim c \leftrightarrow (g \lor k)$

    $\sim a$
    _____

    k

a	b	c	g	k

10. $\sim(w \lor p)$ valid   or

    $(m \to n) \to p$

    $\sim n \to k$
    _____

    $k \leftrightarrow \sim p$

k	m	n	p	w

2.3 Testing for Validity Using the *TF* Method–Equivalences

Introduction

When testing the validity of an argument, it is often useful to replace a statement used in that argument with another, logically equivalent statement. As we discussed in Chapter One, if two statements are equivalent their truth values are always the same, regardless of the truth values of the variables involved. That is, they are either both true or both false. In Sections 1.6, 1.7, and 1.8 we examined a number of equivalences. For instance, we saw that the statement $p \to q$ is equivalent to both $\sim p \vee q$ and $\sim q \to \sim p$. Similarly, $\sim(p \vee q)$ is equivalent to $\sim p \wedge \sim q$, and $\sim(p \to q)$ is equivalent to $p \wedge \sim q$.

Be aware, however, that the fact that two statements are equivalent does not by itself reveal any information about the individual truth values of the variables that are involved in statements.

Using the Conditional Equivalence

Recall that the conditional equivalence allows us to express any conditional statement of the form $p \to q$ as a disjunction of the form $\sim p \vee q$, to which it is logically equivalent. Symbolically, the conditional equivalence is written as $(p \to q) \Leftrightarrow (\sim p \vee q)$. In the following example, the conditional equivalence will prove useful in determining whether an argument is valid.

Example 1

Test the following argument for validity. If it is invalid, produce a counterexample chart.

$$(p \vee \sim q) \to r$$
$$\underline{\sim p \to \sim q}$$
$$r$$

Solution

A close inspection of the argument reveals that the variables in the second premise appear in the LHS of the first premise. We note that the second premise is an implication and the LHS of the first premise is a disjunction. Since the conditional equivalence can be used to express any implication as a disjunction, we suspect that this equivalence may be helpful. Applying the conditional equivalence to the second premise, we express the implication as $p \vee \sim q$. Notice that we still do not know the individual truth values for p and q, but it is of no consequence, since $p \vee \sim q$ is exactly the LHS of the first premise. Hence, r must be true. The argument is valid. ■

Example 2

Test the following argument for validity. If it is invalid, produce a counterexample chart.

$$(p \rightarrow q) \rightarrow r$$
$$r \rightarrow \sim s$$
$$\underline{\sim p \vee q}$$
$$\sim s$$

Solution

If we apply the conditional equivalence to the third premise, $\sim p \vee q$, we may rewrite it as $p \rightarrow q$, which is the LHS of the first premise, $(p \rightarrow q) \rightarrow r$. Since $\sim p \vee q$ is true, $p \rightarrow q$ is also true. Notice again that it is not necessary to know the individual truth values for p and q. We know $(p \rightarrow q) \rightarrow r$ is true and its LHS, $p \rightarrow q$, is also true. Therefore, we can conclude that r must be true. Since both $r \rightarrow \sim s$ and r are true, $\sim s$ is forced to be true. Thus the argument is valid. ■

Using the Contrapositive Equivalence

The contrapositive equivalence states that any conditional statement of the form $p \rightarrow q$ is logically equivalent to the conditional statement $\sim q \rightarrow \sim p$. Symbolically, the contrapositive equivalence is expressed as $(p \rightarrow q) \Leftrightarrow (\sim q \rightarrow \sim p)$. In the next example, the contrapositive equivalence will be useful in determining whether an argument is valid.

Example 3

Test the following argument for validity.

$$p \rightarrow (\sim q \rightarrow r)$$
$$(\sim p \vee s) \rightarrow w$$
$$\underline{\sim (\sim r \rightarrow q)}$$
$$w$$

Solution

The conditional statement within the parentheses in the third premise is the contrapositive of the RHS of the first premise $p \rightarrow (\sim q \rightarrow r)$. Since $\sim (\sim r \rightarrow q)$ is true, the contrapositive equivalence tells us that $\sim (\sim q \rightarrow r)$ must also be true, so the RHS of $p \rightarrow (\sim q \rightarrow r)$ must be false. Therefore, p is false, so $\sim p$ is true. This means that the LHS of $(\sim p \vee s) \rightarrow w$ must be true. But since the LHS of this true conditional statement is true, its RHS must also be true. Therefore, the argument is valid. ∎

Using DeMorgan's Equivalence

As we discussed in Section 1.8, DeMorgan's equivalences allow us to express $\sim (p \vee q)$ as $(\sim p \wedge \sim q)$ and $\sim (p \wedge q)$ as $(\sim p \vee \sim q)$. We can express these two equivalences symbolically by writing $\sim (p \vee q) \Leftrightarrow (\sim p \wedge \sim q)$ and $\sim (p \wedge q) \Leftrightarrow (\sim p \vee \sim q)$. DeMorgan's equivalence will be used in the next two examples.

Example 4

Test the following argument for validity. If it is invalid, produce a counterexample chart.

$$\sim (p \wedge q) \rightarrow r$$
$$\underline{\sim p \vee \sim q}$$
$$r$$

Solution

The second premise, $\sim p \vee \sim q$, is equivalent to $\sim (p \wedge q)$ by DeMorgan's equivalence. Consequently, $\sim (p \wedge q)$ is true. The first premise has $\sim (p \wedge q)$ as its LHS. Therefore, r must be true, and the argument is valid. Again, notice that we still do not know, nor do we need to know, the truth values for p and q. ∎

Example 5

Test the following argument for validity. If it is invalid, produce a counterexample chart.

$$\sim p \rightarrow (q \wedge \sim r)$$
$$\sim q \vee r$$
$$\underline{p \rightarrow (s \wedge h)}$$
$$h$$

Solution

The appearance of the variables q and r in both the first and second premises may provide a hint as to how we should proceed. The RHS of the first premise is the negation of the second premise, because $\sim(\sim q \vee r)$ is equivalent to $q \wedge \sim r$ by DeMorgan's equivalence. Since $\sim q \vee r$ is true, $q \wedge \sim r$ must be false. In order for the first premise to be true, $\sim p$ must be false, making p true. Turning to the third premise, $s \wedge h$ must be true since p is true. The only way a conjunction can be true is if each of its components is true. Therefore, h must be true. The argument is valid.

■

Using the Conditional Negation Equivalence

When we negate the conditional statement $p \rightarrow q$, we obtain $p \wedge \sim q$. If both of these statements appear in an argument, it is important to recognize that one is the negation of the other.

Example 6

Test the following argument for validity. If it is invalid, produce a counterexample chart.

$$r \rightarrow (d \wedge \sim e)$$

$$r \vee s$$

$$\underline{d \rightarrow e}$$

$$s$$

Solution

The negation of the third premise, $d \rightarrow e$, is $d \wedge \sim e$, which is the RHS of the first premise. Since $d \rightarrow e$ is true, $d \wedge \sim e$ must be false. Since $r \rightarrow (d \wedge \sim e)$ is true but its RHS is false, r must be false. However, $r \vee s$ is true, so s must be true. The argument is valid.

■

Example 7

Test the following argument for validity. If it is invalid, produce a counterexample chart.

$$(p \wedge \sim q) \rightarrow (r \vee s)$$

$$\underline{\sim(p \rightarrow q)}$$

$$r$$

Solution

The conditional negation equivalence tells us that the second premise is exactly the LHS of the first premise. Since $\sim(p \to q)$ is true, so is $p \wedge \sim q$. Therefore, $r \vee s$ must be true. However, there are three ways a disjunction can be true, so r is not necessarily true. Since the argument is clearly invalid, we now produce a counterexample chart. Notice that since $\sim(p \to q)$ is true, $p \to q$ must be false. This means that p is true and q is false. We know that the RHS of the first premise must be true. This suggests we construct the counterexample chart as follows:

p	q	r	s
T	F	F	T

With these truth values, every premise is true, but the conclusion is false. The argument is invalid. ∎

In-Class Exercises and Problems for Section 2.3

In-Class Exercises

For Exercises 1–10, use your knowledge of equivalences to test the validity of each argument. If the argument is invalid, produce a counterexample chart.

1. $\sim p \vee s$
   $\sim p \to (q \wedge r)$
   $\dfrac{\sim q \vee \sim r}{s}$

2. $\sim p \vee s$
   $\sim q \to r$
   $\dfrac{(q \vee r) \leftrightarrow s}{\sim p}$

3. $(\sim r \to q) \to s$
   $\sim q \to r$
   $\dfrac{s \to (r \wedge w)}{\sim w}$

4. $\sim p \vee \sim q$
   $(p \to \sim q) \to a$
   $\dfrac{a \to (b \wedge d)}{d \vee c}$

5. $(m \rightarrow \sim n) \rightarrow (q \vee n)$

$\sim (\sim q \rightarrow \sim r)$

$n \rightarrow \sim m$

$\overline{\qquad \sim n \vee p \qquad}$

6. $a \wedge b$

$(m \wedge \sim n) \rightarrow \sim b$

$(\sim m \vee n) \rightarrow p$

$\overline{\qquad p \vee r \qquad}$

7. $(\sim p \rightarrow r) \leftrightarrow b$

$b \rightarrow (c \rightarrow d)$

$\sim d$

$p \vee r$

$\overline{\qquad w \rightarrow \sim c \qquad}$

8. $w \rightarrow \sim s$

$(p \vee \sim q) \rightarrow (w \wedge s)$

$r \vee p$

$r \rightarrow (\sim p \rightarrow \sim q)$

$\overline{\qquad p \qquad}$

9. $(p \wedge \sim q) \rightarrow s$

$(s \vee w) \rightarrow (\sim r \vee m)$

$\sim (p \rightarrow q)$

$c \rightarrow (r \wedge \sim m)$

$\overline{\qquad \sim c \qquad}$

10. $(r \rightarrow \sim w) \rightarrow (\sim s \vee p)$

$(s \rightarrow p) \rightarrow b$

$q \wedge (w \rightarrow \sim r)$

$\overline{\qquad b \qquad}$

Problems

For Problems 11–20, test each argument for validity. If the argument is invalid, produce a counterexample chart.

11. $q \rightarrow w$

$q \vee (d \wedge c)$

$\sim d \vee \sim c$

$\overline{\qquad w \qquad}$

12. $\sim d \rightarrow c$

$(\sim c \rightarrow d) \rightarrow w$

$w \rightarrow (c \wedge s)$

$\overline{\qquad \sim s \qquad}$

13. $g \wedge m$

$(b \wedge \sim e) \rightarrow \sim m$

$(\sim b \vee e) \rightarrow q$

$\overline{\qquad c \vee q \qquad}$

14. $q \rightarrow w$

$\sim d \rightarrow c$

$(d \vee c) \leftrightarrow w$

$\overline{\qquad q \qquad}$

15. $(\sim b \rightarrow e) \rightarrow \sim c$
$\quad \sim (b \rightarrow \sim d) \vee c$
$\quad \dfrac{\sim e \rightarrow b}{\sim d}$

16. $\sim g \rightarrow (m \wedge k)$
$\quad g \rightarrow (q \wedge d)$
$\quad \dfrac{\sim q \vee \sim d}{k \leftrightarrow \sim g}$

17. $m \rightarrow d$
$\quad \sim (q \rightarrow r) \rightarrow p$
$\quad p \rightarrow (m \wedge \sim d)$
$\quad \dfrac{(w \vee p) \vee (q \wedge \sim r)}{w}$

18. $b \leftrightarrow (q \vee m)$
$\quad \sim q \rightarrow m$
$\quad b \rightarrow (p \rightarrow c)$
$\quad \dfrac{\sim c}{s \rightarrow \sim p}$

19. $m \vee q$
$\quad \sim (s \rightarrow \sim w)$
$\quad (\sim q \rightarrow \sim r) \rightarrow m$
$\quad \dfrac{w \leftrightarrow (q \vee \sim r)}{q}$

20. $(\sim w \vee q) \rightarrow \sim (m \rightarrow \sim s)$
$\quad (s \rightarrow \sim m) \wedge \sim r$
$\quad \dfrac{\sim b \rightarrow (w \rightarrow q)}{b \vee p}$

DID YOU KNOW?

The book *Gödel, Escher, Bach: an Eternal Golden Braid* (GEB), a Pulitzer prize-winning exposition on genius written by Douglas Hofstadter, explores the workings of brilliant people's brains with the help of historical examples and brainteaser puzzles. The book touches on math, computers, literature, music, and artificial intelligence, while exploring how math, art, music, and language are interrelated. Hofstadter tackles questions such as "How does intelligent behavior arise from its component parts? Can computers think? Can brains compute?"

This book shows what it means to see symbols and patterns where others see only the chaos. GEB discusses some very serious matters, and it is not an easy book to tackle. However, it is very enjoyable to read, because Hofstadter illustrates each subject with diverse, easily understood examples.

Check Your Understanding

For questions 1–5, answer true, false or can't be determined because of insufficient information.

1. If $\sim a \vee b$ is true, then $\sim(a \wedge \sim b)$ is _____ .

2. If $\sim(q \rightarrow \sim s)$ is true $q \wedge s$ is _____ .

3. If $q \rightarrow (a \wedge \sim b)$ is false then $q \rightarrow (\sim a \vee b)$ is _____ .

4. If $q \rightarrow \sim w$ and $p \rightarrow \sim(w \rightarrow \sim q)$ are both true then p is _____ .

5. If $\sim q$ is false and $(\sim j \rightarrow \sim k) \leftrightarrow q$ is true and then $j \vee \sim k$ is _____ .

For questions 6–8, determine if the truth value of s is true, false or can't be determined because of insufficient information.

6. $(\sim p \vee r) \rightarrow s$
$$\frac{p \rightarrow r}{s}$$

7. $\sim p \vee r$
$$\frac{s \rightarrow (p \wedge \sim r)}{s}$$

8. $\sim(p \rightarrow r)$
$$\frac{s \rightarrow (p \wedge \sim r)}{s}$$

6. _____   7. _____   8. _____

For questions 9–10, use your knowledge of equivalences to determine whether each argument is valid or invalid. If it is valid, circle the word valid. If it is invalid, produce a counterexample chart.

9. $m \vee n$ valid   or

a	k	m	n	p	r	s	w

$m \rightarrow \sim(\sim k \vee s)$

$a \wedge (\sim p \rightarrow w)$

$$\frac{(\sim w \rightarrow p) \leftrightarrow (k \rightarrow s)}{n \vee r}$$

10. $p \rightarrow \sim w$ valid   or

a	b	p	r	w

$\sim p \vee r$

$(\sim b \rightarrow \sim a) \vee w$

$$\frac{(a \rightarrow b) \rightarrow (p \wedge \sim r)}{r}$$

2.4 Testing for Validity Using the *TF* Method–Conditional Conclusions

Introduction

If an argument contains a conditional conclusion, we can use a clever bit of reasoning to determine if the argument is valid or invalid. Recall that the *only* way a conditional statement can be false is if its LHS is true and its RHS is false. Therefore, if we can show that when the LHS of a conditional statement is true its RHS must also be true, the conditional statement must be true. This line of reasoning is the basis for the technique known as the *conditional proof.*

The Conditional Proof Strategy

When an argument contains a conditional conclusion, we can employ the method of conditional proof to test its validity. Note that if the LHS of a conditional conclusion is false, the argument is valid, since when the LHS of an implication is false, the implication must be true, thus making the argument valid. The only condition we have to consider is the case when the LHS of the conclusion is true.

The conditional strategy is to *assume* the LHS of the conclusion is true and show that under this assumption, the RHS of the conditional statement *must* be true, thus making the argument valid.

Example 1

Use a conditional strategy to show that the following argument is valid.

$$a \rightarrow b$$
$$b \rightarrow \sim c$$
$$\overline{a \rightarrow \sim c}$$

Solution

Since the conclusion, $a \rightarrow \sim c$, is a conditional statement, we can assume that a, its LHS, is true. We now proceed as we did previously. Since the first premise is true and since a is assumed to be true, b must be true. The second premise, $b \rightarrow \sim c$, is true and since b is true, $\sim c$ must be true. Thus, when the LHS of the conclusion is true, its RHS must be true as well. Therefore $a \rightarrow \sim c$ is true and the argument is valid.

∎

If we are testing for validity in an argument that contains a conditional conclusion, the conditional strategy does not have to be the first step in our reasoning. We can assume the LHS of the conclusion is true at any step of the reasoning process.

Example 2

Use a conditional strategy to show that the following argument is valid.

$$\sim p \vee q$$
$$\sim r \rightarrow s$$
$$\sim q$$
$$\sim p \rightarrow (s \rightarrow \sim w)$$
$$\overline{\quad w \rightarrow r \quad}$$

Solution

Since the conclusion of this argument is a conditional statement, we try to use the conditional strategy. If we assume w is true, we can proceed no further. While w appears in the premise $\sim p \rightarrow (s \rightarrow \sim w)$, we cannot reason further since the truth values of $\sim p$ and s are as yet unknown. Rather, we choose to begin with the first and third premises. Since $\sim q$ is true q is false. The first premise is a disjunction and one of its parts is false, so $\sim p$ must be true. The fourth premise $\sim p \rightarrow (s \rightarrow \sim w)$ has $\sim p$ as its LHS, so $s \rightarrow \sim w$ must be true. Now use the fact that the argument has a conditional conclusion. We assume its LHS, w, is true. If w is true $\sim w$ is false so s must be false. Now we turn our attention to the second premise. We know $\sim r \rightarrow s$ must be true and s is false. This means that $\sim r$ must be false, so r is true. But r is the RHS of the conclusion. Therefore we have shown that whenever the LHS of the conclusion is true, its RHS must be true. Therefore, the argument is valid.

■

Example 3

Use a conditional strategy to test the following argument for validity.

$$p \rightarrow \sim w$$
$$\sim p \rightarrow (s \vee q)$$
$$\overline{\quad w \rightarrow s \quad}$$

Solution

Since the conclusion of this argument is a conditional statement, we can try to use the conditional strategy. We assume w is true and consider the first premise $p \rightarrow \sim w$. Since premises are true and since w is assumed true, $\sim w$ must be false. Therefore, p must be false. The second premise, $\sim p \rightarrow (s \vee q)$ is true, and since p is false, we know $\sim p$ is true. Therefore $(s \vee q)$ is true. Since the disjunction is true, at least one of its parts must be true, but we can't be sure that s, the RHS of the conclusion is true. Since s is not *forced* to be true, we try to construct a counterexample chart for this argument, i.e. a set of truth values that makes every premise true and the conclusion false. Since we have assumed w is true, in order for the conclusion to be false, s must be false. The counterexample chart is shown below.

p	q	s	w
F	T	F	T

It is worth restating the strategy for a conditional approach. To use this approach, we must be presented with an argument whose conclusion is a conditional statement. If so, we can assume that the LHS of the conclusion is true and try to show that the RHS must be true. If we can, the argument is valid.

There are two common mistakes students make when trying to use this strategy. The first is assuming the LHS of *any* conclusion is true. The conditional proof strategy can be used only if the conclusion is a conditional statement. The other error is assuming the *entire* conclusion is true. This assumption is incorrect, since you would be assuming what you want to prove.

In-Class Exercises and Problems for Section 2.4

In-Class Exercises

For Exercises 1–12, test each argument for validity using the conditional proof technique. If the argument is invalid, produce a counterexample chart.

1. $\dfrac{p \rightarrow (q \wedge r)}{p \rightarrow r}$

2. $\begin{array}{l} \sim a \rightarrow \sim b \\ \sim a \vee c \\ \hline b \rightarrow c \end{array}$

3. $w \rightarrow \sim s$
   $\dfrac{b \rightarrow \sim w}{s \rightarrow b}$

4. $r \vee q$
   $\dfrac{\sim b \vee \sim q}{\sim r \rightarrow \sim b}$

5. $\sim (h \vee p) \vee w$
   $\dfrac{s \rightarrow \sim w}{h \rightarrow \sim s}$

6. $\sim k \rightarrow (p \wedge c)$
   $\dfrac{\sim c \vee a}{\sim k \rightarrow a}$

7. $r \rightarrow \sim p$
   $\sim r \vee (q \rightarrow s)$
   $\dfrac{q}{p \rightarrow \sim s}$

8. $(p \wedge a) \rightarrow m$
   $\sim m \vee (c \wedge d)$
   $\dfrac{p}{a \rightarrow c}$

9. b
   $\sim v \vee s$
   $(q \vee r) \rightarrow x$
   $\dfrac{b \rightarrow (x \rightarrow v)}{q \rightarrow s}$

10. $\sim p \rightarrow \sim q$
    $(p \wedge r) \rightarrow w$
    $\dfrac{(r \vee s) \rightarrow q}{r \rightarrow w}$

11. $\sim d \rightarrow p$
    $p \rightarrow (\sim m \rightarrow n)$
    $\sim d$
    $\dfrac{(m \vee d) \rightarrow r}{\sim n \rightarrow (r \wedge s)}$

12. $(d \wedge a) \rightarrow z$
    d
    $a \wedge (b \rightarrow p)$
    $\dfrac{z \rightarrow (x \vee c)}{\sim p \rightarrow c}$

For Exercises 13–16, express each argument symbolically. Then determine if the argument is valid or invalid. If it is invalid, produce a counterexample chart.

13. If I get married, I go on a honeymoon. If I go to Anguilla or St. Barts then I will be a June bride. If I do not go to Hawaii, then I will not be a June bride. If I go on a honeymoon, I go to Anguilla. Therefore, if I get married, I go to Hawaii.

14. It is not the case that I buy a flat screen television or go on a cruise. If I go on a cruise or buy a computer then if I get a raise I buy a flat screen television. Therefore, if I buy a computer then I get a raise.

15. If I lift weights, then I will be healthy and gain muscle mass. If I do not exercise, I will not be healthy. I either lift weights or do not hire a personal trainer. Therefore, if I hire a personal trainer then either I exercise or overeat.

16. You are not a member of the chess team, or you either do Sudoku number puzzles or surf the web. If you do Sudoku number puzzles then you have a lot of patience. You do not surf the web. Therefore, if you are a member of the chess team you do not have a lot of patience.

Problems

For Problems 17–28, test each argument for validity using the conditional proof technique. If the argument is invalid, produce a counterexample chart.

17. $\dfrac{\sim s \lor \sim p}{s \to \sim p}$

18. $\dfrac{(a \land \sim b) \lor c}{\sim (a \to \sim b) \to c}$

19. $k \lor \sim (a \lor \sim b)$
$\dfrac{b \to c}{\sim k \to (c \lor n)}$

20. $\sim d \lor \sim x$
$\dfrac{b \lor x}{\sim b \to \sim d}$

21. $\sim (a \lor m)$
$\dfrac{(m \lor d) \to (j \to a)}{d \to j}$

22. $(s \to \sim w) \to \sim p$
$\dfrac{s \to a}{p \to (a \land q)}$

23. $\sim (\sim p \lor s)$
$p \to (q \to l)$
$\dfrac{q \lor m}{\sim l \to m}$

24. $d \to \sim a$
$r \to (a \land \sim b)$
$\dfrac{d \lor \sim q}{q \to \sim r}$

25. $(p \wedge a) \to m$
    $\sim m \vee (c \wedge d)$
    $\underline{p}$
    $a \to \sim c$

26. $(p \vee \sim q) \to \sim r$
    $\underline{\sim m \to q}$
    $r \to m$

27. $n \to \sim g$
    $(p \vee q) \to r$
    $\sim w \to \sim r$
    $\underline{\sim g \to p}$
    $n \to w$

28. $\sim q$
    $(a \vee b) \to q$
    $\sim c \vee n$
    $\underline{c \to \sim (b \vee w)}$
    $w \to n$

For Problems 29–31, express each argument symbolically. Then determine if the argument is valid or invalid. If it is invalid, produce a counterexample chart.

29. If the comptroller mails out dividend checks, it is June 30 or December 31. If it is December 31, then tax statements are not mailed. The comptroller mails monthly statements or tax statements. The comptroller is not mailing monthly statements. Therefore, if the comptroller is mailing dividend checks then it is not June 30.

30. If I speak Spanish and Portuguese, I will take Italian lessons. Either I do not take Italian lessons or I visit Rome and Venice. I speak Spanish. Therefore, if I speak Portuguese, I visit Rome.

31. If I do not take singing lessons then I cannot be a thespian. If I take singing lessons and play the piano, I can be a rock singer. If I play the piano or the guitar, I can be a thespian. Therefore, if I play the piano, I can be a rock singer.

JUST FOR FUN
What is the next number in the sequence?

$$61, 52, 63, 94, 46, \underline{\quad\quad}$$

2.5 Testing for Validity Using the *TF* Method–The Indirect Approach

Introduction

We have seen that an argument can be either valid or invalid. There are no other possibilities. If an argument is invalid, it can be shown that the conclusion may be false even though the premises are true.

If we are unable to directly deduce an argument's validity because there does not seem to be a relationship between the premises, we must resort to other methods.

Indirect Approach

One of the alternate methods for testing for validity is the *indirect approach*. It is important to understand the strategy for an indirect approach. If we assume that the argument is *invalid*, we are assuming that it is possible for the conclusion to be *false* while every premise is true. If we are able to provide truth values for the variables such that the conclusion is false while all premises are true, then we will have proven that our argument is invalid. If we are unable to do this, because of any contradiction to the laws of logic, our assumption must have been incorrect, and thus the argument must be valid.

Example 1

Use an indirect approach to show that the following argument is valid.

$$p \rightarrow q$$
$$q \rightarrow r$$
$$\overline{p \rightarrow r}$$

Solution

Using the indirect approach, we begin by assuming the conclusion is *false*. The conclusion is an implication. The only way it can be false is if p is true and r is false. The first premise, $p \rightarrow q$, must be true, and since we deduced p was true from our indirect approach, q must be true. The second premise, $q \rightarrow r$, is true, and since q was deduced true, r must be true as well. However, our assumption led us to the fact that if the conclusion was assumed false, r had to be false. We now have r is true and r is false, clearly a contradiction. Therefore, the assumption must have been incorrect. Since the assumption was that the argument was invalid, and this led to a contradiction, the argument must be valid.

■

In the previous example, it would be impossible to produce a counterexample chart, since the argument was valid. If our indirect approach does not lead to a contradiction, we would strongly suspect that the argument was indeed invalid, and we would try to produce a counterexample chart.

Example 2

Use the indirect approach to test the argument below for validity.

$$p \rightarrow q$$
$$\underline{r \rightarrow q}$$
$$p \vee \sim r$$

Solution

Using the indirect approach, we assume that the conclusion is *false*. The conclusion is a disjunction. The only way a disjunction can be false is if both of its components are false. This means that p is false and $\sim r$ is false, which implies that r is true. Since the second premise, $r \rightarrow q$, is assumed true, and r is true, q must be true. We know p is false and q is true. This confirms the fact that the first premise, $p \rightarrow q$, is indeed true. We have *not* produced a contradiction. That is, it is possible to have true premises and a false conclusion. Therefore, our argument is invalid. We now produce a counterexample chart based on the discussion above.

p	q	r
F	T	T

∎

Example 3

Test the given argument for validity.

$$m \vee \sim n$$
$$n$$
$$r \rightarrow \sim p$$
$$\underline{m \rightarrow (\sim p \rightarrow q)}$$
$$\sim q \rightarrow \sim r$$

Solution

It appears that the direct *TF* approach is the method of choice. We begin with the first two premises. Knowing that n and $m \vee \sim n$ are both true implies that m must be true. Since m is the antecedent of the premise $m \to (\sim p \to q)$, we know that $\sim p \to q$ must be true. There are three ways for $\sim p \to q$ to be true, so we are at an impasse. This suggests considering an indirect approach. We therefore assume the conclusion, $\sim q \to \sim r$, is *false*. Since it is an implication, $\sim q$ must be true and $\sim r$ must be false. We know both n and m are true from our attempt at a direct approach. We also know $\sim p \to q$ is true. Since $\sim q$ was assumed true, q is false. The truth value of $\sim p$ must be false in order for the premise $\sim p \to q$ to remain true. Applying this fact to the third premise, $r \to \sim p$, r must be false. This means $\sim r$ is true. However, $\sim r$ was assumed false. Since $\sim r$ cannot be both true and false, we have arrived at a contradiction. Therefore, the argument is valid.

■

Example 4

Test the given argument for validity.

$$d \to r$$
$$(b \wedge c) \to d$$
$$\underline{a \wedge b}$$
$$r$$

Solution

Since we can begin our reasoning with the conjunction $a \wedge b$, it again appears that the direct *TF* approach is the method of choice. The conjunction $a \wedge b$ is true, so both a and b must be true. We try to incorporate the truth value of b into the second premise. But we can go no further since we do not know the truth value of c or d. Once again, this suggests considering an indirect approach. Therefore, we assume the conclusion, r, is *false*. For the first premise to remain true, d must be false. If the second premise $(b \wedge c) \to d$, is to remain true, $b \wedge c$ must be false. Since b is true, c is forced to be false. Using the indirect method, we have assigned truth values to all the variables such that the premises are true and the conclusion is false, and produced no contradictions in doing so. Hence, argument is invalid. Constructing a counterexample chart using the truth values obtained, we write:

a	b	c	d	r
T	T	F	F	F

■

In-Class Exercises and Problems for Section 2.5

In-Class Exercises

For Exercises 1–10, use the indirect approach to test each argument for validity. If the argument is invalid, produce a counterexample chart.

1. $(\sim m \rightarrow p) \rightarrow r$
   $\underline{\sim r \vee q}$
   q

2. $a \rightarrow b$
   $\underline{m \leftrightarrow a}$
   $\sim m \vee b$

3. $r \vee (\sim p \rightarrow s)$
   $\underline{\sim r \rightarrow (\sim s \vee p)}$
   $p \vee r$

4. p
   $\sim p \vee (\sim a \rightarrow b)$
   $\underline{\sim a \vee c}$
   c

5. $(p \vee q) \rightarrow r$
   $\sim r \vee s$
   $\underline{\sim p \rightarrow w}$
   $s \rightarrow w$

6. $(c \vee d) \leftrightarrow g$
   $a \rightarrow (b \vee c)$
   $\underline{a}$
   $\sim d \rightarrow g$

7. $w \wedge a$
   $\sim a \vee (c \rightarrow s)$
   $\underline{w \rightarrow (\sim s \vee n)}$
   $c \rightarrow n$

8. $w \rightarrow s$
   $(s \vee r) \rightarrow q$
   $\underline{\sim q \vee p}$
   $p \vee (w \rightarrow r)$

9. $w \wedge a$
   $r \rightarrow (\sim p \vee s)$
   $s \vee (w \rightarrow r)$
   $\underline{p \leftrightarrow a}$
   s

10. $\sim p \vee s$
    p
    $c \rightarrow (r \vee w)$
    $\underline{w \rightarrow (s \rightarrow r)}$
    $c \rightarrow r$

In Exercises 11–13, express each argument symbolically. Then determine if it is valid or invalid. If it is invalid, produce a counterexample chart.

11. If the dog barks, then she does not get a treat. If the dog rolls over, then she gets petted and praised. The dog barks or rolls over. Therefore, if you give the dog a treat, she gets praised.

12. If you e-mail the invitations, you get quick responses. Either you do not get quick responses, or you have time to plan for the party and call the caterers two weeks before the party. If you do not have to make telephone calls to guests then you have time to plan the party. Therefore, if you e-mail the invitations, you do not have to make telephone calls to the guests.

13. Jill can't afford to go to college. Here are the scholarship offers she knows she can count on. If college A gives a scholarship then college B will also give a scholorship but college C won't. If college C gives a scholarship, then college A will too. College B decided to give a scholarship. Can Jill determine that either college A or college C will also offer a scholarship?

Problems

For Problems 14–21, use the indirect approach to test each argument for validity. If the argument is invalid, produce a counterexample chart.

14. $s \rightarrow (c \vee m)$

$$\frac{c \wedge h}{s \rightarrow \sim b}$$

15. $p \vee q$

$$\frac{\sim r \vee \sim q}{\sim p \rightarrow \sim r}$$

16. $(a \vee b) \rightarrow c$

$$\frac{}{b \rightarrow (c \vee d)}$$

17. $g \rightarrow (w \wedge x)$

$$\frac{}{g \rightarrow x}$$

18. $p \rightarrow k$
    $\sim p \rightarrow (w \wedge s)$
    $\dfrac{\sim s \vee p}{k}$

19. $a \rightarrow d$
    $\sim d \vee p$
    $\dfrac{p \rightarrow \sim c}{(a \wedge b) \rightarrow c}$

20. $p \rightarrow (d \rightarrow \sim z)$
    $p \rightarrow q$
    $c \vee d$
    $\dfrac{\sim q}{z \vee c}$

21. $\sim p \rightarrow \sim n$
    $r \rightarrow \sim s$
    n
    $\dfrac{\sim p \vee (\sim s \rightarrow q)}{q \vee r}$

In Problems 22–25, express each argument symbolically. Then determine its validity. If it is invalid, produce a counterexample chart.

22. If you pay your taxes and obey the law then you are a good citizen. If you are a good citizen, you drive within the speed limit. If you do not drive within the speed limit, then either you are not a good citizen or you didn't obey the law. Therefore, if you pay your taxes, you drive within the speed limit.

23. If you check your luggage then if you go on a long trip you must arrive at the airport two hours before the flight departs. You check your luggage or you leave your car in long-term parking. It is not the case that you go on a long trip and leave your car in long-term parking. Therefore, you arrive at the airport two hours before the flight departs or you do not go on a long trip.

24. If Grandma makes lasagna or ziti then Grandpa is happy. If Grandpa is not happy or Grandma makes meatballs, then Grandpa complains. Either Grandpa does not complain or Grandma makes ziti. Therefore, if Grandma makes meatballs, Grandpa is happy.

25. If I buy a new car, we carpool to work. If we carpool to work then I pick you up but you eat breakfast at home. If you take the train to work then either you do not eat breakfast at home or you wake up an hour earlier. Either I buy a new car and you wake up an hour earlier, or I do not buy a new car but you take the train to work. Therefore, either I pick you up or you wake up an hour earlier.

Check Your Understanding

For questions 1–4, use the indirect approach or the conditional proof technique to determine whether each argument is valid or invalid. If it is valid, circle the word valid. If it is invalid, produce a counterexample chart.

1. $p \wedge z$
   $w \rightarrow (p \vee k)$
   _____
   $w \rightarrow \sim k$

   valid   or $\begin{array}{c|c|c|c} k & p & w & z \\ \hline & & & \end{array}$

2. $\sim h \vee s$
   h
   $m \rightarrow (r \vee w)$
   $w \rightarrow (s \rightarrow r)$
   _____
   $\sim m \vee r$

   valid   or $\begin{array}{c|c|c|c|c} h & m & r & s & w \\ \hline & & & & \end{array}$

3. $a \rightarrow s$
   $(s \vee r) \rightarrow n$
   $\sim n \vee p$
   _____
   $p \vee (a \rightarrow r)$

   valid   or $\begin{array}{c|c|c|c|c} a & n & p & r & s \\ \hline & & & & \end{array}$

4. $(r \vee \sim m) \rightarrow \sim d$
   $(r \rightarrow n) \leftrightarrow m$
   $\sim a \vee d$
   _____
   $a \rightarrow (m \rightarrow \sim n)$

   valid   or $\begin{array}{c|c|c|c|c} a & d & m & n & r \\ \hline & & & & \end{array}$

For question 5, express the argument symbolically. Then determine if it is valid or invalid. If it is valid, write the word valid. If it is invalid, produce a counterexample chart.

5. If you graduate from <u>h</u>igh school either you will go to <u>c</u>ollege or go to <u>w</u>ork. If you go to work, you can either buy season tickets to the <u>Y</u>ankees or buy a <u>n</u>ew car. You do not buy a new car. Therefore, either you go to college or buy season tickets to the Yankees.

Chapter 2 Review

For questions 1–10, answer true, false or can't be determined.

1. Let s be a premise. What is the truth value of $s \lor w$?

2. Suppose $p \leftrightarrow q$ is true and p is a premise. What is the truth value of $\sim q \to w$?

3. If $\sim p \lor (r \land s)$ is a premise and $w \to s$ is false, what is the truth value of p?

4. If both $p \to q$ and $(p \to q) \to \sim r$ are premises, what is the truth value of r?

5. If $p \land (q \lor r)$ is a premise, what is the truth value of $p \land q$?

6. Suppose $(r \lor w) \land s$ and $s \to \sim w$ are both premises, what is the truth value of r?

7. If both $s \leftrightarrow (q \lor r)$ and $\sim s$ are premises, what is the truth value of $r \to p$?

8. If $(r \lor s) \to \sim p$ and $\sim r \to s$ are both premises, what is the truth value of p?

9. Suppose $\sim (s \to w) \to (q \lor a)$ and $s \land \sim w$ are premises. What is the truth value of a?

10. Let $r \to s$ and $\sim r \lor \sim s$ be premises. What is the truth value of r?

For question 11–20, use the *TF* method to determine if each argument is valid or invalid. If it is invalid, produce a counterexample chart.

11. $w \to \sim p$	12. $q \land \sim a$
$\quad \sim s \to p$	$\quad a \lor \sim p$
$\quad \sim (\sim r \lor s)$	$\quad \sim p \to (q \lor b)$
$\quad\quad\quad \sim w$	$\quad\quad\quad b$

13. $\sim p$	14. $\sim p \lor c$
$\quad \sim (p \leftrightarrow w)$	$\quad q \to \sim c$
$\quad s \to w$	$\quad p$
$\quad\quad\quad s$	$\quad \sim q \to (w \land r)$
	$\quad\quad\quad s \to r$

15. $b \rightarrow (a \wedge \sim c)$

$\sim a \vee p$

$p \rightarrow d$

b

———————

$d \wedge \sim c$

16. $(m \vee n) \rightarrow p$

$\sim p$

$n \leftrightarrow a$

———————

$m \vee a$

17. $(w \vee s) \rightarrow \sim q$

$q \wedge p$

$a \rightarrow w$

$\sim a \rightarrow (b \vee p)$

———————

b

18. $s \wedge w$

$(d \rightarrow h) \leftrightarrow \sim s$

$\sim a \vee h$

$a \vee (b \wedge c)$

———————

$c \vee z$

19. $b \wedge c$

$(p \vee q) \rightarrow \sim w$

$(\sim p \vee r) \rightarrow (s \vee \sim c)$

$(b \vee d) \rightarrow w$

———————

s

20. a

$\sim r \vee (w \wedge s)$

$(a \vee b) \rightarrow r$

$(m \wedge \sim s) \leftrightarrow \sim w$

———————

m

In questions 21–28, use your knowledge of equivalences to test the validity of each argument. If the argument is invalid, produce a counterexample chart.

21. $(s \rightarrow w) \rightarrow r$

$r \rightarrow (a \wedge b)$

$\sim s \vee w$

———————

$c \rightarrow b$

22. $a \wedge r$

$(p \wedge \sim q) \rightarrow \sim r$

$(\sim p \vee q) \rightarrow w$

———————

$w \vee k$

23. $(\sim w \rightarrow p) \wedge \sim q$

$b \rightarrow (\sim q \vee s)$

$(\sim p \rightarrow w) \rightarrow b$

———————

s

24. $b \rightarrow (a \wedge \sim c)$

$\sim a \vee c$

$p \rightarrow \sim b$

———————

p

25. $\sim (a \rightarrow b)$

    $c \leftrightarrow (\sim b \vee p)$

    $\dfrac{(w \vee a) \rightarrow c}{w}$

26. $m \rightarrow n$

    $(\sim m \vee n) \rightarrow (\sim a \wedge s)$

    $(a \vee \sim s) \vee w$

    $\dfrac{(w \vee b) \rightarrow c}{c}$

27. $(a \wedge b) \rightarrow (\sim p \vee q)$

    $(p \rightarrow q) \leftrightarrow w$

    $\sim (a \rightarrow \sim b)$

    $\dfrac{c \rightarrow \sim (w \vee s)}{\sim c}$

28. $\sim (p \vee \sim s)$

    $(\sim c \rightarrow k) \rightarrow d$

    $d \rightarrow (a \wedge \sim b)$

    $\dfrac{(s \vee r) \leftrightarrow (\sim k \rightarrow c)}{\sim b}$

In questions 29–36, use the indirect approach or the conditional proof approach (when appropriate) to test the validity of each argument. If the argument is invalid, produce a counterexample chart.

29. $s \vee p$

    $w \vee q$

    $\dfrac{s \vee \sim w}{p \vee q}$

30. $a \vee s$

    $w \rightarrow (p \rightarrow a)$

    $\dfrac{\sim s \vee (p \wedge w)}{a}$

31. $p \wedge q$

    $s \rightarrow (\sim q \vee r)$

    $\dfrac{p \vee s}{\sim s}$

32. $(\sim c \vee q) \rightarrow a$

    $\sim a \vee p$

    $\dfrac{p \rightarrow r}{\sim c \rightarrow r}$

33. $(m \vee p) \rightarrow s$

    $s \rightarrow (q \wedge w)$

    $\sim a \rightarrow p$

    $\dfrac{m \leftrightarrow a}{q}$

34. $p \wedge \sim s$

    $\sim r \rightarrow \sim q$

    $\sim a \leftrightarrow (c \vee q)$

    $\dfrac{p \rightarrow (a \rightarrow s)}{c \vee r}$

35. $p \rightarrow (\sim q \rightarrow r)$

    $\sim r$

    $r \vee s$

    $\dfrac{(s \vee a) \leftrightarrow (p \vee w)}{\sim w \rightarrow q}$

36. $(a \wedge b) \wedge (c \rightarrow \sim b)$

    $(p \leftrightarrow r) \rightarrow w$

    $\sim c \rightarrow (p \vee q)$

    $\dfrac{w \rightarrow (\sim a \vee c)}{p \vee w}$

In questions 37–48, represent each argument symbolically. Then determine if it is valid or invalid. If it is invalid, produce a counterexample chart.

37. We will go to Disneyworld or to Las Vegas. If we go to Disneyworld we will stand on long lines. If we go to Las Vegas we stay up late. We don't stay up late. Therefore, we stand on long lines.

38. Callie is a good worker and she is a whiz at the computer. If Callie is a whiz at the computer then she can either diagnose and fix any problem or reprogram the computer. If she can fix any problem then her reports are not late. Therefore, if Callie can reprogram the computer, her reports are not late.

39. If your alarm clock rings you will not be late for work. If either your car does not start or you lose your electricity, you will be late for work. It is not true that both your car starts and you do not lose electricity. Therefore, your alarm clock does not ring.

40. It is not the case that if I receive a paycheck then I spend it all. I go to the bank during my lunch hour if I receive a paycheck. If I go to the bank during my lunch hour then either I put money in my saving account or my checking account. I don't put money in my saving account. Therefore, I put money into my checking account if and only if I don't spend it all.

41. Either you stop at a service station or your car runs properly. If you either need gas or the gas gauge is broken, you have to stop at a service station. If you do not need gas then your gas gauge is broken. Therefore, your car does not run properly.

42. If Noah takes the day off, then he either goes out with Averi or goes out on his boat. If Noah works overtime, he does not go out with Averi. Either Noah works overtime, or he takes the day off but makes less money. Therefore, if Noah goes out on his boat, he makes less money.

43. Either Scott did not go to work today or it is not the case that if Scott goes to work early then the job was done on time. If Scott's entire department is out sick, then he did go to work today. Therefore, if Scott did not go to work early his entire department is not out sick.

44. If the neighborhood market closes then I drive further and shop in the next town. If I drive further then I spend more money on gas and cannot buy all the groceries I need. Either the neighborhood market closes or I get home early. I did not get home early. Therefore, I spend more money on gas.

45. If Gary puts his white clothes in the washing machine then he uses bleach. If he puts white clothes and red clothes in the washing machine, then some of his clothes turn pink. If some of his clothes turn pink then he uses bleach. Either Gary puts his red clothes in the washing machine or he does not use bleach. Therefore, Gary does not put his white clothes in the washing machine.

46. If the President comes to town, then there will be gridlock and the National Guard will be called. If the National Guard is called it will take twice as long to get home. If it takes twice as long to get home then either you will be very hungry or your dinner will get cold. Therefore, if the President comes to town you will be very hungry.

47. The computer program either does a payroll update or an inventory update. If the program does a payroll update it calculates federal tax and state tax. If the program does an inventory update, we cannot use the computer for three hours. We can't use the computer for three hours and the program did not compute federal tax. Therefore, the program computes state tax.

48. She is a Democrat or a Republican. If she is a Republican, she wants less government control and a revised tax plan. She does not want less government control. Therefore, she is a Democrat.

Sample Exam: Chapter 2

For questions 1–5, answer true, false or can't be determined.

1. If $\sim(\sim r \rightarrow w)$ is a premise, then what is the truth value of $r \rightarrow (s \vee w)$?
2. If $\sim(r \vee \sim w)$ is a premise, then what is the truth value of $\sim(\sim r \leftrightarrow w)$?
3. If $\sim r \wedge w$ is a premise, then what is the truth value of $\sim w \vee p$?
4. If $(w \rightarrow \sim r) \vee p$ is false, what is the truth value of $w \rightarrow p$?
5. If $\sim[(r \wedge w) \leftrightarrow s]$ is a premise and r is false, then what is the truth value of $s \vee p$?

For questions 6–9, construct a counterexample chart for each invalid argument.

6. $(a \wedge b) \vee c$
   $\sim c$
   $\dfrac{(a \vee d) \rightarrow (e \vee g)}{d \rightarrow g}$

7. $(p \vee q) \rightarrow r$
   $\sim r \wedge s$
   $\dfrac{\sim p \rightarrow (a \vee b)}{s \leftrightarrow b}$

8. p
   $p \rightarrow (q \vee r)$
   $\dfrac{q \rightarrow s}{s \vee a}$

9. $p \rightarrow q$
   $(p \rightarrow q) \rightarrow (r \vee s)$
   $\dfrac{r \rightarrow w}{q \rightarrow w}$

For questions 10–21, use the *TF* method to determine if each argument is valid or invalid. If it is invalid, produce a counterexample chart.

10. $a \vee (b \rightarrow c)$
    $c \rightarrow (\sim w \vee r)$
    $\dfrac{\sim a \wedge b}{w \rightarrow r}$

11. $(m \rightarrow n) \rightarrow q$
    $\sim q$
    $\dfrac{(\sim n \vee w) \rightarrow s}{a \rightarrow s}$

12. $(z \vee m) \rightarrow \sim n$
    $\sim z \rightarrow \sim (p \rightarrow q)$
    $\underline{n \wedge y}$
    $\sim p$

13. $p \wedge q$
    $(q \vee r) \rightarrow \sim s$
    $\underline{s \rightarrow w}$
    w

14. $(\sim q \wedge \sim r) \rightarrow a$
    $(a \vee b) \rightarrow c$
    $\underline{\sim (q \vee r)}$
    c

15. $\sim p \vee w$
    $(p \rightarrow w) \rightarrow \sim r$
    $\underline{r \vee s}$
    $s \rightarrow w$

16. $(p \rightarrow q) \rightarrow \sim a$
    $q \vee s$
    $a \wedge b$
    $\underline{(s \vee w) \rightarrow r}$
    $r \leftrightarrow b$

17. $\sim a \vee b$
    $\sim c \vee d$
    $(a \rightarrow b) \rightarrow c$
    $\underline{b \rightarrow d}$
    $\sim b$

18. $p \rightarrow (s \wedge r)$
    $a \leftrightarrow b$
    $\underline{(a \wedge c) \vee p}$
    $s \vee b$

19. $(\sim p \vee \sim s) \wedge r$
    $w \vee q$
    $q \rightarrow (p \wedge s)$
    $\underline{(c \wedge d) \rightarrow \sim w}$
    $\sim c$

20. a
    $(b \vee d) \rightarrow (p \rightarrow \sim q)$
    $\sim a \leftrightarrow (b \rightarrow c)$
    $\underline{(q \rightarrow \sim p) \rightarrow m}$
    m

21. $s \rightarrow \sim p$
    $r \rightarrow (s \vee q)$
    $(r \rightarrow s) \leftrightarrow \sim c$
    $\underline{(\sim s \vee \sim p) \rightarrow c}$
    q

In questions 22–26, express each argument symbolically. Then use the *TF* method to determine if it is valid or invalid. If it is invalid, produce a counterexample chart.

22. If the butler was working the night of the crime, then he was a suspect. If he was not working the night of the crime, then the victim was not poisoned. The detective found his girlfriend locked in a closet and ruled the butler was not a suspect. Therefore, the victim was not poisoned.

23. If the Yankees win the baseball game then either the pitcher had nine strikeouts or the defense was outstanding. If the defense was not outstanding the crowd did not cheer. The crowd cheered. Therefore, the Yankees won.

24. If I go to college I will get a job. If I get a job then the company will pay for my master's degree. If the company pays for my master's degree then I can't work overtime. I worked overtime. Therefore, I didn't go to college.

25. It is not true that if you live in Idaho then you like potatoes. Either you like potatoes or you do not live on a farm. If you like horseback riding then you live on a farm. Therefore, you either don't like horseback riding or you like milking cows.

26. If you live in a glass house you shouldn't throw stones. If you throw stones then either you will break a window or the neighbors will call the police. If the neighbors call the police you will have to do community service. Therefore, if you live in a glass house you have to do community service.

JUST FOR FUN

I went to visit my friend Melissa. Melissa has three children and challenged me to guess their ages. I said that I needed some information.

First, Melissa said, "The product of their ages is 36."

I said, "I don't have enough information to know their ages."

Melissa then said, "The sum of their ages is the number on the house next door."

I went outside and looked at the house number next door. When I returned, I said, "I still don't know their ages."

Then Melissa said, "My oldest child, Averi, has brown hair."

I said, "Ah ha! Now I know their ages!"

How old are Melissa's children?

CHAPTER THREE

FORMAL PROOFS

In Chapter Two, we explored arguments in an intuitive way. We now turn our attention to a more formal presentation that is used to prove an argument's validity.

Objectives

After completing Chapter Three, the student should be able to:

- Identify and apply valid argument forms.

- Construct a formal proof of a valid argument utilizing the applicable equivalences and valid argument forms.

- Use conditional proof methods and indirect methods to prove that an argument is valid.

3.1 Valid Argument Forms–Part I

Introduction

To formally prove that arguments are valid, we use tautologies extensively. While there are a rather large number of tautologies at our disposal, it turns out that only a few of them are frequently used when constructing formal proofs.

We now turn our attention to a group of tautologies that are implications. These conditional tautologies are often called *valid argument forms*. A concise listing of all the valid argument forms we study in this chapter can be found on page 228.

Modus Ponens

The first valid argument form we will consider is called *modus ponens,* abbreviated as MP. This tautology can be represented symbolically as $[(p \rightarrow q) \wedge p] \rightarrow q$.

We may also express modus ponens in its argument form. If we think of the two components of the antecedent, $p \rightarrow q$ and $p,$ as premises of an argument, and q as the conclusion, we can express the tautology as:

$$p \rightarrow q$$
$$\underline{p}$$
$$q$$

Examining this tautology, we see that if one premise of an argument is an implication and the other premise is the LHS of this implication, we can conclude that the RHS of the implication must be true. It is critical to recognize the pattern for modus ponens. Modus ponens deals only with an argument whose premises are an implication and its LHS. When this occurs, we are guaranteed that the RHS of the implication must be true.

To verify that modus ponens is indeed a tautology, we will construct a truth table.

p	q	$p \rightarrow q$	$(p \rightarrow q) \wedge p$	$[(p \rightarrow q) \wedge p] \rightarrow q$
T	T	T	T	T
T	F	F	F	T
F	T	T	F	T
F	F	T	F	T

Example 1

Is the argument an illustration of modus ponens?

$$q \rightarrow p$$

$$q$$

$$p$$

Solution

The premises of the argument are an implication and its LHS. The conclusion is the RHS of the implication. This is exactly the pattern for modus ponens. ∎

Example 2

Is the given argument an illustration of modus ponens?

$$\sim p \rightarrow q$$

$$\sim p$$

$$q$$

Solution

Do not let the appearance of the negation symbol lead you astray. The premises of the argument are an implication and its LHS. The conclusion is the RHS of the implication. This is exactly the pattern for modus ponens. ∎

Example 3

Is the following argument an illustration of modus ponens?

$$p \rightarrow q$$

$$q$$

$$p$$

Solution

Since the premises of the argument are an implication and its *RHS*, this argument is not an illustration of modus ponens. ∎

Example 4

Is the given argument an illustration of modus ponens?

$$(\sim r \vee s) \rightarrow (w \rightarrow b)$$

$$\sim r \vee s$$

$$w \rightarrow b$$

Solution

Do not let the appearance of compound statements confuse you. The premises of the argument are an implication and its LHS. The conclusion is the RHS of the implication. This is exactly the pattern for modus ponens.

∎

Example 5

Is the given argument an illustration of modus ponens?

$$(\sim r \vee s) \to (w \to b)$$
$$\underline{r \vee \sim s}$$
$$w \to b$$

Solution

In this case, one of the premises is an implication. However, the second premise is not its LHS. Therefore, this is not an illustration of modus ponens.

∎

Modus Tollens

The *modus tollens* tautology, abbreviated as MT, can be represented symbolically as $[(p \to q) \wedge \sim q] \to \sim p$. We may also express modus tollens in its argument form. If we think of the two components of the antecedent, $p \to q$ and $\sim q$, as premises of an argument, and $\sim p$ as the conclusion, we can express the tautology as:

$$p \to q$$
$$\underline{\sim q}$$
$$\sim p$$

Examining this tautology, we see that if one premise of an argument is an implication and the other premise is the negation of its RHS, we can conclude that the negation of the LHS of the implication must be true. Again, recognition of the modus tollens pattern is crucial.

As with modus ponens, we will use a truth table to verify that modus tollens is a tautology.

p	q	$p \to q$	$(p \to q) \wedge \sim q$	$[(p \to q) \wedge \sim q] \to \sim p$
T	T	T	F	T
T	F	F	F	T
F	T	T	F	T
F	F	T	T	T

Example 6

Is the argument below an illustration of modus tollens?

$$\sim p \rightarrow q$$
$$\underline{\sim q}$$
$$p$$

Solution

The premises of the argument are an implication and the negation of its RHS. The conclusion of the argument is the negation of the LHS of the implication. Therefore, the argument is an illustration of modus tollens.

■

Example 7

Is the argument below an illustration of modus tollens?

$$p \rightarrow (q \wedge r)$$
$$\underline{\sim (q \wedge r)}$$
$$\sim p$$

Solution

The premises of the argument are an implication and the negation of its RHS. We are not concerned that the RHS is a conjunction. The conclusion of the argument is the negation of the LHS of the implication. This is exactly the pattern required for us to conclude that the argument is valid because it is an illustration of the modus tollens tautology.

■

Example 8

Is the argument shown below an illustration of modus ponens, modus tollens or neither of these tautologies?

$$\sim p \rightarrow \sim q$$
$$\underline{\sim q}$$
$$\sim p$$

Solution

For the argument to have the modus ponens form, we must have an implication and its LHS as premises. This is not the case here. If the tautology is of the modus tollens form, it must have an implication and the negation of its RHS as premises. This is not the case either. This argument form is neither modus ponens nor modus tollens. In fact, the argument presented is not a tautology, and thus, is invalid.

■

To show a formal proof for an argument's validity, we construct a sequence of *true* statements, as well as reasons for their truth, in table form. A formal proof for the argument in Example 7 would be:

Statement	Reason
1. $p \rightarrow (q \wedge r)$	Premise
2. $\sim(q \wedge r)$	Premise
3. $\sim p$	Modus Tollens (1)(2)

The numbers after the words modus tollens refer to the statements that were used to deduce ~p.

Example 9

Construct a formal proof for the following valid argument.

$$r \rightarrow s$$

$$\sim s$$

$$\sim r \rightarrow p$$

$$\overline{\qquad p \qquad}$$

Solution

We record only true statements, beginning with the simplest premise.

Statement	Reason
1. $\sim s$	Premise

Since the first premise is an implication containing an *s*, we try to use either modus ponens or modus tollens in combination with ~*s*. To use modus tollens, we need an implication and the negation of its RHS to be true. We know that ~*s* is true, so we can continue our formal proof and show that ~*r* is true.

Statement	Reason
1. $\sim s$	Premise
2. $r \rightarrow s$	Premise
3. $\sim r$	Modus Tollens (1)(2)

Now that we know ~*r* is true, we combine it with the third premise ~ *r* → *p* and use modus ponens to conclude that *p* is true.

Statement	Reason
1. ~ *s*	Premise
2. *r* → *s*	Premise
3. ~ *r*	Modus Tollens (1)(2)
4. ~ *r* → *p*	Premise
5. *p*	Modus Ponens (3)(4)

This completes the proof.   ∎

In-Class Exercises and Problems for Section 3.1

In-Class Exercises

In Exercises 1–8, state the valid argument form shown.

1. ~ *s* → ~ *k*

   k

   ─────

   s

2. ~ *r* → *w*

   ~ *r*

   ─────

   w

3. *d* → (*w* ∨ *b*)

   ~ (*w* ∨ *b*)

   ──────────

   ~ *d*

4. (*c* ∧ *h*) → ~ *l*

   l

   ──────────

   ~ (*c* ∧ *h*)

5. (*d* → *c*) → ~ *p*

   d → *c*

   ──────────

   ~ *p*

6. ~ *j* → ~ (*s* ∨ ~ *p*)

   ~ *j*

   ──────────

   ~ (*s* ∨ ~ *p*)

$$7.\ (m \lor n) \to \sim (r \to s)$$
$$\frac{r \to s}{\sim (m \lor n)}$$

$$8.\ \sim (w \to c) \to (a \land b)$$
$$\frac{\sim (a \land b)}{w \to c}$$

In Exercises 9–18, use the *TF* method to test each argument for validity. If it is valid, show a formal proof. If invalid, provide a counterexample chart.

$$9.\ \sim a$$
$$\sim a \to b$$
$$\frac{c \to \sim b}{\sim c}$$

$$10.\ r \to p$$
$$s \to w$$
$$\sim r \to s$$
$$\frac{\sim w}{p}$$

$$11.\ \sim r \to (a \lor b)$$
$$\sim (a \lor b) \to w$$
$$\frac{\sim w}{r}$$

$$12.\ \sim q \to s$$
$$s \to (p \lor r)$$
$$\frac{\sim (p \lor r)}{q}$$

$$13.\ (p \to q) \to (r \to s)$$
$$(r \to s) \to (m \to b)$$
$$\frac{p \to q}{m \to b}$$

$$14.\ \sim (w \land s)$$
$$k \to (w \land s)$$
$$\frac{p \to \sim k}{\sim p}$$

$$15.\ \sim w \to (c \lor s)$$
$$\sim r \to \sim p$$
$$p \to \sim w$$
$$\frac{\sim (c \lor s)}{r}$$

$$16.\ (r \lor \sim s) \to p$$
$$c \to (r \lor \sim s)$$
$$\sim p$$
$$\frac{\sim c \to d}{d}$$

17. $\sim(p \vee r) \rightarrow s$
    $\sim(a \rightarrow c)$
    $s \rightarrow (k \wedge g)$
    $\dfrac{(p \vee r) \rightarrow (a \rightarrow c)}{k \wedge g}$

18. $k \rightarrow (l \vee p)$
    $(r \rightarrow w) \rightarrow q$
    $\sim(l \vee p)$
    $\dfrac{(r \rightarrow w) \rightarrow k}{\sim q}$

Problems

For each formal proof in Problems 19–23, supply the reason that justifies each statement.

19. $g \rightarrow \sim p$
    g
    $\dfrac{\sim p \rightarrow n}{n}$

Statement	Reason
1. $g \rightarrow \sim p$	1.
2. g	2.
3. $\sim p$	3.
4. $\sim p \rightarrow n$	4.
5. n	5.

20. $r \rightarrow (p \vee \sim s)$
    $q \rightarrow r$
    $\dfrac{\sim(p \vee \sim s)}{\sim q}$

Statement	Reason
1. $\sim(p \vee \sim s)$	1.
2. $r \rightarrow (p \vee \sim s)$	2.
3. $\sim r$	3.
4. $q \rightarrow r$	4.
5. $\sim q$	5.

21. $(s \wedge b) \rightarrow k$
    $\sim(s \wedge b) \rightarrow c$
    $\dfrac{\sim k}{c}$

Statement	Reason
1. $(s \wedge b) \rightarrow k$	1.
2. $\sim k$	2.
3. $\sim(s \wedge b)$	3.
4. $\sim(s \wedge b) \rightarrow c$	4.
5. c	5.

22. $(a \rightarrow b) \rightarrow \sim w$
    $p \rightarrow \sim (r \vee s)$
    $a \rightarrow b$
    $\sim w \rightarrow (r \vee s)$
    _____
    $\sim p$

Statement	Reason
1. $a \rightarrow b$	1.
2. $(a \rightarrow b) \rightarrow \sim w$	2.
3. $\sim w$	3.
4. $\sim w \rightarrow (r \vee s)$	4.
5. $r \vee s$	5.
6. $p \rightarrow \sim (r \vee s)$	6.
7. $\sim p$	7.

23. $(s \vee w) \rightarrow \sim d$
    $g \rightarrow (c \rightarrow h)$
    d
    $\sim (s \vee w) \rightarrow g$
    _____
    $c \rightarrow h$

Statement	Reason
1. d	1.
2. $(s \vee w) \rightarrow \sim d$	2.
3. $\sim (s \vee w)$	3.
4. $\sim (s \vee w) \rightarrow g$	4.
5. g	5.
6. $g \rightarrow (c \rightarrow h)$	6.
7. $c \rightarrow h$	7.

In Problems 24–29, construct a formal proof for each valid argument.

24. $r \rightarrow \sim s$
    r
    _____
    $\sim s$

25. $\sim r \rightarrow s$
    $\sim s$
    _____
    r

26. $p \rightarrow (q \vee r)$
    p
    _____
    $q \vee r$

27. $s \rightarrow (p \rightarrow r)$
    $\sim (p \rightarrow r)$
    _____
    $\sim s$

28. $p \rightarrow r$

    $r \rightarrow q$

    $\underline{p}$

    q

29. $p \rightarrow r$

    $r \rightarrow q$

    $\underline{\sim q}$

    $\sim p$

In Problems 30–39, use the *TF* method to test each argument for validity. If it is valid, construct a formal proof. If invalid, produce a counterexample chart.

30. $p \rightarrow r$

    $q \rightarrow \sim r$

    $\underline{p}$

    $\sim q$

31. $p \rightarrow q$

    $\sim q$

    $\underline{p \rightarrow r}$

    r

32. $(p \rightarrow q) \rightarrow r$

    $\sim s \rightarrow \sim r$

    $\underline{p \rightarrow q}$

    s

33. $w \rightarrow (p \wedge a)$

    $\sim (p \wedge a)$

    $\underline{\sim w \rightarrow k}$

    k

34. $(p \wedge q) \rightarrow (r \vee s)$

    $\sim (r \vee s)$

    $\underline{\sim (p \wedge q) \rightarrow m}$

    m

35. $a \rightarrow (d \vee g)$

    $(d \vee g) \rightarrow \sim m$

    a

    $\underline{\sim r \rightarrow \sim m}$

    r

36. $\sim a$

    $\sim s \rightarrow (p \leftrightarrow q)$

    $s \rightarrow (r \vee w)$

    $\underline{(r \vee w) \rightarrow a}$

    $p \leftrightarrow q$

37. $(c \vee d) \rightarrow q$

    $q \rightarrow \sim r$

    $a \rightarrow r$

    $b \rightarrow a$

    $\underline{c \vee d}$

    $\sim b$

38. $r \rightarrow w$

$\quad p \rightarrow \sim(r \rightarrow w)$

$\quad (s \vee q) \rightarrow b$

$\quad \underline{\sim p \rightarrow (s \vee q)}$

$\quad\quad\quad b$

39. $(m \vee n) \rightarrow w$

$\quad \sim(a \rightarrow b) \rightarrow (p \wedge q)$

$\quad \sim(m \vee n) \rightarrow (p \wedge q)$

$\quad \underline{\sim w}$

$\quad\quad\quad a \rightarrow b$

In Problems 40–43, each set of premises is followed by four possible conclusions. State the conclusion(s) that makes the argument valid. There may be more than one correct conclusion. Be sure to state all that apply.

40. p

$\quad \sim q \rightarrow (w \vee r)$

$\quad p \rightarrow \sim(a \leftrightarrow b)$

$\quad \underline{q \rightarrow (a \leftrightarrow b)}$

$\quad$ a. q $\quad$ b. $w \vee r$

$\quad$ c. w $\quad$ d. $a \leftrightarrow b$

41. $m \rightarrow p$

$\quad \sim n \rightarrow \sim(m \rightarrow p)$

$\quad n \rightarrow (s \vee q)$

$\quad \underline{\sim w \rightarrow \sim(s \vee q)}$

$\quad$ a. m $\quad$ b. $s \vee q$

$\quad$ c. w $\quad$ d. $\sim n$

42. If we invite Joshua we can't invite Shira. If we don't invite Shira, then we can't invite Reuben. If we don't invite Reuben, then we can invite Maxine. We will invite Joshua. Therefore,

$\quad$ a. we can't invite Shira. $\quad\quad$ b. we can't invite Reuben.

$\quad$ c. we can't invite Maxine. $\quad\quad$ d. we can invite Maxine.

43. If it rains, the crops will grow. If the winds increase, it will not be sunny. If the crops grow we will have a good harvest. It will be sunny if it doesn't rain. We did not have a good harvest. Therefore,

$\quad$ a. the winds did not increase. $\quad\quad$ b. the crops grew.

$\quad$ c. it did not rain. $\quad\quad$ d. it was not sunny.

In Problems 44–47, express each argument symbolically. If the argument is valid, construct a formal proof. If not, produce a counterexample chart.

44. If there is a south wind, there will not be rain. If there is no increase in barometric pressure, there will be rain. There is a south wind. Therefore, there is an increase in barometric pressure.

45. If the butler or the cook is awake then the kitchen light is on. If the kitchen light is on, then the dog does not bark. If there is a prowler, the dog barks. If there is no prowler, the maid is in the hall and she stole the silver. The butler or the cook is awake. Therefore, the maid is in the hall and she stole the silver.

46. If my alarm doesn't ring in the morning I will have to take a later train to work. If my alarm does ring then I will get up early and not oversleep. If I get up early and do not oversleep then I will have time for my morning coffee. I didn't have time for my morning coffee. Therefore, I took the later train to work.

47. If Harvey is your friend then your friend is a rabbit. If Harvey is your friend, you watch too much television. If your friend is a rabbit, then you have an active imagination. You don't have an active imagination. Therefore, you don't watch too much television.

DID YOU KNOW?

Modus Ponens, or MP, is the abbreviated form of modus ponendo ponens. In logic, this means "the mode that affirms by affirming." This is the valid form of an argument in which the antecedent (LHS) of a conditional statement is true, thereby concluding the truth of the consequent (RHS).

Modus Tollens, or MT, is the abbreviated form of modus tollendo tollens. It means "the mode that denies by denying." This is the valid form of an argument in which the consequent (RHS) of a conditional statement is false thus implying the negation of the antecedent (LHS).

Check Your Understanding

For question 1, supply the reason that justifies each statement.

1. $(w \vee a) \rightarrow \sim p$

    $s \rightarrow \sim (c \rightarrow d)$

    $w \vee a$

    $\sim p \rightarrow (c \rightarrow d)$

    ―――――――――

    $\quad \sim s$

Statement	Reason
1. $w \vee a$	1._____
2. $(w \vee a) \rightarrow \sim p$	2._____
3. $\sim p$	3._____
4. $\sim p \rightarrow (c \rightarrow d)$	4._____
5. $c \rightarrow d$	5._____
6. $s \rightarrow \sim (c \rightarrow d)$	6._____
7. $\sim s$	7._____

For question 2, construct a formal proof for the valid argument shown below.

2. $a \rightarrow b$

    $p \rightarrow \sim n$

    $w \rightarrow \sim (a \rightarrow b)$

    $\sim w \rightarrow p$

    ―――――――――

    $\quad \sim n$

Statement	Reason

3.2 Valid Argument Forms-Part II

Disjunctive Syllogism

The third valid argument form we will consider is called *disjunctive syllogism*, abbreviated as DS. This tautology can be represented symbolically as $[(p \vee q) \wedge \sim p] \to q$ or $[(p \vee q) \wedge \sim q] \to p$.

In its argument form, we can present the disjunctive syllogism tautology as:

$$\begin{array}{ccc} p \vee q & & p \vee q \\ \underline{\sim p} & \text{or} & \underline{\sim q} \\ q & & p \end{array}$$

Examining this tautology, we see that if one of the premises of an argument is a disjunction and the other premise is the negation of one of its parts, we can conclude that the remaining part of the disjunction must be true. To verify that disjunctive syllogism is a tautology, we will construct a truth table.

p	q	$p \vee q$	$(p \vee q) \wedge \sim p$	$[(p \vee q) \wedge \sim p] \to q$
T	T	T	F	T
T	F	T	F	T
F	T	T	T	T
F	F	F	F	T

Again, recognition of the disjunctive syllogism pattern is crucial. The name of the tautology should provide a reminder that this tautology involves a disjunction, not an implication, as one of its premises.

Example 1

Is the argument below an illustration of disjunctive syllogism?

$$\begin{array}{c} r \vee \sim s \\ \underline{s} \\ r \end{array}$$

Solution

The premises are a disjunction and the negation of one of its parts. Therefore, we are guaranteed that the remaining part of the disjunction must be true. This is exactly the disjunctive syllogism pattern. ∎

Example 2

Is the following argument an illustration of disjunctive syllogism?

$$r \rightarrow\ \sim s$$

$$\frac{s}{r}$$

Solution

This argument cannot be an illustration of disjunctive syllogism since a disjunction does not appear as one of the premises. ∎

Example 3

Is the following argument an illustration of disjunctive syllogism?

$$p \vee q$$

$$\frac{q}{p}$$

Solution

Although one of the premises is a disjunction, the other is not the negation of one of its parts. Therefore, this is not an illustration of disjunctive syllogism. ∎

Example 4

Is the argument below an illustration of disjunctive syllogism?

$$(p \rightarrow r) \vee q$$

$$\frac{\sim (p \rightarrow r)}{q}$$

Solution

The premises consist of a disjunction and the negation of one of its parts. Therefore, we are guaranteed that the remaining part of the disjunction must be true. This is the disjunctive syllogism pattern. ∎

Example 5

Express the given valid argument symbolically and construct a formal proof.

If it is Halloween I will buy candy.
It is Halloween or Christmas.
I will not buy candy.
Therefore, it is Christmas.

Solution

The argument can be represented symbolically as:

$$h \rightarrow b$$
$$h \vee c$$
$$\underline{\sim b}$$
$$c$$

We begin the formal proof with the simplest premise, ~*b*.

Statement	Reason
1. ~*b*	Premise

Since the statement *b* appears in the first premise, we proceed to:

Statement	Reason
1. ~*b*	Premise
2. $h \rightarrow b$	Premise

Which tautology should we use to combine the two premises? Since we have an implication and the negation of its RHS as premises, modus tollens is applicable.

Statement	Reason
1. ~*b*	Premise
2. $h \rightarrow b$	Premise
3. ~*h*	Modus Tollens (1) (2)

We now combine this result with the second premise of the argument.

Statement	Reason
1. ~*b*	Premise
2. $h \rightarrow b$	Premise
3. ~*h*	Modus Tollens (1) (2)
4. $h \vee c$	Premise

Statements three and four tell us that the disjunction is true and the negation of one of its parts is also true. Therefore, disjunctive syllogism guarantees that the remaining part of the disjunction is true.

Statement	Reason
1. $\sim b$	Premise
2. $h \rightarrow b$	Premise
3. $\sim h$	Modus Tollens (1) (2)
4. $h \vee c$	Premise
5. c	Disjunctive Syllogism (3)(4)

Thus, we have proven that the argument is valid.   ∎

Example 6

Use the *TF* method to determine whether the given argument is valid. If so, construct a formal proof of its validity. If it is not valid, produce a counterexample chart.

$$q \vee \sim r$$
$$\sim p$$
$$\underline{p \vee \sim q}$$
$$\sim r$$

Solution

Since ~p is true, p is false. We know that $p \vee \sim q$ is true and p is false, so ~q must be true. This means that q is false. But $q \vee \sim r$ is true, hence ~r must be true. Therefore, the argument is valid. Now we proceed with a formal proof, following the same line of reasoning. Begin by combining the second and third premises.

Statement	Reason
1. $\sim p$	Premise
2. $p \vee \sim q$	Premise
3. $\sim q$	Disjunctive Syllogism (1) (2)

Now use the first premise.

Statement	Reason
1. $\sim p$	Premise
2. $p \vee \sim q$	Premise
3. $\sim q$	Disjunctive Syllogism (1) (2)
4. $q \vee \sim r$	Premise
5. $\sim r$	Disjunctive Syllogism (3) (4)

∎

Example 7

Use the *TF* method to determine whether the given argument is valid. If so, provide a formal proof of the validity. If not, produce a counterexample chart.

$$b \vee \sim c$$

$$c$$

$$\frac{\sim a \rightarrow b}{\sim a}$$

Solution

Since c is true, $\sim c$ is false. We know $b \vee \sim c$ is true and $\sim c$ is false, so b must be true. But $\sim a \rightarrow b$ is true, so $\sim a$ may be true or false. Since the conclusion cannot be forced to be true, the argument is invalid and a counterexample must be produced. The counterexample chart is constructed using truth values for a, b, and c, as per our discussion above.

a	b	c
T	T	T

That is, when a is true, b is true, and c is true, every premise is true, but the conclusion is false. Thus, we have shown that the argument is invalid.

■

Disjunctive Addition

Another argument form that involves disjunctions is called *disjunctive addition* abbreviated as DA. This tautology can be represented symbolically as $p \rightarrow (p \vee z)$. The tautology tells us that if we know that a statement, p, is true, we can join any other statement to p using the disjunction connective, and the resulting disjunction must be true.

In its argument form, the disjunctive addition tautology may be expressed as:

$$\frac{p}{p \vee z}$$

Notice that in each of the three previous valid argument forms we have studied, there were two premises from which we drew a conclusion. The disjunctive addition tautology has only one premise.

Example 8

Use the *TF* method to determine whether the given argument is valid. If so, construct a formal proof. If not, produce a counterexample chart.

$$g$$
$$(g \vee h) \to w$$
$$\overline{}$$
$$w$$

Solution

Since g is true, $g \vee h$ must be true. We know $(g \vee h) \to w$ is also true, so w must be true. Since the conclusion is forced to be true, the argument is valid. We now supply a formal proof. The first premise is the simple statement g. The statement g also appears as part of the antecedent of the second premise. Therefore, we want to establish the fact that the antecedent is true.

Statement	Reason
1. g	Premise
2. $g \vee h$	Disjunctive Addition (1)

The disjunction $g \vee h$ is true because we have joined h to a known true statement, g. Now that we know that the LHS of the second premise is true, the proof can be completed.

Statement	Reason
1. g	Premise
2. $g \vee h$	Disjunctive Addition (1)
3. $(g \vee h) \to w$	Premise
4. w	Modus Ponens (2) (3)

Notice that in step three it was necessary to state that $(g \vee h) \to w$ was true in order to apply the modus ponens tautology.   ■

Example 9

Construct a formal proof for the following valid argument.

$$p$$
$$s \to \sim r$$
$$r \vee \sim p$$
$$(\sim s \vee q) \to z$$
$$\overline{}$$
$$z$$

Solution

It appears that the simple statement p can be combined with the disjunction $r \vee \sim p$. Here we are combining a disjunctive statement and the negation of one of its parts to deduce a conclusion.

Statement	Reason
1. p	Premise
2. $r \vee \sim p$	Premise
3. r	Disjunctive Syllogism (1) (2)

The true statement r can be combined with $s \rightarrow \sim r$.

Statement	Reason
1. p	Premise
2. $r \vee \sim p$	Premise
3. r	Disjunctive Syllogism (1) (2)
4. $s \rightarrow \sim r$	Premise
5. $\sim s$	Modus Tollens (3) (4)

If we could show that the LHS of the last premise, $(\sim s \vee q) \rightarrow z$, is true, we could use modus ponens to deduce z. We just showed that $\sim s$ is true. Since disjunctive addition allows us to join any statement to a true statement using the disjunctive connective, we proceed by stating that $\sim s \vee q$ is true. Finally, we apply modus ponens.

Statement	Reason
1. p	Premise
2. $r \vee \sim p$	Premise
3. r	Disjunctive Syllogism (1) (2)
4. $s \rightarrow \sim r$	Premise
5. $\sim s$	Modus Tollens (3) (4)
6. $\sim s \vee q$	Disjunctive Addition (5)
7. $(\sim s \vee q) \rightarrow z$	Premise
8. z	Modus Ponens (6) (7) ∎

Example 10

Use the *TF* method to determine whether the given argument is valid. If so, construct a formal proof. If not, produce a counterexample chart.

$$p \to q$$
$$p$$
$$\underline{q \to (r \vee s)}$$
$$s$$

Solution

We know p and $p \to q$ are true. Therefore, q is true. Since $q \to (r \vee s)$ is true, $r \vee s$ must be true. However, we cannot discern which component of this disjunction is true. In particular, we cannot force s to be true. Thus, the argument is invalid. The counterexample will show that when r is true and s is false, all the premises will be true, but the conclusion will be false.

p	q	r	s
T	T	T	F

■

In-Class Exercises and Problems for Section 3.2

In-Class Exercises

In Exercises 1–10, state the illustrated valid argument form.

1. $\sim a \vee h$
   $\underline{\sim h}$
   $\sim a$

2. $(p \to q) \vee \sim b$
   $\underline{b}$
   $p \to q$

3. $\underline{\sim c}$
   $\sim c \vee s$

4. $\underline{m}$
   $m \vee (a \to c)$

5. $(k \wedge p) \vee \sim d$
   $\underline{\sim (k \wedge p)}$
   $\sim d$

6. $(n \leftrightarrow s)$
   $\overline{(n \leftrightarrow s) \vee \sim r}$

7. $(a \vee \sim b) \rightarrow \sim d$
   d
   _____
   $\sim (a \vee \sim b)$

8. $r \vee \sim (g \rightarrow \sim k)$
   $\sim r$
   _____
   $\sim (g \rightarrow \sim k)$

9. $w \rightarrow (s \wedge \sim p)$
   w
   _____
   $s \wedge \sim p$

10. $(s \rightarrow p) \wedge k$
    _____
    $[(s \rightarrow p) \wedge k] \vee (m \wedge n)$

In Exercises 11–28, test each argument for validity. If it is valid, provide a formal proof. If it is invalid, provide a counterexample chart.

11. a
    $b \rightarrow c$
    $\sim a \vee b$
    _____
    c

12. r
    $r \rightarrow (s \vee q)$
    $\sim s$
    _____
    q

13. $p \rightarrow \sim q$
    q
    $\sim p \vee \sim r$
    _____
    r

14. $w \rightarrow c$
    w
    $(c \vee p) \rightarrow b$
    _____
    b

15. $p \vee (q \wedge r)$
    $\sim (q \wedge r)$
    $p \rightarrow s$
    _____
    s

16. $p \rightarrow w$
    $\sim w$
    $(\sim p \vee s) \rightarrow k$
    _____
    k

17. $(a \rightarrow b) \vee c$
    $\sim c$
    $\sim b$
    _____
    $\sim a \vee p$

18. $m \vee p$
    $\sim p$
    $w \rightarrow (m \vee l)$
    _____
    $\sim w$

19. $a \rightarrow \sim b$
    b
    $c \rightarrow a$
    $c \lor d$
    $\overline{\quad d \quad}$

20. p
    $s \rightarrow \sim r$
    $\sim p \lor r$
    $\sim s \rightarrow w$
    $\overline{\quad w \quad}$

21. $\sim p \lor d$
    $\sim a$
    $p \rightarrow \sim s$
    $\sim a \rightarrow s$
    $\overline{\quad d \quad}$

22. $(w \lor z) \rightarrow k$
    $p \rightarrow q$
    $\sim q$
    $\sim p \rightarrow w$
    $\overline{\quad k \quad}$

23. $\sim c \rightarrow \sim d$
    $d \lor q$
    $\sim q$
    $\overline{c \lor (p \rightarrow s)}$

24. $(r \rightarrow w) \rightarrow (s \lor g)$
    $r \rightarrow w$
    $\sim s$
    $\overline{\quad g \lor p \quad}$

25. $a \lor (b \rightarrow c)$
    $\sim a \rightarrow \sim d$
    $\sim (b \rightarrow c)$
    $\overline{\quad d \quad}$

26. $m \lor \sim s$
    s
    $n \rightarrow \sim (m \lor q)$
    $\overline{\quad \sim n \quad}$

27. $r \lor \sim s$
    s
    $(r \lor w) \rightarrow p$
    $q \rightarrow \sim p$
    $\overline{\quad \sim q \quad}$

28. $\sim p \rightarrow q$
    $\sim (s \lor c)$
    $(\sim r \lor w) \rightarrow d$
    $\sim (\sim p \rightarrow q) \lor [r \rightarrow (s \lor c)]$
    $\overline{\quad d \quad}$

For Exercises 29–31, symbolize each argument. If the argument is valid, provide a formal proof. If not, provide a counterexample chart.

29. If I travel to the southwest United States, I will see the Indian ruins. If I see either the Indian ruins or the Painted Desert, then I will not go to San Francisco. Either I go to San Francisco or my trip was cancelled. I travel to the southwest United States. Therefore, my trip was cancelled.

30. I will either save my money or buy a new computer. If I buy a new computer, I will buy either a PC or a Mac. I didn't save my money. I bought a Mac. Therefore I didn't buy a PC.

31. If your major is engineering then you must take calculus. If you take calculus, you will not be able to work after school. You either work after school or take out a loan. You do not take out a loan. Therefore, your major is not engineering.

Problems

For each formal proof in Problems 32–36, supply the reason that justifies each statement.

32. $c \vee \sim d$

d

$p \rightarrow \sim c$

———————

$\sim p$

Statement	Reason
1. d	1.
2. $c \vee \sim d$	2.
3. c	3.
4. $p \rightarrow \sim c$	4.
5. $\sim p$	5.

33. $w \rightarrow s$

w

$(s \vee m) \rightarrow \sim n$

———————

$\sim n$

Statement	Reason
1. w	1.
2. $w \rightarrow s$	2.
3. s	3.
4. $s \vee m$	4.
5. $(s \vee m) \rightarrow \sim n$	5.
6. $\sim n$	6.

34. $a \vee b$
$(g \vee b) \rightarrow c$
$\sim q \rightarrow \sim c$
$\dfrac{\sim (a \vee b) \vee g}{q}$

Statement	Reason
1. $a \vee b$	1.
2. $\sim (a \vee b) \vee g$	2.
3. g	3.
4. $g \vee b$	4.
5. $(g \vee b) \rightarrow c$	5.
6. c	6.
7. $\sim q \rightarrow \sim c$	7.
8. q	8.

35. $\sim (p \rightarrow r)$
$w \vee q$
$\sim (p \rightarrow r) \rightarrow h$
$\dfrac{q \rightarrow \sim h}{w \vee a}$

Statement	Reason
1. $\sim (p \rightarrow r)$	1.
2. $\sim (p \rightarrow r) \rightarrow h$	2.
3. h	3.
4. $q \rightarrow \sim h$	4.
5. $\sim q$	5.
6. $w \vee q$	6.
7. w	7.
8. $w \vee a$	8.

36. p

   $p \to (a \lor \sim c)$

   $(w \to b) \to \sim s$

   $(a \lor g) \to s$

   c

   $\overline{\sim (w \to b) \lor z}$

Statement	Reason
1. p	1.
2. $p \to (a \lor \sim c)$	2.
3. $a \lor \sim c$	3.
4. c	4.
5. a	5.
6. $a \lor g$	6.
7. $(a \lor g) \to s$	7.
8. s	8.
9. $(w \to b) \to \sim s$	9.
10. $\sim (w \to b)$	10.
11. $\sim (w \to b) \lor z$	11.

In Problems 37–42, construct a formal proof for each valid argument.

37. $q \lor \sim p$

   $\sim q$

   $\sim r \to p$

   $\overline{r}$

38. $d \to \sim e$

   e

   $d \lor g$

   $\overline{g}$

39. $(m \lor p) \to s$

   $c \to m$

   c

   $\overline{s}$

40. $(c \to k) \lor (k \to p)$

   $\sim (c \to k)$

   $\sim p$

   $\overline{\sim k}$

41. s

   $(a \lor b) \to c$

   $s \to a$

   $\overline{c}$

42. $r \to (s \lor g)$

   r

   $\sim s$

   $\overline{g \lor p}$

In Problems 43–56, test each argument for validity. If it is valid, construct a formal proof. If it is invalid, produce a counterexample chart.

43. w

$\quad w \rightarrow (p \rightarrow q)$

$\quad p$

$\quad\overline{\quad\quad\quad\quad\quad}$

$\quad\quad q \vee r$

44. $r \rightarrow q$

$\quad \sim q$

$\quad \sim w \vee \sim r$

$\quad\overline{\quad\quad\quad\quad}$

$\quad\quad w$

45. $m \vee \sim n$

$\quad n$

$\quad m \rightarrow (p \vee r)$

$\quad \sim r$

$\quad\overline{\quad\quad\quad\quad}$

$\quad\quad p$

46. $s \vee p$

$\quad b \vee a$

$\quad p \rightarrow b$

$\quad \sim s$

$\quad\overline{\quad\quad\quad}$

$\quad\quad a$

47. a

$\quad (s \vee q) \rightarrow w$

$\quad (a \vee p) \rightarrow k$

$\quad k \rightarrow s$

$\quad\overline{\quad\quad\quad\quad}$

$\quad\quad w$

48. $a \vee c$

$\quad p \rightarrow r$

$\quad \sim r$

$\quad \sim p \rightarrow \sim a$

$\quad\overline{\quad\quad\quad\quad}$

$\quad\quad c \vee q$

49. $(b \rightarrow c) \vee a$

$\quad \sim a$

$\quad (c \vee w) \rightarrow p$

$\quad b$

$\quad\overline{\quad\quad\quad}$

$\quad\quad p$

50. $\sim r \rightarrow w$

$\quad \sim s \vee p$

$\quad \sim w$

$\quad (r \vee q) \rightarrow p$

$\quad\overline{\quad\quad\quad\quad}$

$\quad\quad s$

51. $s \vee w$
    $w \to a$
    $(a \vee b) \to p$
    $\sim s$
    _____
    $p \vee q$

52. $k \leftrightarrow l$
    $\sim (k \leftrightarrow l) \vee p$
    $\sim s \vee (a \to c)$
    $(p \vee w) \to s$
    _____
    $a \to c$

53. $\sim p \to (a \vee b)$
    $\sim s$
    $a \to q$
    $\sim p$
    $b \to s$
    _____
    q

54. $(a \to b) \to (c \vee d)$
    $\sim c$
    $a \to b$
    $(d \vee k) \to \sim b$
    _____
    $\sim a$

55. $(m \leftrightarrow n) \vee r$
    a
    $\sim (m \leftrightarrow n)$
    $r \to p$
    $\sim (a \to m) \to \sim p$
    _____
    m

56. w
    $(p \vee k) \to c$
    $w \to (p \vee s)$
    $\sim z \vee c$
    $\sim s$
    _____
    z

In Problems 57–63, each set of premises is followed by four possible conclusions. State the conclusion(s) that makes the argument valid. There may be more than one correct conclusion. Be sure to state all that apply.

57. $(f \vee g) \to \sim h$
    $f \vee \sim k$
    m
    $\sim k \to \sim m$
    _____
    a. $\sim k$ b. f
    c. g d. $\sim h$

58. $f \to (h \vee j)$
    $\sim f \vee \sim h$
    $(n \vee \sim s) \to f$
    $\sim s$
    _____
    a. f b. j
    c. $n \vee \sim s$ d. n

59. a

$p \to {\sim}(a \lor b)$

$p \lor (q \to c)$

$\underline{{\sim}(q \to c) \lor k}$

  a. $p \lor k$ b. $q \to c$

  c. ${\sim}k \lor b$ d. ${\sim}p \lor q$

60. ${\sim}({\sim}p \lor q)$

$({\sim}p \lor q) \lor r$

$r \to {\sim}w$

$\underline{{\sim}w \to (m \lor {\sim}r)}$

  a. ${\sim}w \lor {\sim}p$ b. $w \lor c$

  c. $m \lor k$ d. ${\sim}p \lor q$

61. If he exercises he will not get ill. He will exercise or go on a diet. He will not go on a diet if he is thin. He is thin. Therefore,

  a. he goes on a diet.   b. he doesn't exercise.

  c. he gets ill.   d. he is strong or not ill.

62. Today is Sunday or Monday. If today is Sunday, either Hiram doesn't get up early or he attends church. Today is Monday if Hiram doesn't get up early. It is not Monday. Therefore,

  a. today is Wednesday or Sunday. b Hiram attends church.

  c. Hiram does not get up early.   d. today is not Sunday.

63. If it rains or snows, the game will not be played. Either the game is played or the team cannot be in the finals. The temperature is increasing. Either it will rain or the temperature will not increase. Therefore,

  a. it is snowing.

  b. the team cannot be in the finals.

  c. the temperature increases or remains constant.

  d. it is not raining.

For Problems 64–67, express each argument symbolically. If the argument is valid, construct a formal proof. If not, produce a counterexample chart.

64. If Emma invites Pat or Jean to her party then she invites Max. Emma does not invite Shawn to her party if she invites Max. Either Shawn is invited to her party or Emma's party is cancelled. Emma invites Pat to her party. Therefore, Emma's party is cancelled.

65. If you like baseball, then you watch the World Series at home. If you have season tickets, then you do not watch the World Series at home. Either you have season tickets or you invite friends to your home. You like baseball. Therefore, you invite friends to your home.

66. It will snow if the temperature is below freezing. Either it doesn't snow or I shovel my driveway. If the temperature is not below freezing or it rains, the roads become slippery. I don't shovel my driveway. Therefore, the roads become slippery.

67. If a murder is committed then either the FBI or the CIA is called in to investigate. If the FBI is called in to investigate then Roberts is in charge of the case. Roberts is not in charge of the case. A murder is committed. Therefore, either the CIA is called in to investigate or the NYPD solved the case.

DID YOU KNOW?

Disjunctive Syllogism, which is abbreviated to DS, had been referred to as modus tollendo ponens. The reason this line of reasoning is called disjunctive syllogism is that, first, it is a syllogism (a three-step argument) and second, it contains a disjunction, which means an "or" statement.

Check Your Understanding

For questions 1–4, state the illustrated valid argument form.

1. $c \rightarrow (d \rightarrow p)$	2. $c \vee (d \rightarrow p)$	3. $\sim(d \rightarrow p) \rightarrow c$	4. $\sim(d \rightarrow p)$
$\underline{\sim(d \rightarrow p)}$	$\underline{\sim(d \rightarrow p)}$	$\underline{\sim(d \rightarrow p)}$	$\underline{\sim(d \rightarrow p) \vee c}$
$\sim c$	c	c	

1. _____   2. _____   3. _____   4. _____

In questions 5–7, each argument is missing its second premise. Supply each second premise so that the argument becomes valid.

5. $(r \wedge \sim s) \rightarrow \sim(p \vee w)$	6. $(r \wedge \sim s) \vee \sim(p \vee w)$	7. $(r \wedge \sim s) \rightarrow \sim(p \vee w)$
$\underline{\text{Premise 2}}$	$\underline{\text{Premise 2}}$	$\underline{\text{Premise 2}}$
$\sim(r \wedge \sim s)$	$r \wedge \sim s$	$\sim(p \vee w)$

5._____   6._____   7. _____

For question 8, supply the reason that justifies each statement.

8. $\sim(m \rightarrow n) \vee h$
   $h \rightarrow \sim(a \rightarrow w)$
   $m \rightarrow n$
   $\underline{\sim p \rightarrow (a \rightarrow w)}$
   $p \vee s$

Statement	Reason
1. $m \rightarrow n$	1. _____
2. $\sim(m \rightarrow n) \vee h$	2. _____
3. h	3. _____
4. $h \rightarrow \sim(a \rightarrow w)$	4. _____
5. $\sim(a \rightarrow w)$	5. _____
6. $\sim p \rightarrow (a \rightarrow w)$	6. _____
7. p	7. _____
8. $p \vee s$	8. _____

3.3 Valid Argument Forms–Part III

Conjunctive Addition

There are two commonly used argument forms that involve conjunctions. The first one that we will consider is *conjunctive addition,* abbreviated as CA. This is expressed symbolically as:

$$p$$
$$\underline{q\qquad}$$
$$p \wedge q$$

which states that if we know that two statements are true, we can join them with the conjunction connective and the resulting conjunction will be true.

This argument form is similar to disjunctive addition. However, with disjunctive addition we only needed to know that one statement was true in order to produce a true disjunctive statement. Since the only way a conjunction can be true is if both of its components are true, we need to know that two statements are true in order to state that their conjunction is true.

Example 1

Construct a formal proof for the following valid argument.

$$(\sim r \wedge p) \rightarrow s$$
$$p$$
$$\underline{r \rightarrow \sim p\qquad}$$
$$s \vee b$$

Solution

We begin with the second premise, p, and the third premise, $r \rightarrow \sim p$.

Statement	Reason
1. p	Premise
2. $r \rightarrow \sim p$	Premise
3. $\sim r$	Modus Tollens (1) (2)

Now that we know $\sim r$ is true, we can apply the conjunctive addition argument form to produce $\sim r \wedge p$, the LHS of the first premise.

Statement	Reason
1. p	Premise
2. $r \rightarrow \sim p$	Premise
3. $\sim r$	Modus Tollens (1) (2)
4. $\sim r \wedge p$	Conjunctive Addition (1) (3)

Since $\sim r \wedge p$ and $(\sim r \wedge p) \rightarrow s$ are both true, we apply modus ponens to deduce s.

Statement	Reason
1. p	Premise
2. $r \rightarrow \sim p$	Premise
3. $\sim r$	Modus Tollens (1) (2)
4. $\sim r \wedge p$	Conjunctive Addition (1) (3)
5. $(\sim r \wedge p) \rightarrow s$	Premise
6. s	Modus Ponens (4) (5)

Since we know s is true, disjunctive addition allows us to join any statement to it and the resulting disjunction will be true as well.

Statement	Reason
1. p	Premise
2. $r \rightarrow \sim p$	Premise
3. $\sim r$	Modus Tollens (1) (2)
4. $\sim r \wedge p$	Conjunctive Addition (1) (3)
5. $(\sim r \wedge p) \rightarrow s$	Premise
6. s	Modus Ponens (4) (5)
7. $s \vee b$	Disjunctive Addition (6)

Thus, we have shown that the argument is valid.   ■

Example 2

Write a formal proof for the following valid argument.

$$\sim s$$

$$\sim s \rightarrow q$$

$$(p \vee q) \rightarrow r$$
$$\overline{}$$
$$q \wedge r$$

Solution

Statement	Reason
1. $\sim s$	Premise
2. $\sim s \rightarrow q$	Premise
3. q	Modus Ponens (1) (2)
4. $p \vee q$	Disjunctive Addition (3)
5. $(p \vee q) \rightarrow r$	Premise
6. r	Modus Ponens (4) (5)
7. $q \wedge r$	Conjunctive Addition (3) (6)

Observe the use of both disjunctive addition and conjunctive addition in this proof. When disjunctive addition was applied in step four, it was only necessary to know one statement was true, namely, q. When conjunctive addition was applied in step seven, we had to know that both r and q were true. ∎

Conjunctive Simplification

The second argument form that involves a conjunction is *conjunctive simplification,* abbreviated as CS. This argument form can be expressed symbolically as:

$$\frac{p \wedge q}{p} \quad \text{or} \quad \frac{p \wedge q}{q}$$

which means that if a conjunction is true, each of its components must be true. Conjunctive simplification may be thought of as conjunctive addition "in reverse".

Notice that there is no analogous argument for the disjunction. That is, knowing that a disjunction is true does not provide us with information as to which specific component of the disjunction is true.

Example 3

Construct a formal proof for the following valid argument.

$$p \wedge q$$

$$\sim r \to \sim p$$

$$\overline{r \wedge q}$$

Solution

We begin by applying conjunctive simplification to $p \wedge q$.

Statement	Reason
1. $p \wedge q$	Premise
2. p	Conjunctive Simplification (1)
3. q	Conjunctive Simplification (1)

Since p is the negation of the RHS of $\sim r \to \sim p$, we can use modus tollens to deduce that r is true.

Statement	Reason
1. $p \wedge q$	Premise
2. p	Conjunctive Simplification (1)
3. q	Conjunctive Simplification (1)
4. $\sim r \to \sim p$	Premise
5. r	Modus Tollens (2) (4)

We now know that both r and q are true. Therefore, $r \wedge q$ is true by conjunctive addition.

Statement	Reason
1. $p \wedge q$	Premise
2. p	Conjunctive Simplification (1)
3. q	Conjunctive Simplification (1)
4. $\sim r \rightarrow \sim p$	Premise
5. r	Modus Tollens (2) (4)
6. $r \wedge q$	Conjunctive Addition (3) (5)

Thus, we have shown that the argument is valid. ■

Example 4

Use the *TF* method to determine whether the argument below is valid. If so, construct a formal proof. If not, produce a counterexample chart.

$$(p \vee q) \rightarrow r$$
$$(w \wedge r) \rightarrow m$$
$$p \vee q$$
$$(r \vee s) \rightarrow w$$
$$\sim m \vee b$$
$$\overline{}$$
$$b$$

Solution

Since $p \vee q$ and $(p \vee q) \rightarrow r$ are premises and thus both true, r is true by modus ponens. By disjunctive addition, $r \vee s$ is true. Since $(r \vee s) \rightarrow w$ is a premise and its LHS is true, w, its RHS is true by modus ponens. Now, $w \wedge r$ is true by conjunctive addition. Since this is the LHS of the premise $(w \wedge r) \rightarrow m$, m is true by modus ponens. The disjunction $\sim m \vee b$ is a premise and is therefore true. But m is true, so b must be true by disjunctive syllogism. These steps are shown in the following formal proof.

Statement	Reason
1. $p \vee q$	Premise
2. $(p \vee q) \to r$	Premise
3. r	Modus Ponens (1) (2)
4. $r \vee s$	Disjunctive Addition (3)
5. $(r \vee s) \to w$	Premise
6. w	Modus Ponens (4) (5)
7. $w \wedge r$	Conjunctive Addition (3) (6)
8. $(w \wedge r) \to m$	Premise
9. m	Modus Ponens (7) (8)
10. $\sim m \vee b$	Premise
11. b	Disjunctive Syllogism (9) (10)

Therefore, we have shown that the argument is valid.   ∎

Example 5

Use the *TF* method to determine whether the argument below is valid. If so, construct a formal proof. If not, produce a counterexample chart.

$$(r \vee s) \to (b \vee a)$$
$$\sim b \wedge r$$
$$\underline{(a \wedge c) \to d}$$
$$d$$

Solution

Since the conjunction $\sim b \wedge r$ is true, each of its components, r and $\sim b$, is true. Since r is true, $r \vee s$ is true. Therefore, $b \vee a$ is true. We already know $\sim b$ is true, so a must be true. We are unable to proceed further, since we know nothing about the truth values of c and d in the third premise. This suggests that the argument may be invalid. If so, d would have to be false. If c were false as well, the third premise would be true, and yet the conclusion, d, would be false. We now have enough information to produce a counterexample chart. Notice that since the truth value of s is irrelevant, there are two, equally correct counterexample charts, one for the case when s is true, the other, when s is false. All the other truth values have been deduced from the above line of reasoning.

a	b	c	d	r	s
T	F	F	F	T	T
T	F	F	F	T	F

■

In-Class Exercises and Problems for Section 3.3

In-Class Exercises

In Exercises 1–12, state the illustrated valid argument form.

1. $s \wedge \sim p$

$\overline{\qquad\qquad}$

s

2. k

$\sim d$

$\overline{\qquad\qquad}$

$k \wedge \sim d$

3. $p \vee q$

r

$\overline{\qquad\qquad}$

$(p \vee q) \wedge r$

4. $(\sim c \rightarrow a) \wedge \sim w$

$\overline{\qquad\qquad}$

$\sim c \rightarrow a$

5. $q \wedge \sim s$

$\overline{\qquad\qquad}$

$(q \wedge \sim s) \vee m$

6. j

$p \rightarrow s$

$\overline{\qquad\qquad}$

$j \wedge (p \rightarrow s)$

7. $h \vee (s \wedge \sim n)$

$\sim (s \wedge \sim n)$

$\overline{\qquad\qquad}$

h

8. $p \rightarrow (a \wedge \sim b)$

$\sim (a \wedge \sim b)$

$\overline{\qquad\qquad}$

$\sim p$

9. $(r \rightarrow s) \wedge (a \vee b)$

$\overline{\qquad\qquad}$

$r \rightarrow s$

10. $w \wedge k$

$\overline{\qquad\qquad}$

$(w \wedge k) \vee (d \rightarrow a)$

11. $(s \rightarrow d) \rightarrow \sim (g \rightarrow h)$

$s \rightarrow d$

$\overline{\qquad\qquad}$

$\sim (g \rightarrow h)$

12. $\sim (\sim a \vee b)$

$c \rightarrow (\sim a \vee b)$

$\overline{\qquad\qquad}$

$\sim c$

In Exercises 13–28, test each argument for validity. If it is valid, provide a formal proof. If it is invalid, provide a counterexample chart.

13. $\sim p \vee s$

p

$(p \wedge s) \to q$

_____

q

14. $p \to \sim q$

$q \to r$

$p \wedge b$

_____

$\sim r$

15. $m \wedge n$

$p \to w$

$\sim m \vee p$

_____

w

16. $s \to (a \vee w)$

$s \wedge g$

$d \to \sim (a \vee w)$

_____

$\sim d$

17. $\sim p \vee (\sim s \to \sim w)$

$s \to \sim p$

p

_____

$\sim s \wedge \sim w$

18. $(b \wedge p) \to s$

$a \to p$

$a \wedge b$

_____

s

19. $r \to w$

$p \to (w \wedge s)$

$r \wedge s$

_____

$\sim p$

20. $g \vee \sim h$

$k \to \sim g$

$a \wedge k$

_____

$a \wedge \sim h$

21. $\sim a \wedge b$

$\sim d \to e$

$a \vee c$

$d \to \sim c$

_____

e

22. $s \to b$

$s \vee c$

$(c \wedge \sim s) \to k$

$\sim b$

_____

k

23. $r \vee \sim s$

$\sim w \rightarrow \sim p$

s

$\underline{r \rightarrow (p \wedge q)}$

$w \vee c$

24. $p \rightarrow s$

$c \wedge g$

$j \vee p$

$\underline{c \rightarrow \sim j}$

$s \wedge g$

25. k

$c \rightarrow (a \wedge b)$

$(b \vee z) \rightarrow m$

$\underline{c \vee \sim k}$

m

26. $(s \vee w) \rightarrow m$

$p \wedge b$

$r \rightarrow \sim q$

$s \vee r$

$\underline{p \rightarrow q}$

m

27. $(j \vee c) \rightarrow d$

$d \rightarrow k$

$\sim a \rightarrow j$

$(\sim a \wedge s) \vee p$

$\underline{\sim p}$

$k \wedge s$

28. $a \wedge \sim b$

$a \rightarrow e$

$\sim d \vee (c \vee \sim a)$

$(\sim b \wedge e) \rightarrow d$

$\underline{p \rightarrow c}$

$\sim p$

In Exercises 29–31, symbolize each argument. If the argument is valid, provide a formal proof. If not, provide a counterexample chart.

29. Every winter I go cross-country skiing or snowshoeing. If I go snowshoeing I have to buy new boots. If I go cross-country skiing but I don't go snowshoeing then I can save money. I didn't buy new boots. Therefore, I saved money.

30. If you like to take pictures then you should be a photographer. If you want to be a photographer, then you own a camera and can develop your own film. If you can develop your own film,

then either you have a darkroom in your house or you have a rich uncle. You like to take pictures but don't have a rich uncle. Therefore, you own a camera and have a darkroom in your house.

31. Jay likes to play his guitar or ride his motorcycle, but he doesn't like to get up early. If Jay doesn't wear his helmet, he doesn't ride his motorcycle. If Jay wants to be considered a safe rider he wears his helmet. Jay does not play his guitar. Therefore, Jay is considered to be a safe rider.

Problems

In Problems 32–36, supply the reason that justifies each statement in the given valid argument.

	Statement	Reason
32. $a \wedge \sim c$	1. $a \wedge \sim c$	1.
$\sim s \rightarrow p$	2. $\sim c$	2.
$\underline{c \vee \sim p}$	3. $c \vee \sim p$	3.
s	4. $\sim p$	4.
	5. $\sim s \rightarrow p$	5.
	6. s	6.

	Statement	Reason
33. $\sim s$	1. $\sim s$	1.
$\sim s \rightarrow (q \wedge b)$	2. $\sim s \rightarrow (q \wedge b)$	2.
$\underline{\sim b \vee p}$	3. $q \wedge b$	3.
$p \wedge q$	4. b	4.
	5. q	5.
	6. $\sim b \vee p$	6.
	7. p	7.
	8. $p \wedge q$	8.

34. $(k \vee p) \to s$
$\quad d \to \sim (s \wedge \sim m)$
$\quad k \wedge \sim m$
$\quad \overline{}$
$\quad\quad \sim d$

Statement	Reason
1. $k \wedge \sim m$	1.
2. k	2.
3. $\sim m$	3.
4. $k \vee p$	4.
5. $(k \vee p) \to s$	5.
6. s	6.
7. $s \wedge \sim m$	7.
8. $d \to \sim (s \wedge \sim m)$	8.
9. $\sim d$	9.

35. $(d \wedge p) \vee k$
$\quad \sim a \to \sim p$
$\quad (a \wedge d) \to m$
$\quad \sim k$
$\quad \overline{}$
$\quad\quad d \wedge m$

Statement	Reason
1. $\sim k$	1.
2. $(d \wedge p) \vee k$	2.
3. $d \wedge p$	3.
4. d	4.
5. p	5.
6. $\sim a \to \sim p$	6.
7. a	7.
8. $a \wedge d$	8.
9. $(a \wedge d) \to m$	9.
10. m	10.
11. $d \wedge m$	11.

36. $p \vee r$

$\sim(s \wedge w) \rightarrow \sim(p \vee r)$

$(w \vee a) \rightarrow g$

$\dfrac{\sim(g \wedge s) \vee k}{k}$

Statement	Reason
1. $p \vee r$	1.
2. $\sim(s \wedge w) \rightarrow \sim(p \vee r)$	2.
3. $s \wedge w$	3.
4. s	4.
5. w	5.
6. $w \vee a$	6.
7. $(w \vee a) \rightarrow g$	7.
8. g	8.
9. $g \wedge s$	9.
10. $\sim(g \wedge s) \vee k$	10.
11. k	11.

In Problems 37–42, construct a formal proof for each valid argument.

37. m

n

$\dfrac{\sim(m \wedge n) \vee q}{q}$

38. $b \wedge \sim s$

$\sim b \vee q$

$\dfrac{(q \wedge \sim s) \rightarrow l}{l}$

39. $q \rightarrow s$

$p \wedge q$

$\dfrac{w \rightarrow \sim s}{\sim w}$

40. $\sim a \rightarrow b$

$\sim b$

$\dfrac{(a \wedge \sim b) \rightarrow r}{r}$

41. w

$\sim(a \wedge c) \rightarrow \sim r$

$\dfrac{w \rightarrow r}{a}$

42. $\sim c \vee g$

$c \wedge d$

$\dfrac{(d \wedge g) \rightarrow p}{p}$

In Problems 43–56, test each argument for validity. If it is valid, construct a formal proof. If it is invalid, produce a counterexample chart.

43. s

$(s \vee q) \rightarrow p$

$\sim (p \wedge s) \vee a$
_____
a

44. $r \wedge \sim w$

$\sim (r \vee s) \vee q$

$q \rightarrow \sim d$
_____
$\sim d$

45. q

$(\sim p \rightarrow \sim q) \wedge (\sim r \vee s)$

r
_____
$p \wedge s$

46. $a \rightarrow g$

$\sim (d \wedge h) \rightarrow \sim (a \rightarrow g)$

$(d \vee n) \rightarrow s$
_____
$s \wedge h$

47. $\sim (s \wedge q) \rightarrow a$

$\sim p \vee s$

$p \wedge q$
_____
$\sim a$

48. $(p \wedge s) \rightarrow b$

$(r \leftrightarrow q) \rightarrow p$

s

$r \leftrightarrow q$
_____
b

49. $(p \vee q) \wedge w$

$\sim p$

$r \rightarrow \sim q$

$\sim r \rightarrow z$
_____
z

50. $(p \wedge q) \rightarrow r$

$\sim w \wedge \sim s$

$\sim q \rightarrow w$

$s \vee p$
_____
r

51. $\sim p \rightarrow \sim r$

$q \wedge r$

$(p \vee s) \rightarrow w$
_____
w

52. $k \rightarrow p$

$r \rightarrow s$

$\sim s \vee \sim c$

$r \wedge (p \vee c)$
_____
$\sim k$

53. $\sim w \vee (a \wedge b)$
$(b \wedge s) \rightarrow q$
$w \wedge s$
$(q \vee z) \rightarrow c$
_____
c

54. $\sim s$
$r \vee q$
$\sim w \vee (a \wedge b)$
$r \rightarrow s$
$q \rightarrow w$
_____
$b \vee j$

55. $\sim p \rightarrow q$
$(w \vee b) \rightarrow c$
$\sim p \wedge r$
$(s \wedge r) \rightarrow w$
$\sim q \vee s$
_____
c

56. $(c \vee d) \wedge (p \rightarrow q)$
$d \rightarrow p$
$(d \wedge q) \rightarrow z$
a
$c \rightarrow \sim a$
_____
z

In Problems 57–63, each set of premises is followed by four possible conclusions. State the conclusion(s) that makes the argument valid. There may be more than one correct conclusion. Be sure to state all that apply.

57. $a \wedge b$
$b \rightarrow (p \rightarrow r)$
$(w \vee r) \rightarrow \sim a$
$w \vee (a \rightarrow c)$
_____
a. $c \wedge b$ b. $p \vee w$
c. $\sim w \vee k$ d. $r \wedge a$

58. $m \rightarrow (\sim j \wedge k)$
$\sim w \rightarrow y$
$(w \wedge s) \rightarrow m$
$\sim y \wedge s$
_____
a. $k \wedge m$ b. $y \rightarrow a$
c. $\sim m \vee y$ d. $j \wedge s$

59. $r \vee (s \wedge \sim w)$
$s \rightarrow \sim p$
$\sim q \wedge \sim r$
$\sim m \rightarrow w$
_____
a. $s \wedge m$ b. $\sim m \vee c$
c. $p \wedge w$ d. $\sim r \vee k$

60. $\sim p \wedge \sim q$
$p \vee \sim j$
$s \vee q$
$(s \wedge \sim j) \rightarrow (\sim p \rightarrow y)$
_____
a. $y \wedge s$ b. $s \vee j$
c. $j \vee p$ d. $p \wedge q$

61. He'll attract women if his car is sporty. He'll dine in fancy restaurants if his car is not expensive. He'll put money in the bank or his car is expensive. His car is sporty but not expensive. Therefore,
    a. he will attract women and he will put money in the bank.
    b. he will attract women but doesn't drive an expensive car.
    c. he dines in fancy restaurants or his car is expensive.
    d. he doesn't put money in the bank.

62. If it was poison and there was no break-in then either the maid or the chauffeur was the murderer. A weapon would be found if the chauffeur was the murderer. A weapon was found or it was poison. There was no break-in and a weapon was not found. Therefore,
    a. there was a break-in.
    b. a weapon was found.
    c. the chauffeur was the murderer.
    d. the maid was the murderer and it was poison.

In Problems 63–66, express each argument symbolically. If the argument is valid, construct a formal proof. If not, produce a counterexample chart.

63. If you join the circus you want to be a clown but not a lion tamer. Either you join the circus or become an acrobat. If you want to be a clown then you like to make people laugh. You don't become an acrobat. Therefore, you like to make people laugh or you enjoy standing on your head.

64. If Phil can do card tricks and make quarters disappear, then Phil is a magician. If Phil can't make quarters disappear he needs more practice. Phil can either pull rabbits out of a hat or do card tricks. Phil doesn't need more practice and he can't pull rabbits out of a hat. Therefore, Phil is a magician.

65. If you like downhill skiing then you like cold weather. Either you like downhill skiing or ice hockey. If you like ice skating then you don't like ice hockey. You like both ice skating and ice dancing. Therefore, you like ice dancing and cold weather.

66. If the Giants do not make the playoffs, then the Jets will. The Giants will make the playoffs if the Steelers don't, and either the Dolphins win or the Giants don't make the playoffs. If the Jets and Steelers make the playoffs then so will the Lions. The Dolphins do not win. Therefore, the Lions make the playoffs.

Check Your Understanding

For questions 1–6, each argument is incomplete. Supply the missing component so that the argument becomes valid.

1. $\sim k \vee (a \wedge \sim b)$

   Premise 2
   _____
   $\sim k$

2. $\sim (g \to \sim k) \to \sim b$

   Premise 2
   _____
   $g \to \sim k$

3. $\sim g \wedge (a \to \sim b)$
   _____
   Conclusion

1. _____

2. _____

3. _____

4. $(g \to \sim k)$

   Premise 2
   _____
   $(g \to \sim k) \wedge a$

5. Premise 1
   _____
   $(g \to \sim k) \vee a$

6. $\sim k \to (a \wedge \sim b)$

   Premise 2
   _____
   $a \wedge \sim b$

4. _____

5. _____

6. _____

For question 7, supply the reason that justifies each statement.

7. $\sim m \vee n$

   p

   $s \to \sim (\sim w \vee q)$

   $(\sim s \wedge k) \to m$

   $(p \to \sim w) \wedge k$
   _____
   n

Statement	Reason
1. $(p \to \sim w) \wedge k$	1. _____
2. $p \to \sim w$	2. _____
3. p	3. _____
4. $\sim w$	4. _____
5. $\sim w \vee q$	5. _____
6. $s \to \sim (\sim w \vee q)$	6. _____
7. $\sim s$	7. _____
8. k	8. _____
9. $\sim s \wedge k$	9. _____
10. $(\sim s \wedge k) \to m$	10. _____
11. m	11. _____
12. $\sim m \vee n$	12. _____
13. n	13. _____

3.4 Using Equivalences in Proofs

Introduction

In this section, we will use the conditional equivalence, the contrapositive equivalence, the conditional negation equivalence, the DeMorgan's equivalences and the distributive equivalences in formal proofs, in much the same way we did in Section 2.3.

Using the Conditional Equivalence in Proofs

Recall that the conditional equivalence, abbreviated as CE, allows us to express a conditional statement as a disjunction to which it is logically equivalent. This equivalence is expressed symbolically as $(p \rightarrow q) \Leftrightarrow (\sim p \vee q)$, that is, $p \rightarrow q$ and $\sim p \vee q$ are logically equivalent. Notice that the left side of the disjunction is the negation of the LHS of the original conditional statement.

Example 1

Construct a formal proof for the following valid argument.

$$p \rightarrow (r \rightarrow s)$$

$$p$$

$$(\sim r \vee s) \rightarrow w$$

$$\overline{p \wedge w}$$

Solution

We begin with the premises p and $p \rightarrow (r \rightarrow s)$ to conclude $r \rightarrow s$ by modus ponens.

Statement	Reason
1. p	Premise
2. $p \rightarrow (r \rightarrow s)$	Premise
3. $r \rightarrow s$	Modus Ponens (1) (2)

We continue by recognizing that $r \rightarrow s$ is equivalent to $\sim r \vee s$ by the conditional equivalence.

This is useful since $\sim r \vee s$ is the LHS of the premise $(\sim r \vee s) \rightarrow w$, and thus w is true by modus ponens.

Statement	Reason
1. p	Premise
2. $p \to (r \to s)$	Premise
3. $r \to s$	Modus Ponens (1) (2)
4. $\sim r \lor s$	Conditional Equivalence (3)
5. $(\sim r \lor s) \to w$	Premise
6. w	Modus Ponens (4) (5)

Now, since both p and w are true, their conjunction is true by conjunctive addition.

Statement	Reason
1. p	Premise
2. $p \to (r \to s)$	Premise
3. $r \to s$	Modus Ponens (1) (2)
4. $\sim r \lor s$	Conditional Equivalence (3)
5. $(\sim r \lor s) \to w$	Premise
6. w	Modus Ponens (4) (5)
7. $p \land w$	Conjunctive Addition (1) (6)

Thus, we have shown that the argument is valid.   ■

Example 2

Construct a formal proof for the following valid argument.

$$(p \to q) \to (r \lor w)$$
$$\frac{(\sim p \lor q) \land \sim r}{w \lor z}$$

Solution

Begin with the conjunctive statement.

Statement	Reason
1. $(\sim p \lor q) \land \sim r$	Premise
2. $\sim p \lor q$	Conjunctive Simplification (1)
3. $\sim r$	Conjunctive Simplification (1)

Since $\sim p \vee q$ is equivalent to $p \rightarrow q$ by the conditional equivalence and $p \rightarrow q$ is the LHS of $(p \rightarrow q) \rightarrow (r \vee w)$, we can deduce $r \vee w$ by modus ponens. Notice that we must state both the conditional equivalence and modus ponens to deduce $r \vee w$.

Statement	Reason
1. $(\sim p \vee q) \wedge \sim r$	Premise
2. $\sim p \vee q$	Conjunctive Simplification (1)
3. $\sim r$	Conjunctive Simplification (1)
4. $p \rightarrow q$	Conditional Equivalence (2)
5. $(p \rightarrow q) \rightarrow (r \vee w)$	Premise
6. $r \vee w$	Modus Ponens (4) (5)

We know $\sim r$ is true from statement three. Therefore, w is true by disjunctive syllogism.

Finally, since w is true, we can join any statement to it using disjunctive addition, and the resulting disjunction will also be true. We can then conclude that $w \vee z$ is true.

Therefore, the argument is valid. The completed proof follows. Be sure to understand that statement four is true not because it is a premise, but rather because it is equivalent to statement two.

Statement	Reason
1. $(\sim p \vee q) \wedge \sim r$	Premise
2. $\sim p \vee q$	Conjunctive Simplification (1)
3. $\sim r$	Conjunctive Simplification (1)
4. $p \rightarrow q$	Conditional Equivalence (2)
5. $(p \rightarrow q) \rightarrow (r \vee w)$	Premise
6. $r \vee w$	Modus Ponens (4) (5)
7. w	Disjunctive Syllogism (3) (6)
8. $w \vee z$	Disjunctive Addition (7)

■

Using the Contrapositive Equivalence in Proofs

The contrapositive of a conditional statement is formed by interchanging its LHS with its RHS, and negating both sides of the new implication. In Section 1.7, we learned that any conditional statement is logically equivalent to its contrapositive. Symbolically, we write $(p \rightarrow q) \Leftrightarrow (\sim q \rightarrow \sim p)$. We abbreviate the contrapositive equivalence as CP.

Example 3

Construct a formal proof for the following valid argument.

$$p \rightarrow (\sim q \rightarrow r)$$
$$(\sim p \vee s) \rightarrow w$$
$$\underline{\sim (\sim r \rightarrow q)}$$
$$w$$

Solution

The conditional statement within the parentheses in the third premise is the contrapositive of the RHS of the first premise. Therefore, the third premise is the negation of the RHS of the first premise. We can then deduce $\sim p$ by modus tollens.

Statement	Reason
1. $\sim (\sim r \rightarrow q)$	Premise
2. $\sim (\sim q \rightarrow r)$	Contrapositive Equivalence (1)
3. $p \rightarrow (\sim q \rightarrow r)$	Premise
4. $\sim p$	Modus Tollens (2) (3)

Since $\sim p$ is true, $\sim p \vee s$ is also true by disjunctive addition, and $\sim p \vee s$ is the LHS of $(\sim p \vee s) \rightarrow w$. Hence, we deduce w by modus ponens.

Statement	Reason
1. $\sim(\sim r \rightarrow q)$	Premise
2. $\sim(\sim q \rightarrow r)$	Contrapositive Equivalence (1)
3. $p \rightarrow (\sim q \rightarrow r)$	Premise
4. $\sim p$	Modus Tollens (2) (3)
5. $\sim p \vee s$	Disjunctive Addition (4)
6. $(\sim p \vee s) \rightarrow w$	Premise
7. w	Modus Ponens (5) (6)

This completes the proof. ∎

Using the Conditional Negation Equivalence in Proofs

The conditional negation equivalence, abbreviated as CN, is expressed symbolically as $\sim(p \rightarrow q) \Leftrightarrow (p \wedge \sim q)$. That is, the negation of a conditional statement is logically equivalent to the conjunction of its LHS with the negation of its RHS.

Example 4

Produce a formal proof for the following valid argument.

$$p \wedge (q \vee \sim z)$$
$$(r \wedge \sim s) \rightarrow z$$
$$\underline{(r \rightarrow s) \rightarrow \sim p}$$
$$q$$

Solution

We begin with the conjunctive statement.

Statement	Reason
1. $p \wedge (q \vee \sim z)$	Premise
2. p	Conjunctive Simplification (1)
3. $q \vee \sim z$	Conjunctive Simplification (1)

Now combine p with the third premise, $(r \rightarrow s) \rightarrow \sim p$ to obtain $\sim(r \rightarrow s)$ by modus tollens.

Statement	Reason
1. $p \land (q \lor \sim z)$	Premise
2. p	Conjunctive Simplification (1)
3. $q \lor \sim z$	Conjunctive Simplification (1)
4. $(r \rightarrow s) \rightarrow \sim p$	Premise
5. $\sim (r \rightarrow s)$	Modus Tollens (2) (4)

Now use the conditional negation equivalence to express $\sim (r \rightarrow s)$ as $r \land \sim s$.

Statement	Reason
1. $p \land (q \lor \sim z)$	Premise
2. p	Conjunctive Simplification (1)
3. $q \lor \sim z$	Conjunctive Simplification (1)
4. $(r \rightarrow s) \rightarrow \sim p$	Premise
5. $\sim (r \rightarrow s)$	Modus Tollens (2) (4)
6. $r \land \sim s$	Conditional Negation (5)

Since $r \land \sim s$ is the LHS of the premise $(r \land \sim s) \rightarrow z$, we deduce z by modus ponens.

Notice that in step three we showed $q \lor \sim z$ is true. Thus, q is true by disjunctive syllogism. This completes the proof.

Statement	Reason
1. $p \land (q \lor \sim z)$	Premise
2. p	Conjunctive Simplification (1)
3. $q \lor \sim z$	Conjunctive Simplification (1)
4. $(r \rightarrow s) \rightarrow \sim p$	Premise
5. $\sim (r \rightarrow s)$	Modus Tollens (2) (4)
6. $r \land \sim s$	Conditional Negation (5)
7. $(r \land \sim s) \rightarrow z$	Premise
8. z	Modus Ponens (6) (7)
9. q	Disjunctive Syllogism (3) (8) ∎

Example 5
Construct a formal proof for the following valid argument.

$$p \wedge \sim (q \to r)$$
$$r \vee \sim k$$
$$(b \to w) \to k$$
$$(\sim w \vee z) \to s$$
$$\overline{\qquad p \wedge s \qquad}$$

Solution
We begin with the first premise since it is a conjunction.

Statement	Reason
1. $p \wedge \sim (q \to r)$	Premise
2. p	Conjunctive Simplication (1)
3. $\sim (q \to r)$	Conjunctive Simplication (1)

Next, we negate the conditional statement to produce a conjunctive statement, and then apply conjunctive simplification.

Statement	Reason
1. $p \wedge \sim (q \to r)$	Premise
2. p	Conjunctive Simplication (1)
3. $\sim (q \to r)$	Conjunctive Simplication (1)
4. $q \wedge \sim r$	Conditional Negation (3)
5. q	Conjunctive Simplication (4)
6. $\sim r$	Conjunctive Simplication (4)

Now use the results of statement six with premise two.

Statement	Reason
1. $p \wedge \sim (q \to r)$	Premise
2. p	Conjunctive Simplication (1)
3. $\sim (q \to r)$	Conjunctive Simplication (1)
4. $q \wedge \sim r$	Conditional Negation (3)
5. q	Conjunctive Simplication (4)
6. $\sim r$	Conjunctive Simplication (4)
7. $r \vee \sim k$	Premise
8. $\sim k$	Disjunctive Syllogism (6) (7)

Realizing that $\sim k$ is the negation of the RHS of premise three, we apply modus tollens to obtain $\sim (b \to w)$.

Statement	Reason
1. $p \wedge \sim (q \to r)$	Premise
2. p	Conjunctive Simplication (1)
3. $\sim (q \to r)$	Conjunctive Simplication (1)
4. $q \wedge \sim r$	Conditional Negation (3)
5. q	Conjunctive Simplication (4)
6. $\sim r$	Conjunctive Simplication (4)
7. $r \vee \sim k$	Premise
8. $\sim k$	Disjunctive Syllogism (6) (7)
9. $(b \to w) \to k$	Premise
10. $\sim (b \to w)$	Modus Tollens (8) (9)

We now negate the conditional statement and simplify the resulting conjunction. Once we do this we can use disjunctive addition to produce the LHS of premise four, and then apply modus ponens to deduce s is true. Then, since s and p are both true, their conjunction is true.

Statement	Reason
1. $p \wedge \sim (q \rightarrow r)$	Premise
2. p	Conjunctive Simplication (1)
3. $\sim (q \rightarrow r)$	Conjunctive Simplication (1)
4. $q \wedge \sim r$	Conditional Negation (3)
5. q	Conjunctive Simplication (4)
6. $\sim r$	Conjunctive Simplication (4)
7. $r \vee \sim k$	Premise
8. $\sim k$	Disjunctive Syllogism (6) (7)
9. $(b \rightarrow w) \rightarrow k$	Premise
10. $\sim (b \rightarrow w)$	Modus Tollens (8) (9)
11. $b \wedge \sim w$	Conditional Negation (10)
12. b	Conjunctive Simplication (11)
13. $\sim w$	Conjunctive Simplication (11)
14. $\sim w \vee z$	Disjunctive Addition (13)
15. $(\sim w \vee z) \rightarrow s$	Premise
16. s	Modus Ponens (14) (15)
17. $p \wedge s$	Conjunctive Addition (2)(16) ∎

Using DeMorgan's Equivalences in Proofs

DeMorgan's equivalences, expressed as $\sim (p \wedge q) \Leftrightarrow (\sim p \vee \sim q)$ and $\sim (p \vee q) \Leftrightarrow (\sim p \wedge \sim q),$ allow us to express the negation of a conjunction as a disjunction and the negation of a disjunction as a conjunction. DeMorgan's equivalence is abbreviated as DM.

Example 6

Let's show a formal proof for the valid argument first encountered as Example 5 of Section 2.3.

$$\sim p \to (q \wedge \sim r)$$
$$\sim q \vee r$$
$$\underline{p \to (s \wedge h)}$$
$$h$$

Solution

Note the relationship between the second premise and the RHS of the first premise. Using DeMorgan's equivalence, the second premise may be expressed as $\sim(q \wedge \sim r)$. Thus, the second premise is the negation of the RHS of premise one.

Statement	Reason
1. $\sim q \vee r$	Premise
2. $\sim(q \wedge \sim r)$	DeMorgan's Equivalence (1)
3. $\sim p \to (q \wedge \sim r)$	Premise

Now apply modus tollens to obtain p, which is the LHS of premise three.

Statement	Reason
1. $\sim q \vee r$	Premise
2. $\sim(q \wedge \sim r)$	DeMorgan's Equivalence (1)
3. $\sim p \to (q \wedge \sim r)$	Premise
4. p	Modus Tollens (2) (3)
5. $p \to (s \wedge h)$	Premise

Finally, apply modus ponens to statements four and five to deduce $s \wedge h$, and then use conjunctive simplification to infer the truth of h.

Statement	Reason
1. $\sim q \vee r$	Premise
2. $\sim(q \wedge \sim r)$	DeMorgan's Equivalence (1)
3. $\sim p \to (q \wedge \sim r)$	Premise
4. p	Modus Tollens (2) (3)
5. $p \to (s \wedge h)$	Premise
6. $s \wedge h$	Modus Ponens (4) (5)
7. h	Conjunctive Simplification (6) ∎

Example 7

Use the *TF* method to test the following argument for validity. If it is valid, construct a formal proof. If not, produce a counterexample chart.

$$(q \lor r) \to s$$
$$\sim q \to w$$
$$\underline{\sim s}$$
$$w$$

Solution

The premise $\sim s$ is true, so s is false. Since the premise $(q \lor r) \to s$ is true and s is false, $q \lor r$ is false. That means both q and r are false. Since q is false, $\sim q$, the LHS of the second premise, is true. Therefore, w must be true.

We proceed with the formal proof, following the same strategy outlined above. Begin by using the first and third premises to deduce $\sim (q \lor r)$.

Statement	Reason
1. $\sim s$	Premise
2. $(q \lor r) \to s$	Premise
3. $\sim (q \lor r)$	Modus Tollens (1) (2)

Now, apply DeMorgan's equivalence.

Statement	Reason
1. $\sim s$	Premise
2. $(q \lor r) \to s$	Premise
3. $\sim (q \lor r)$	Modus Tollens (1) (2)
4. $\sim q \land \sim r$	DeMorgan's Equivalence (3)

Since the conjunction is true, each of its components is true.

Statement	Reason
1. $\sim s$	Premise
2. $(q \vee r) \rightarrow s$	Premise
3. $\sim (q \vee r)$	Modus Tollens (1) (2)
4. $\sim q \wedge \sim r$	DeMorgan's Equivalence (3)
5. $\sim q$	Conjunctive Simplification (4)

Since $\sim q$ is the LHS of the second premise, we deduce w by modus ponens.

Statement	Reason
1. $\sim s$	Premise
2. $(q \vee r) \rightarrow s$	Premise
3. $\sim (q \vee r)$	Modus Tollens (1) (2)
4. $\sim q \wedge \sim r$	DeMorgan's Equivalence (3)
5. $\sim q$	Conjunctive Simplification (4)
6. $\sim q \rightarrow w$	Premise
7. w	Modus Ponens (5) (6)

∎

Using the Distributive Equivalences in Proofs

Recall that the distributive equivalence involves two different connectives, and it has two forms. One form tells us that a conjunction distributes over a disjunction. This form can be expressed symbolically as $[p \wedge (q \vee r)] \Leftrightarrow [(p \wedge q) \vee (p \wedge r)]$. We abbreviate the distributive equivalence as DE.

The other form of the distributive equivalence tells us that a disjunction distributes over a conjunction. It is expressed as $[p \vee (q \wedge r)] \Leftrightarrow [(p \vee q) \wedge (p \vee r)]$

Example 8

Show a formal proof for the following valid argument.

$$\sim r \vee \sim s$$
$$w \rightarrow p$$
$$\underline{r \wedge (s \vee w)}$$
$$p$$

Solution

Note that the distributive equivalence allows us to express the third premise as $(r \wedge s) \vee (r \wedge w)$, while DeMorgan's equivalence tells us that the first premise is logically equivalent to $\sim (r \wedge s)$. We begin our proof using these observations.

Statement	Reason
1. $r \wedge (s \vee w)$	Premise
2. $(r \wedge s) \vee (r \wedge w)$	Distributive Equivalence (1)
3. $\sim r \vee \sim s$	Premise
4. $\sim (r \wedge s)$	DeMorgan's Equivalence (3)

We now apply disjunctive syllogism to statements two and four, and then apply conjunctive simplification to the resulting statement.

Statement	Reason
1. $r \wedge (s \vee w)$	Premise
2. $(r \wedge s) \vee (r \wedge w)$	Distributive Equivalence (1)
3. $\sim r \vee \sim s$	Premise
4. $\sim (r \wedge s)$	DeMorgan's Equivalence (3)
5. $r \wedge w$	Disjunctive Syllogism (2) (4)
6. w	Conjunctive Simplification (5)

Finally, we can apply modus ponens after we state $w \rightarrow p$ is true. This will produce the desired conclusion.

Statement	Reason
1. $r \wedge (s \vee w)$	Premise
2. $(r \wedge s) \vee (r \wedge w)$	Distributive Equivalence (1)
3. $\sim r \vee \sim s$	Premise
4. $\sim (r \wedge s)$	DeMorgan's Equivalence (3)
5. $r \wedge w$	Disjunctive Syllogism (2) (4)
6. w	Conjunctive Simplification (5)
7. $w \rightarrow p$	Premise
8. p	Modus Ponens (6) (7)

In-Class Exercises and Problems for Section 3.4

In-Class Exercises

In Exercises 1–14, state the illustrated valid argument form or equivalence.

1. $\dfrac{\sim(a \to \sim q)}{a \wedge q}$

2. $\dfrac{\sim b \vee \sim c}{b \to \sim c}$

3. $\dfrac{\sim(n \wedge \sim p)}{\sim n \vee p}$

4. $\dfrac{\sim(k \to p)}{\sim(k \to p) \vee q}$

5. $\dfrac{r \to \sim d}{d \to \sim r}$

6. $\dfrac{\begin{array}{c} \sim g \\ (\sim h \vee a) \to g \end{array}}{\sim(\sim h \vee a)}$

7. $\dfrac{q \vee (s \wedge \sim r)}{(q \vee s) \wedge (q \vee \sim r)}$

8. $\dfrac{\begin{array}{c} a \\ (s \to c) \vee \sim a \end{array}}{s \to c}$

9. $\dfrac{\sim g \wedge a}{\sim(g \vee \sim a)}$

10. $\dfrac{\begin{array}{c} q \wedge c \\ (q \wedge c) \to \sim s \end{array}}{\sim s}$

11. $\dfrac{(d \to s) \wedge a}{d \to s}$

12. $\dfrac{g \vee \sim m}{\sim(\sim g \wedge m)}$

13. $\dfrac{(a \wedge b) \to \sim c}{\sim(a \wedge b) \vee \sim c}$

14. $\dfrac{\sim[(p \to w) \to \sim k]}{(p \to w) \wedge k}$

In Exercises 15–22, construct a formal proof for each valid argument.

15. $\begin{array}{l} \sim p \to s \\ \underline{\sim(p \vee q)} \\ s \end{array}$

16. $\begin{array}{l} b \to (a \to c) \\ b \\ \underline{\sim(\sim a \vee c) \vee n} \\ n \end{array}$

17. $\sim (s \to w)$

    $w \vee p$

    $\underline{a \to \sim p}$

    $\sim a$

18. $s \to (\sim a \wedge b)$

    $\sim s \to c$

    $\underline{a \vee \sim b}$

    c

19. $(\sim m \to p) \to q$

    $a \vee \sim q$

    $\underline{\sim p \to m}$

    $a \vee b$

20. $(\sim s \vee p) \to k$

    $d \vee c$

    $\underline{s \to \sim (\sim d \to c)}$

    k

21. $s \vee p$

    $p \to (a \to \sim b)$

    $\sim (w \to s)$

    $\underline{r \to \sim (b \to \sim a)}$

    $\sim r$

22. $w \to s$

    $m \vee (q \to d)$

    $\sim (p \vee \sim w)$

    $\underline{m \to \sim (s \vee b)}$

    $\sim q \vee d$

In Exercises 23–34, test each argument for validity. If valid, show a formal proof. If not, produce a counterexample chart.

23. $(p \to q) \to \sim r$

    $r \vee c$

    $\underline{(\sim p \vee q) \wedge s}$

    c

24. $(a \wedge b) \to p$

    $\sim w \to p$

    $\underline{\sim (a \to \sim b)}$

    w

25. $\sim p \vee a$

    $\sim r \to q$

    $\underline{r \to (p \wedge \sim a)}$

    q

26. $\sim (\sim k \to q) \vee p$

    $(p \vee w) \to n$

    $\underline{r \wedge (\sim q \to k)}$

    $r \wedge n$

27. $r \vee (m \to n)$

    $\sim (\sim s \to r)$

    $\underline{\sim (d \vee n)}$

    $\sim m$

28. $(p \to w) \to s$

    $c \wedge (\sim p \vee w)$

    $\underline{\sim r \to \sim (s \wedge c)}$

    r

29. $g \rightarrow \sim k$

$(\sim g \vee \sim k) \rightarrow p$

$(w \vee \sim s) \rightarrow \sim p$

___

s

30. $\sim a \vee (p \wedge r)$

$(a \rightarrow p) \rightarrow (w \wedge \sim r)$

$w \vee \sim b$

___

b

31. $(\sim p \vee q) \rightarrow w$

$\sim r \vee c$

$p \rightarrow q$

$(s \vee \sim r) \rightarrow \sim w$

___

c

32. $\sim (w \rightarrow \sim s)$

$(p \wedge s) \rightarrow q$

$q \rightarrow (\sim a \vee b)$

$\sim w \vee p$

___

$a \rightarrow b$

33. $h \vee a$

$p \rightarrow (a \rightarrow b)$

$s \rightarrow (b \vee c)$

$\sim (h \vee \sim p)$

___

$\sim s$

34. $(b \rightarrow c) \rightarrow q$

$(s \wedge a) \rightarrow p$

$(a \wedge \sim b) \vee (a \wedge c)$

$(w \vee \sim s) \rightarrow \sim q$

___

p

In Exercises 35–37, symbolize each argument. If the argument is valid, provide a formal proof. If not, provide a counterexample chart.

35. It is not true that if you work hard you will not be successful. If you work hard and are successful then you will be able to afford the car of your dreams. Either you cannot afford the car of your dreams or you buy an iPod. Therefore, you buy an iPod.

36. If I read a book I will read a novel but not a biography. If I do not read a book then I will either go to the opera or go to a museum. It is not the case that if I do not go to a museum I will go to the opera. Therefore, I will read a novel or go out to dinner.

37. If you like the outdoors then you should train for the New York Marathon and join a soccer team. It is not true that if you like to run you should become a basketball player. If you like to run then you should train for the New York Marathon. Either you want to become a basketball player or join a soccer team. Therefore, you like the outdoors.

Problems

For each formal proof in Problems 38–43, supply the reason that justifies each statement.

38. $\sim k \vee p$

$\sim (h \vee \sim k)$

$\dfrac{p \to a}{a \wedge \sim h}$

Statement	Reason
1. $\sim (h \vee \sim k)$	1.
2. $\sim h \wedge k$	2.
3. $\sim h$	3.
4. k	4.
5. $\sim k \vee p$	5.
6. p	6.
7. $p \to a$	7.
8. a	8.
9. $a \wedge \sim h$	9.

39. $\sim q \vee b$

$(a \wedge b) \to s$

$\dfrac{\sim (a \to \sim q)}{s}$

Statement	Reason
1. $\sim (a \to \sim q)$	1.
2. $a \wedge q$	2.
3. a	3.
4. q	4.
5. $\sim q \vee b$	5.
6. b	6.
7. $a \wedge b$	7.
8. $(a \wedge b) \to s$	8.
9. s	9.

40. $(s \rightarrow c) \rightarrow d$

$h \rightarrow \sim (d \vee p)$

$\dfrac{\sim s \vee c}{\sim h}$

Statement	Reason
1. $\sim s \vee c$	1.
2. $s \rightarrow c$	2.
3. $(s \rightarrow c) \rightarrow d$	3.
4. d	4.
5. $d \vee p$	5.
6. $h \rightarrow \sim (d \vee p)$	6.
7. $\sim h$	7.

41. $s \rightarrow \sim (\sim m \rightarrow p)$

$s \vee (n \rightarrow c)$

$\dfrac{\sim p \rightarrow m}{\sim n \vee c}$

Statement	Reason
1. $\sim p \rightarrow m$	1.
2. $\sim m \rightarrow p$	2.
3. $s \rightarrow \sim (\sim m \rightarrow p)$	3.
4. $\sim s$	4.
5. $s \vee (n \rightarrow c)$	5.
6. $n \rightarrow c$	6.
7. $\sim n \vee c$	7.

42. $p \wedge (a \vee b)$

$(c \vee \sim w) \rightarrow \sim b$

$\dfrac{\sim (p \wedge a)}{w}$

Statement	Reason
1. $p \wedge (a \vee b)$	1.
2. $(p \wedge a) \vee (p \wedge b)$	2.
3. $\sim (p \wedge a)$	3.
4. $p \wedge b$	4.
5. b	5.
6. $(c \vee \sim w) \rightarrow \sim b$	6.
7. $\sim (c \vee \sim w)$	7.
8. $\sim c \wedge w$	8.
9. w	9.

	Statement	Reason
43. $\sim g$	1. $\sim g$	1.
$(c \rightarrow \sim g) \rightarrow \sim (a \rightarrow \sim b)$	2. $\sim g \vee \sim c$	2.
$s \vee \sim (a \vee p)$	3. $g \rightarrow \sim c$	3.
$\overline{\qquad s \wedge b \qquad}$	4. $c \rightarrow \sim g$	4.
	5. $(c \rightarrow \sim g) \rightarrow \sim (a \rightarrow \sim b)$	5.
	6. $\sim (a \rightarrow \sim b)$	6.
	7. $a \wedge b$	7.
	8. a	8.
	9. b	9.
	10. $a \vee p$	10.
	11. $s \vee \sim (a \vee p)$	11.
	12. s	12.
	13. $s \wedge b$	13.

In Problems 44–51, construct a formal proof for each valid argument.

44. $\sim (a \rightarrow c)$
$$\frac{q \vee c}{q}$$

45. $(\sim m \rightarrow q) \rightarrow p$
$n \rightarrow \sim p$
$$\frac{m \vee q}{\sim n}$$

46. $s \rightarrow n$
$\sim n \vee p$
$$\frac{\sim (\sim s \vee c)}{p \wedge \sim c}$$

47. $\sim (p \rightarrow \sim q) \vee w$
$\sim r \rightarrow \sim w$
$$\frac{q \rightarrow \sim p}{r}$$

48. $(p \rightarrow w) \rightarrow s$
$\sim s$
$$\frac{(\sim p \vee w) \vee a}{a}$$

49. $r \rightarrow (p \wedge w)$
$\sim p \vee \sim w$
$$\frac{r \vee s}{s}$$

50. $s \vee a$

  $\sim c \rightarrow r$

  $\sim (\sim r \rightarrow s)$

  $(\sim s \rightarrow a) \rightarrow b$

  ────────

  $b \wedge c$

51. $(p \vee \sim r) \rightarrow \sim (a \rightarrow \sim g)$

  $r \rightarrow s$

  $(s \vee w) \rightarrow m$

  $g \rightarrow \sim a$

  ────────

  m

In Problems 52–65, test each argument for validity. If valid, construct a formal proof. If not, produce a counterexample chart.

52. $\sim s \vee q$

  $q \rightarrow \sim b$

  $\sim (s \rightarrow \sim p)$

  ────────

  $\sim b \wedge p$

53. $\sim (a \rightarrow b)$

  $w \vee (\sim a \vee b)$

  $(w \vee p) \rightarrow c$

  ────────

  c

54. p

  $p \rightarrow \sim (c \vee \sim n)$

  $(w \rightarrow \sim r) \rightarrow c$

  ────────

  r

55. $m \rightarrow (p \wedge \sim s)$

  $(\sim m \vee a) \rightarrow w$

  $\sim p \vee s$

  ────────

  w

56. $p \vee \sim s$

  $\sim g \vee k$

  $s \rightarrow \sim (g \rightarrow k)$

  ────────

  $\sim p$

57. $d \rightarrow \sim n$

  $w \rightarrow (a \wedge b)$

  $\sim (n \rightarrow \sim d) \vee w$

  ────────

  $b \vee c$

58. $\sim (a \wedge \sim b)$

  $\sim r \vee p$

  $(a \rightarrow b) \rightarrow (q \wedge r)$

  ────────

  p

59. $r \wedge (\sim a \rightarrow q)$

  $\sim r \vee s$

  $w \rightarrow (a \vee q)$

  ────────

  $s \wedge w$

60. $p \vee r$

$\quad (d \vee h) \rightarrow \sim p$

$\quad \sim h \rightarrow q$

$\quad \dfrac{\sim (s \rightarrow r)}{q}$

61. $\sim (c \wedge p) \rightarrow (s \rightarrow w)$

$\quad s \wedge \sim w$

$\quad \dfrac{(c \wedge s) \rightarrow n}{n}$

62. $(\sim r \wedge s) \rightarrow \sim (\sim p \vee q)$

$\quad \sim (w \vee b) \vee c$

$\quad p \rightarrow q$

$\quad \dfrac{(r \vee \sim s) \rightarrow w}{c}$

63. $p \wedge q$

$\quad (r \vee s) \rightarrow (w \rightarrow \sim a)$

$\quad (a \rightarrow \sim w) \rightarrow \sim b$

$\quad \dfrac{\sim r \rightarrow s}{p \wedge \sim b}$

64. $(\sim a \vee c) \wedge (p \rightarrow s)$

$\quad \sim w \rightarrow (a \wedge \sim c)$

$\quad \sim (\sim p \vee s) \vee k$

$\quad \dfrac{(w \wedge k) \rightarrow q}{q}$

65. $(a \wedge \sim b) \rightarrow c$

$\quad (c \vee q) \rightarrow (w \rightarrow \sim s)$

$\quad \sim (a \rightarrow b)$

$\quad \dfrac{s}{a \wedge \sim w}$

In Problems 66–71, each set of premises is followed by four possible conclusions. State the conclusion(s) that makes the argument valid. There may be more than one correct conclusion. Be sure to state all that apply.

66. $\sim g \wedge k$

$\quad (g \vee \sim k) \vee \sim l$

$\quad \sim (y \wedge \sim l)$

$\quad \dfrac{\sim x \rightarrow y}{}$

a. $\sim y$ b. $\sim l$

c. x d. $\sim (y \vee l)$

67. $\sim (m \vee \sim r)$

$\quad k \rightarrow m$

$\quad (w \rightarrow \sim p) \vee k$

$\quad \dfrac{\sim (p \rightarrow \sim w) \vee s}{}$

a. $r \wedge w$ b. $s \wedge k$

c. $\sim m \vee p$ d. $k \wedge r$

68. $\sim p \lor q$

    $(p \to q) \to (r \lor \sim s)$

    $(\sim r \land s) \lor k$

    $w \to \sim k$

    ────────────────

    a. $r \land k$ b. $p \lor s$

    c. $w \to r$ d. $q \to \sim k$

69. $a \lor (p \land s)$

    $(a \lor s) \to j$

    $\sim j \lor \sim (c \to d)$

    $d \lor w$

    ────────────────

    a. $a \lor j$ b. $w \to d$

    c. $j \land d$ d. $w \land c$

70. It is not true that if today is Monday then classes meet on a Tuesday schedule. It is not the case that it is Wednesday or Friday. There will be an exam, if it is not Wednesday and classes do not meet on a Tuesday schedule. Therefore,
    a. there will be an exam.
    b. today is not Monday.
    c. today is Wednesday.
    d. today is Friday.

71. If he invests wisely, he is rich. He will be successful, if he does not invest wisely or he is rich. If he is successful or not unemployed, then it is not true that he is successful but not married. Therefore,
    a. he is married.
    b. he is not successful.
    c. he is not rich.
    d. he does not invest wisely.

In Problems 72–75, express each argument symbolically. If the argument is valid, construct a formal proof. If not, produce a counterexample chart.

72. If Sam looks for a new job in another state then he wants to work part-time and learn to play tennis. Sam will either look for a new job in another state or not sell his house. If he decides to move to Arizona or Florida, he must sell his house. If he does not move to Arizona then he will definitely move to Florida. Therefore, Sam learns to play tennis.

73. Harry relaxes or he is hyperactive. It's not true that if Harry goes to the movies then he does not like videos. If Harry goes to the movies or eats popcorn, he relaxes. Therefore, Harry is hyperactive.

74. Kenny goes biking and follows the stock market. If Kenny goes biking it is not true that he either rides when the temperature is above 60° or he doesn't ride alone. Either he rides when the temperature is above 60° or, if he follows the stock market then his computer is on. Therefore, Kenny's computer is on and he rides his bike alone.

75. Oleg either boots up his computer and plays games, or boots up his computer and does homework. If he boots up his computer, then he reads his e-mail if he accesses the internet. Oleg accesses the internet and does not do homework. Therefore, Oleg reads his e-mail and plays games.

DID YOU KNOW?

A disarmingly uncomplicated, yet powerful concept in higher mathematics is called the pigeonhole principle. The idea is simple to understand. If three pigeons are placed into two pigeonholes, one of the pigeonholes must have more than one pigeon. This principle can be generalized to say that if more than n pigeons are placed into n pigeonholes, some pigeonhole must contain more than one pigeon. While the principle seems obvious, its implications are amazing. For instance, each of the following can be shown to be true using the principle. Can you see how?

- Among 366 people, there are at least two people who have the exact same birthday.
- In London, there are at least two people who have exactly the same number of hairs on their head.
- If a college has 6,000 American students, at least one from each of the 50 states, then there must be a group of 120 students who come from same state.
- If there are twenty-five students in a class and each student received an A, B, or C on an exam, then at least nine students received the same grade.

Check Your Understanding

In questions 1–8, the first two steps of a formal proof are shown. The second step is missing either the statement or its corresponding reason. Supply the missing element.

1. Statement   Reason

   1. $\sim(m \vee \sim n)$ Premise

   2. $\sim m \wedge n$ _____

2. Statement   Reason

   1. $g \rightarrow \sim d$ Premise

   2. $\sim g \vee \sim d$ _____

3. Statement   Reason

   1. $d \wedge (c \vee \sim g)$ Premise

   2. _____ Distributive Eq.(1)

4. Statement   Reason

   1. $\sim(j \rightarrow b)$ Premise

   2. $j \wedge \sim b$ _____

5. Statement   Reason

   1. $\sim b \vee \sim k$ Premise

   2. _____ Conditional Eq. (1)

6. Statement   Reason

   1. $\sim l \vee n$ Premise

   2. _____ DeMorgan's Eq. (1)

7. Statement   Reason

   1. $p \rightarrow \sim m$ Premise

   2. _____ Contrapositive Eq. (1)

8. Statement   Reason

   1. $r \wedge s$ Premise

   2. _____ Cond. Neg. Eq. (1)

For questions 9–10, the first few steps of a formal proof are shown, however, some of steps are missing. Supply the missing statements and reasons.

9. Statement   Reason

   1. $\sim(b \rightarrow a)$ Premise

   2. _____ _____

   3. _____ DM (2)

   4. $d \rightarrow (\sim b \vee a)$ Premise

10. Statement   Reason

   1. $\sim(\sim s \wedge \sim w)$ Premise

   2. _____ _____

   3. _____ _____

   4. $(\sim s \rightarrow w) \vee k$ Premise

3.5 Conditional Proofs

In Section 2.4, we saw that if the conclusion of an argument was a conditional statement, we could often employ a subtle line of reasoning to construct a proof. Recall that if the conclusion of an argument is a conditional statement and its LHS is false, the argument is valid, regardless of the truth value of the RHS. However, if the LHS of the conclusion is true, then in order to show that the argument is valid, we must prove that the RHS is true as well.

Therefore, if we *assume* the LHS of the conclusion is true and we are able to show that under this assumption, its RHS is true, we will have shown that the argument is valid. This sort of strategy is called a *conditional proof.* It is important to realize that this strategy may only be used if the conclusion is a conditional statement.

When we assume that the LHS of a conclusion is true, the reason we supply in our formal proof is called the *assumption for a conditional proof,* abbreviated *ACP.*

Example 1

Write a conditional proof for the following valid argument.

$$p \rightarrow q$$
$$q \rightarrow \sim r$$
$$\overline{p \rightarrow \sim r}$$

Solution

Since the conclusion of the argument is a conditional statement, we can use a conditional proof and assume that its LHS is true. We then proceed as we did previously, employing all the tools at our disposal.

Statement	Reason
1. p	ACP
2. $p \rightarrow q$	Premise
3. q	Modus Ponens (1) (2)
4. $q \rightarrow \sim r$	Premise
5. $\sim r$	Modus Ponens (3) (4)
6. $p \rightarrow \sim r$	Conditional Proof (1) (5)

Notice that in statement six, the reason we use to show that the conclusion, $p \rightarrow \sim r$, is true is "Conditional Proof." ∎

When constructing a formal proof of an argument whose conclusion is a conditional statement, the assumption for a conditional proof need not be the first step in our proof. We may assume the LHS of the conclusion is true at any step in a proof, as our next example illustrates.

Example 2

Write a conditional proof for the following valid argument.

$$\sim a \lor b$$
$$\sim c \to d$$
$$\sim b$$
$$\sim a \to (d \to \sim e)$$
$$\overline{e \to c}$$

Solution

Since the conclusion of this argument is a conditional statement, we will attempt to use a conditional proof. However, if we inspect the premises, we note that the assumption that e is true will not allow us to proceed. While e appears in the premise $\sim a \to (d \to \sim e)$, we cannot reason further, since the truth values of $\sim a$ and d are as yet unknown.

We choose to begin with the first and third premises.

Statement	Reason
1. $\sim a \lor b$	Premise
2. $\sim b$	Premise
3. $\sim a$	Disjunctive Syllogism (1) (2)

Now we proceed to the fourth premise, knowing its LHS is true.

Statement	Reason
1. $\sim a \lor b$	Premise
2. $\sim b$	Premise
3. $\sim a$	Disjunctive Syllogism (1) (2)
4. $\sim a \to (d \to \sim e)$	Premise
5. $d \to \sim e$	Modus Ponens (3) (4)

Now it seems reasonable to assume the LHS of the conclusion, *e,* is true, since it is the negation of the RHS of $d \rightarrow\, \sim e$. We then proceed in the usual way.

Statement	Reason
1. $\sim a \vee b$	Premise
2. $\sim b$	Premise
3. $\sim a$	Disjunctive Syllogism (1) (2)
4. $\sim a \rightarrow (d \rightarrow\, \sim e)$	Premise
5. $d \rightarrow\, \sim e$	Modus Ponens (3) (4)
6. e	ACP
7. $\sim d$	Modus Tollens (5) (6)
8. $\sim c \rightarrow d$	Premise
9. c	Modus Tollens (7) (8)
10. $e \rightarrow c$	Conditional Proof (6) (9)

Thus, we have shown that the argument is valid.   ■

In-Class Exercises and Problems for Section 3.5

In-Class Exercises

In Exercises 1–12, test each argument for validity. If valid, construct a formal proof. If not, produce a counterexample chart.

1. $c \rightarrow (d \wedge g)$
$$\frac{}{c \rightarrow g}$$

2. $\sim p \rightarrow\, \sim r$
$$\frac{\sim p \vee q}{r \rightarrow q}$$

3. $r \vee s$
$$\frac{\sim a \vee \sim s}{\sim r \rightarrow\, \sim a}$$

4. $s \rightarrow\, \sim w$
$$\frac{h \rightarrow\, \sim s}{w \rightarrow h}$$

5. $\sim k \rightarrow (p \wedge b)$

$\dfrac{\sim b \vee a}{\sim k \rightarrow a}$

6. $\sim (h \vee p) \vee l$

$\dfrac{s \rightarrow \sim l}{h \rightarrow \sim s}$

7. p

$(p \wedge a) \rightarrow n$

$\dfrac{\sim n \vee (c \wedge d)}{a \rightarrow c}$

8. $r \rightarrow \sim p$

$\sim r \vee (q \rightarrow s)$

$\dfrac{q}{p \rightarrow \sim s}$

9. $\sim p \rightarrow \sim q$

$(p \wedge r) \rightarrow w$

$\dfrac{(r \vee s) \rightarrow q}{r \rightarrow w}$

10. a

$\sim s \vee r$

$(p \vee q) \rightarrow w$

$\dfrac{a \rightarrow (w \rightarrow s)}{p \rightarrow r}$

11. $\sim d \vee p$

$p \rightarrow (\sim m \rightarrow n)$

d

$\dfrac{(m \wedge d) \rightarrow r}{\sim n \rightarrow (r \vee a)}$

12. $(a \wedge d) \rightarrow w$

$b \vee d$

$a \wedge (b \rightarrow p)$

$\dfrac{c \rightarrow (w \vee s)}{\sim p \rightarrow c}$

Problems

In Problems 13–14, supply the reason that justifies each statement in the formal proof.

13. $\sim p \vee q$

$r \rightarrow \sim q$

$\dfrac{\sim r \rightarrow s}{p \rightarrow s}$

Statement	Reason
1. p	1.
2. $\sim p \vee q$	2.
3. q	3.
4. $r \rightarrow \sim q$	4.
5. $\sim r$	5.
6. $\sim r \rightarrow s$	6.
7. s	7.
8. $p \rightarrow s$	8.

	Statement	Reason
14. $a \lor \sim b$	1. b	1.
b	2. $a \lor \sim b$	2.
$c \to \sim d$	3. a	3.
$a \to (\sim d \to q)$	4. $a \to (\sim d \to q)$	4.
$\sim q \to \sim c$	5. $\sim d \to q$	5.
	6. $\sim q$	6.
	7. d	7.
	8. $c \to \sim d$	8.
	9. $\sim c$	9.
	10. $\sim q \to \sim c$	10.

In Problems 15–20, construct a formal proof for each valid argument.

15. $(c \lor p) \to s$
$$\frac{}{p \to (s \lor b)}$$

16. $a \lor b$

$a \to c$
$$\frac{}{\sim b \to c}$$

17. $\sim a \to \sim w$

$(a \lor d) \to \sim p$
$$\frac{}{w \to \sim p}$$

18. $(w \lor \sim s) \to p$

$\sim a \to w$
$$\frac{}{\sim p \to (a \lor q)}$$

19. $\sim p$

$(\sim p \land s) \to q$

$\sim q \lor \sim r$
$$\frac{}{s \to \sim r}$$

20. $\sim (\sim p \lor s)$

$p \to (q \to l)$

$q \lor m$
$$\frac{}{\sim l \to m}$$

In Problems 21–30, test each argument for validity. If valid, construct a formal proof. If not, produce a counterexample chart.

21. $d \to \sim a$

$r \to (a \land \sim b)$

$d \lor \sim q$
$$\frac{}{r \to \sim q}$$

22. $s \land k$

$k \to (\sim d \lor c)$

$b \to c$
$$\frac{}{d \to b}$$

23. $\sim s \rightarrow (\sim c \vee q)$
    $\sim d$
    $d \vee \sim s$
    $\underline{q \rightarrow p}$
    $c \rightarrow p$

24. $g \rightarrow \sim n$
    $(p \vee q) \rightarrow r$
    $r \rightarrow w$
    $\underline{g \vee p}$
    $n \rightarrow w$

25. $p \wedge \sim w$
    $\sim w \rightarrow (b \vee c)$
    $\underline{p \rightarrow (a \rightarrow c)}$
    $a \rightarrow b$

26. a
    $\sim a \vee b$
    $(b \vee d) \rightarrow (p \vee s)$
    $\underline{\sim k \rightarrow \sim s}$
    $\sim p \rightarrow k$

27. $\sim w \rightarrow \sim a$
    $\sim (w \wedge p) \vee s$
    $s \rightarrow b$
    $\underline{\sim (a \wedge \sim p)}$
    $a \rightarrow b$

28. $\sim r$
    $(b \vee c) \rightarrow r$
    $\sim d \vee m$
    $\underline{d \rightarrow \sim (c \vee p)}$
    $p \rightarrow m$

29. $(a \wedge \sim b) \rightarrow (c \vee d)$
    $g \rightarrow (w \wedge \sim d)$
    $\sim (a \rightarrow b)$
    $\underline{\sim c \vee z}$
    $g \rightarrow z$

30. $(p \rightarrow a) \rightarrow (\sim m \vee c)$
    $r \rightarrow \sim (c \vee q)$
    $\sim r \rightarrow k$
    $\underline{\sim p \vee a}$
    $m \rightarrow k$

In Problems 31–36, each set of premises is followed by four possible conclusions. State the conclusion(s) that makes the argument valid. There may be more than one correct conclusion. Be sure to state all that apply.

31. $(x \rightarrow y) \rightarrow g$

$w \vee \sim x$

$s \rightarrow \sim w$

----

a. $\sim g \rightarrow y$

b. $\sim g \rightarrow \sim s$

c. $\sim g \rightarrow \sim x$

d. $\sim g \rightarrow \sim w$

32. $m \wedge \sim n$

$m \rightarrow (j \vee \sim k)$

$(y \wedge \sim j) \rightarrow n$

----

a. $\sim j \rightarrow \sim m$

b. $\sim j \rightarrow \sim y$

c. $\sim j \rightarrow n$

d. $\sim j \rightarrow \sim k$

33. $(r \rightarrow \sim s) \rightarrow p$

$(x \vee y) \rightarrow \sim z$

$(w \vee p) \rightarrow z$

----

a. $x \rightarrow (\sim s \wedge \sim w)$

b. $x \rightarrow \sim r$

c. $x \rightarrow (\sim p \vee a)$

d. $x \rightarrow y$

34. $\sim s \rightarrow (e \wedge \sim g)$

$g \vee (\sim w \wedge k)$

$\sim p \rightarrow \sim k$

----

a. $\sim s \rightarrow (g \wedge p)$

b. $\sim s \rightarrow (\sim g \wedge \sim w)$

c. $\sim s \rightarrow w$

d. $\sim s \rightarrow (p \vee m)$

35. She will fly to Rome or she won't fly to Paris. If she flies to London then she won't fly to Rome. She won't need an umbrella if she doesn't fly to London. If she doesn't have the cash then she won't fly to Paris. Therefore,
    a. if she flies to Paris she will fly to Rome.
    b. if she flies to Paris she will fly to London.
    c. if she flies to Paris, then either she has the cash or she has frequent flier points.
    d. if she flies to Paris she will need an umbrella.

36. Alice will not get to Wonderland if she doesn't follow the rabbit. If Alice follows the rabbit, she will meet the Queen or attend a mad tea party. Alice does not attend a mad tea party. Therefore,
    a. if Alice gets to Wonderland then she followed the rabbit.
    b. if Alice gets to Wonderland then she does not meet the Queen.
    c. if Alice gets to Wonderland, then she either meets the Queen or she attends a mad tea party.
    d. if Alice gets to Wonderland then she meets the Queen.

In Problems 37–39, symbolize each argument. If the argument is valid, construct a formal proof. If not, produce a counterexample chart.

37. Allen is having a surprise party for Eric. If Lou is unable to come to the party then Allen will not have the party. If Emad comes to the party then Jane won't come. If Lou comes to the party then either Jane or Ken will attend. Therefore, if Ken doesn't come to the party then either Emad won't come or Bette will stay home.

38. Maria goes swimming and skydiving. If she rides her bike and goes Rollerblading then she won't go swimming. If Maria doesn't ride her bike, then if she goes skydiving she has to buy a parachute. Therefore, if Maria goes Rollerblading she buys a parachute.

39. Tom will either work on his new computer program or he will build his daughter a dollhouse. If he works on his new computer program then he won't have time to go boating. If he stays home to finish his model airplane then, he won't go boating or he won't go horseback riding. Therefore, if Tom doesn't build his daughter a dollhouse then he will finish his model airplane.

3.6 Indirect Proofs

As we saw in Section 2.5, an alternate method of testing for validity is the *indirect approach* or *indirect method.* Recall that the strategy for an indirect approach is to assume that the argument is invalid, i.e., that the premises are all true but the conclusion is false. If, in truth, the argument is valid, a contradiction will arise.

In practical terms, this means that we assume the negation of the conclusion is true. Then, if we are able to show that any statement and its negation are both true, we will have arrived at a contradiction. This means that the assumption was incorrect and thus, the conclusion is true, making the argument valid.

This technique can be employed regardless of the form of the conclusion; that is, the conclusion of the argument may be a simple sentence, or any compound statement, including a conditional statement.

Example 1

Construct a formal proof for the following valid argument using the indirect method.

$$p \vee r$$

$$r \rightarrow s$$

$$\underline{p \rightarrow q}$$

$$q \vee s$$

Solution

Begin by assuming $\sim (q \vee s)$ is true. This is the assumption for the indirect proof, abbreviated AIP. We next apply DeMorgan's equivalence and then conjunctive simplification.

Statement	Reason
1. $\sim (q \vee s)$	AIP
2. $\sim q \wedge \sim s$	DeMorgan's Equivalence (1)
3. $\sim q$	Conjunctive Simplification (2)
4. $\sim s$	Conjunctive Simplification (2)

Now combine $\sim q$ with the third premise using modus tollens to conclude $\sim p$. (We could just as easily have used $\sim s$ with the second premise.)

Statement	Reason
1. $\sim (q \vee s)$	AIP
2. $\sim q \wedge \sim s$	DeMorgan's Equivalence (1)
3. $\sim q$	Conjunctive Simplification (2)
4. $\sim s$	Conjunctive Simplification (2)
5. $p \rightarrow q$	Premise
6. $\sim p$	Modus Tollens (3) (5)

Using the first premise, $p \vee r$, and the result from step six, we can reason r by disjunctive syllogism.

Statement	Reason
1. $\sim (q \vee s)$	AIP
2. $\sim q \wedge \sim s$	DeMorgan's Equivalence (1)
3. $\sim q$	Conjunctive Simplification (2)
4. $\sim s$	Conjunctive Simplification (2)
5. $p \rightarrow q$	Premise
6. $\sim p$	Modus Tollens (3) (5)
7. $p \vee r$	Premise
8. r	Disjunctive Syllogism (6) (7)

Combining r with the second premise produces s by modus ponens. However, in step four above, we showed that $\sim s$ was true. This is a contradiction. Therefore, the assumption in step one must have been incorrect. Hence, $q \vee s$ is true.

Statement	Reason
1. $\sim(q \vee s)$	AIP
2. $\sim q \wedge \sim s$	DeMorgan's Equivalence (1)
3. $\sim q$	Conjunctive Simplification (2)
4. $\sim s$	Conjunctive Simplification (2)
5. $p \rightarrow q$	Premise
6. $\sim p$	Modus Tollens (3) (5)
7. $p \vee r$	Premise
8. r	Disjunctive Syllogism (6) (7)
9. $r \rightarrow s$	Premise
10. s	Modus Ponens (8) (9)
11. $q \vee s$	Contradiction (4) (10)

We have shown that the argument is valid.   ∎

Example 2

Write a formal proof for the following valid argument using the indirect approach.

$$m \rightarrow (\sim p \rightarrow q)$$
$$n$$
$$r \rightarrow \sim p$$
$$\underline{m \vee \sim n}$$
$$q \vee \sim r$$

Solution

As in a conditional proof, the assumption need not be stated in the first line of an indirect proof. We choose to begin this proof by combining the second and fourth premises, to deduce m by disjunctive syllogism.

Statement	Reason
1. $m \vee \sim n$	Premise
2. n	Premise
3. m	Disjunctive Syllogism (1) (2)

Combining this result with the first premise will produce $\sim p \rightarrow q$ by modus ponens.

Statement	Reason
1. $m \vee \sim n$	Premise
2. n	Premise
3. m	Disjunctive Syllogism (1) (2)
4. $m \rightarrow (\sim p \rightarrow q)$	Premise
5. $\sim p \rightarrow q$	Modus Ponens (3) (4)

It is at this point that we choose to state our assumption for the indirect proof, and assume that the negation of $q \vee \sim r$ is true. We next apply DeMorgan's equivalence followed by conjunctive simplification.

Statement	Reason
1. $m \vee \sim n$	Premise
2. n	Premise
3. m	Disjunctive Syllogism (1) (2)
4. $m \rightarrow (\sim p \rightarrow q)$	Premise
5. $\sim p \rightarrow q$	Modus Ponens (3) (4)
6. $\sim (q \vee \sim r)$	AIP
7. $\sim q \wedge r$	DeMorgan's Equivalence (6)
8. $\sim q$	Conjunctive Simplification (7)
9. r	Conjunctive Simplification (7)

We now combine r with the third premise to deduce $\sim p$. Then using $\sim p$ in combination with statement five we deduce q. However, in statement eight we showed $\sim q$ is true. Hence, we have a contradiction based on our incorrect assumption in step six.

Statement	Reason
1. $m \vee \sim n$	Premise
2. n	Premise
3. m	Disjunctive Syllogism (1) (2)
4. $m \rightarrow (\sim p \rightarrow q)$	Premise
5. $\sim p \rightarrow q$	Modus Ponens (3) (4)
6. $\sim (q \vee \sim r)$	AIP
7. $\sim q \wedge r$	DeMorgan's Equivalence (6)
8. $\sim q$	Conjunctive Simplification (7)
9. r	Conjunctive Simplification (7)
10. $r \rightarrow \sim p$	Premise
11. $\sim p$	Modus Ponens (9) (10)
12. q	Modus Ponens (5) (11)
13. $q \vee \sim r$	Contradiction (8) (12)

Therefore, the conclusion is true and the argument is valid. ∎

Example 3

Construct an indirect proof for the following valid argument.

$$p \rightarrow q$$
$$q \rightarrow \sim r$$
$$\overline{p \rightarrow \sim r}$$

Solution

We already have shown that this argument is valid using a conditional proof in Example 1 of Section 3.5. Since the indirect method can be used for *any* valid argument, we should be able to construct a formal proof using an indirect approach as well.

First, assume the negation of the conclusion is true, and then apply the conditional negation equivalence followed by conjunctive simplification.

Statement	Reason
1. $\sim(p \to \sim r)$	AIP
2. $p \wedge r$	Conditional Negation (1)
3. p	Conjunctive Simplification (2)
4. r	Conjunctive Simplification (2)

Using p with premise one produces q by modus ponens. Then, using q with premise two produces $\sim r$, again by modus ponens.

Statement	Reason
1. $\sim(p \to \sim r)$	AIP
2. $p \wedge r$	Conditional Negation (1)
3. p	Conjunctive Simplification (2)
4. r	Conjunctive Simplification (2)
5. $p \to q$	Premise
6. q	Modus Ponens (3) (5)
7. $q \to \sim r$	Premise
8. $\sim r$	Modus Ponens (6) (7)

Steps four and eight are contradictory. Therefore, our assumption was incorrect, so $p \to \sim r$ is true.

Statement	Reason
1. $\sim(p \to \sim r)$	AIP
2. $p \wedge r$	Conditional Negation (1)
3. p	Conjunctive Simplification (2)
4. r	Conjunctive Simplification (2)
5. $p \to q$	Premise
6. q	Modus Ponens (3) (5)
7. $q \to \sim r$	Premise
8. $\sim r$	Modus Ponens (6) (7)
9. $p \to \sim r$	Contradiction (4) (8)

■

In-Class Exercises and Problems for Section 3.6

In-Class Exercises

In Exercises 1–12, test the each argument for validity. If valid, construct a formal proof. If not, produce a counterexample chart.

1. $m \to n$

   $m \lor n$
   _____

   n

2. $a \to (c \to r)$
   _____

   $(a \land c) \to r$

3. $c \to g$

   $l \to m$

   $c \lor l$
   _____

   $g \lor m$

4. $p \to s$

   $(d \lor r) \to k$

   $p \lor \sim k$
   _____

   $d \lor s$

5. $(p \lor n) \to a$

   $\sim a \lor (b \land c)$

   $b \to (c \land n)$
   _____

   $g \lor n$

6. $r \to \sim p$

   $\sim (s \lor \sim n)$

   $\sim s \to (\sim p \to w)$
   _____

   $\sim w \to \sim r$

7. $\sim a \lor w$

   $d \to s$

   $(a \to w) \to (p \lor d)$
   _____

   $p \lor s$

8. $(r \lor p) \to m$

   $(p \lor s) \to w$

   $\sim m \lor (w \to \sim s)$
   _____

   $\sim (r \land s)$

9. $\sim s \lor h$

   $d \to a$

   $\sim s \to (a \to \sim c)$

   $\sim h$
   _____

   $c \to \sim d$

10. $\sim c$

    $p \lor q$

    $\sim b \to (q \to c)$

    $(\sim p \to q) \to (a \lor \sim b)$
    _____

    a

11. $\sim z \vee (\sim s \to w)$

$a \to \sim s$

$\sim z \to \sim k$

k

———

$w \vee \sim a$

12. r

$\sim (p \wedge r) \vee s$

$(k \vee c) \to \sim w$

$s \to (c \to w)$

———

$c \to \sim p$

Problems

Each argument in Problems 13–16 is valid. In each problem, supply the reasons that justify the statements made.

13. $a \vee (s \wedge q)$

$q \to p$

$(s \vee w) \to \sim p$

———

a

Statement	Reason
1. $\sim a$	1.
2. $a \vee (s \wedge q)$	2.
3. $s \wedge q$	3.
4. s	4.
5. q	5.
6. $q \to p$	6.
7. p	7.
8. $(s \vee w) \to \sim p$	8.
9. $\sim (s \vee w)$	9.
10. $\sim s \wedge \sim w$	10.
11. $\sim s$	11.
12. a	12.

14. $(\sim c \vee r) \rightarrow a$

    $\sim a \vee p$

    $\underline{p \rightarrow q}$

    $\sim c \rightarrow q$

Statement	Reason
1. $\sim (\sim c \rightarrow q)$	1.
2. $\sim c \wedge \sim q$	2.
3. $\sim c$	3.
4. $\sim q$	4.
5. $\sim c \vee r$	5.
6. $(\sim c \vee r) \rightarrow a$	6.
7. a	7.
8. $\sim a \vee p$	8.
9. p	9.
10. $p \rightarrow q$	10.
11. q	11.
12. $\sim c \rightarrow q$	12.

15. $n \vee (\sim b \wedge s)$

    $(c \vee p) \rightarrow (a \rightarrow b)$

    $\underline{s \rightarrow a}$

    $n \vee \sim c$

Statement	Reason
1. $\sim (n \vee \sim c)$	1.
2. $\sim n \wedge c$	2.
3. $\sim n$	3.
4. c	4.
5. $n \vee (\sim b \wedge s)$	5.
6. $\sim b \wedge s$	6.
7. $\sim b$	7.
8. s	8.
9. $s \rightarrow a$	9.
10. a	10.
11. $c \vee p$	11.
12. $(c \vee p) \rightarrow (a \rightarrow b)$	12.
13. $a \rightarrow b$	13.
14. b	14.
15. $n \vee \sim c$	15.

16. $p \wedge \sim s$

  $\sim s \rightarrow (w \rightarrow k)$

  $\underline{\sim w \rightarrow \sim r}$

  $k \vee \sim r$

Statement	Reason
1. $p \wedge \sim s$	1.
2. p	2.
3. $\sim s$	3.
4. $\sim s \rightarrow (w \rightarrow k)$	4.
5. $w \rightarrow k$	5.
6. $\sim (k \vee \sim r)$	6.
7. $\sim k \wedge r$	7.
8. $\sim k$	8.
9. r	9.
10. $\sim w$	10.
11. $\sim w \rightarrow \sim r$	11.
12. $\sim r$	12.
13. $k \vee \sim r$	13.

In Problems 17–22, construct a formal proof for each valid argument.

17. $\sim h \vee k$

  $\underline{\sim k \rightarrow h}$

  k

18. $a \rightarrow (p \wedge s)$

  $\overline{\sim a \vee p}$

19. $(p \rightarrow q) \rightarrow s$

  $\sim s \vee k$

  $\underline{\sim k \vee r}$

  $\sim p \rightarrow r$

20. $b \vee a$

  $(p \rightarrow b) \rightarrow a$

  $\underline{a \rightarrow c}$

  c

21. $\sim n \rightarrow s$

  $(n \vee b) \rightarrow w$

  $\underline{\sim w \vee \sim q}$

  $\sim q \vee s$

22. r

  $r \rightarrow (g \vee c)$

  $\underline{(c \vee p) \rightarrow k}$

  $\sim g \rightarrow k$

In Problems 23–32, test each argument for validity. If valid, construct a formal proof. If not, produce a counterexample chart.

23. $a \vee (c \to \sim p)$
    $\sim a \to (\sim p \vee c)$
    _____
    $\sim p \vee a$

24. $(\sim s \wedge p) \vee b$
    $(m \vee p) \to s$
    _____
    b

25. $(\sim c \vee d) \to e$
    $\sim e \vee d$
    $d \to z$
    _____
    z

26. $p \vee h$
    $w \to (k \to p)$
    $\sim h \vee (k \wedge w)$
    _____
    p

27. $(s \vee a) \to q$
    $(c \vee \sim d) \to \sim q$
    $d \to p$
    _____
    $s \to \sim p$

28. $(a \to h) \vee m$
    $(p \to m) \to \sim a$
    $\sim h \to \sim m$
    _____
    $h \vee p$

29. $\sim (d \to l)$
    $d \to (b \vee s)$
    $s \to (m \wedge \sim c)$
    _____
    $b \vee \sim c$

30. $(p \to c) \to a$
    $(w \vee p) \to \sim b$
    $\sim a \vee s$
    _____
    $w \to s$

31. $\sim r$
    $s \to r$
    $\sim s \to (d \to l)$
    $c \vee d$
    _____
    $\sim l \to c$

32. $(p \vee q) \to \sim a$
    $a \vee \sim s$
    $\sim w \to (p \wedge s)$
    $w \to r$
    _____
    r

In Problems 33–35, symbolize each argument. If the argument is valid, construct a formal proof. If not, produce a counterexample chart.

33. You either fly with a charter airline or you can't go on a vacation. If you pay full price for your airline ticket or your flight departs between 9:00 a.m. and 5:00 p.m., then you didn't fly with a charter airline. If you didn't pay full price for your airline ticket you can stay in a better hotel. Therefore, if you go on a vacation, then you can stay in a better hotel.

34. If Mike doesn't play nine holes of golf then he will play in his jazz band. If Mike goes to Virginia to visit his friend, then if the weather is nice he will play nine holes of golf. Either Mike will not play in his jazz band, or the weather is nice and he will go to Virginia to visit his friend. Therefore, Mike plays nine holes of golf.

35. Rich will either create a crossword puzzle or a cryptic crossword puzzle. If Rich goes to Connecticut then either he will not judge a crossword puzzle contest or he will create a crossword puzzle. Rich will either go to Connecticut and judge a crossword puzzle contest or not create a cryptic crossword puzzle. Therefore, Rich creates a crossword puzzle.

DID YOU KNOW?

Since ancient times, logic has contributed to the study of foundations of mathematics. In fact, the logic developed by Aristotle predominated until the mid-nineteenth century. The modern study of logic began in the late 19th century with the development of axiomatic approaches for geometry, arithmetic, and analysis. In 1903, the British mathematicians Alfred North Whitehead (1861–1947) and Bertrand Russell (1872–1970) published *Principia Mathematica,* which is a text on the foundations of mathematics. They attempted to derive mathematical truths from axioms and inference rules in logic.

In the early 20th century, logic was shaped by David Hilbert (1862–1943), who was a German mathematician. He has been recognized as one of the most influential mathematicians of the last one hundred years. It is a little known fact that Hilbert supplied significant portions of the mathematics required for Einstein's General Theory of Relativity.

Check Your Understanding

A valid argument is shown below. Two formal proofs verifying the validity of this argument are given. Supply the reason that justifies each statement in each proof.

1. $(\sim s \vee g) \rightarrow b$

   $\sim b \vee w$

   $\dfrac{w \rightarrow k}{\sim s \rightarrow k}$

Proof A	Proof B

Proof A

Statement	Reason
1. $\sim s$	1. _____
2. $\sim s \vee g$	2. _____
3. $(\sim s \vee g) \rightarrow b$	3. _____
4. b	4. _____
5. $\sim b \vee w$	5. _____
6. w	6. _____
7. $w \rightarrow k$	7. _____
8. k	8. _____
9. $\sim s \rightarrow k$	9. _____

Proof B

Statement	Reason
1. $\sim(\sim s \rightarrow k)$	1. _____
2. $\sim s \wedge \sim k$	2. _____
3. $\sim s$	3. _____
4. $\sim k$	4. _____
5. $w \rightarrow k$	5. _____
6. $\sim w$	6. _____
7. $\sim b \vee w$	7. _____
8. $\sim b$	8. _____
9. $(\sim s \vee g) \rightarrow b$	9. _____
10. $\sim(\sim s \vee g)$	10. _____
11. $s \wedge \sim g$	11. _____
12. s	12. _____
13. $\sim s \rightarrow k$	13. _____

Chapter 3 Review

In questions 1–10, match each valid argument to the equivalence or valid argument form it illustrates.

1. $\dfrac{\sim[r \rightarrow \sim(w \vee s)]}{r \wedge (w \vee s)}$

2. $\dfrac{\sim[r \wedge (w \rightarrow s)]}{\sim r \vee \sim(w \rightarrow s)}$

3. $r \rightarrow \sim(w \vee s)$
   $\dfrac{w \vee s}{\sim r}$

4. $r \vee \sim(w \vee s)$
   $\dfrac{w \vee s}{r}$

5. $\dfrac{r}{r \vee (w \vee s)}$

6. $\dfrac{r \wedge \sim(w \vee s)}{\sim(w \vee s)}$

7. $r \rightarrow \sim(w \vee s)$
   $\dfrac{r}{\sim(w \vee s)}$

8. $\sim(w \vee s)$
   $\dfrac{r}{r \wedge \sim(w \vee s)}$

9. $\dfrac{r \rightarrow \sim(w \vee s)}{(w \vee s) \rightarrow \sim r}$

10. $\dfrac{r \rightarrow \sim(w \vee s)}{\sim r \vee \sim(w \vee s)}$

a. Conjunctive Addition
b. Modus Tollens
c. Disjunctive Syllogism
d. Conditional Negation
e. Conditional Equivalence
f. Modus Ponens
g. DeMorgan's Equivalence
h. Conjunctive Simplification
i. Disjunctive Addition
j. Contrapositive Equivalence

Questions 11–13 each show the proof of a valid argument. Supply the reason that justifies each statement.

11. $b \wedge c$

$(r \vee s) \rightarrow \sim w$

$\sim r \rightarrow (p \vee \sim c)$

$(b \vee d) \rightarrow w$

_____

p

Statement	Reason
1. $b \wedge c$	1.
2. b	2.
3. c	3.
4. $b \vee d$	4.
5. $(b \vee d) \rightarrow w$	5.
6. w	6.
7. $(r \vee s) \rightarrow \sim w$	7.
8. $\sim (r \vee s)$	8.
9. $\sim r \wedge \sim s$	9.
10. $\sim r$	10.
11. $\sim r \rightarrow (p \vee \sim c)$	11.
12. $p \vee \sim c$	12.
13. p	13.

12. $s \rightarrow p$

$\sim c \rightarrow (d \wedge r)$

$\sim p \vee b$

$c \rightarrow \sim b$

_____

$s \rightarrow r$

Statement	Reason
1. s	1.
2. $s \rightarrow p$	2.
3. p	3.
4. $\sim p \vee b$	4.
5. b	5.
6. $c \rightarrow \sim b$	6.
7. $\sim c$	7.
8. $\sim c \rightarrow (d \wedge r)$	8.
9. $d \wedge r$	9.
10. r	10.
11. $s \rightarrow r$	11.

13. $r \rightarrow (a \rightarrow b)$

$(c \vee q) \rightarrow a$

$\underline{(c \rightarrow b) \rightarrow w}$

$\sim r \vee w$

Statement	Reason
1. $\sim (\sim r \vee w)$	1.
2. $r \wedge \sim w$	2.
3. r	3.
4. $\sim w$	4.
5. $r \rightarrow (a \rightarrow b)$	5.
6. $a \rightarrow b$	6.
7. $(c \rightarrow b) \rightarrow w$	7.
8. $\sim (c \rightarrow b)$	8.
9. $c \wedge \sim b$	9.
10. c	10.
11. $\sim b$	11.
12. $c \vee q$	12.
13. $(c \vee q) \rightarrow a$	13.
14. a	14.
15. $\sim a$	15.
16. $\sim r \vee w$	16.

In questions 14–21, construct a formal proof for each valid argument. Where appropriate, you may use either a direct proof, conditional proof or indirect proof.

14. $(\sim p \vee q) \rightarrow r$

$\sim p$

$\underline{a \rightarrow \sim r}$

$\sim a$

15. $(c \vee d) \wedge a$

$b \rightarrow \sim (c \vee d)$

$\underline{b \vee \sim s}$

$\sim s$

16. $(\sim d \vee a) \rightarrow s$

$d \vee \sim s$

$\underline{\sim p \rightarrow \sim d}$

p

17. $\sim (p \wedge s) \rightarrow q$

$a \wedge \sim q$

$\underline{a \rightarrow (\sim s \vee \sim c)}$

$p \wedge \sim c$

18. $b \to c$

$\sim c$

$(\sim b \lor \sim p) \to s$

_____

$s \lor q$

19. $\sim d \to c$

$k \to p$

$(d \lor c) \to k$

_____

p

20. $m \to \sim n$

$\sim b \lor s$

$m \land p$

$\sim n \to (a \land b)$

_____

s

21. $a \to (p \to \sim s)$

$l \lor a$

p

$(\sim s \lor k) \to b$

_____

$\sim l \to b$

In questions 22–35, test each argument for validity. If valid, construct a formal proof. Where appropriate, you may use either a direct proof, a conditional proof or an indirect proof. If invalid, produce a counterexample chart.

22. $\sim a \to g$

$a \to \sim g$

_____

g

23. $p \lor (\sim r \to s)$

$\sim (p \lor a) \land \sim s$

_____

r

24. $k \lor \sim (a \lor \sim b)$

$b \to c$

_____

$\sim k \to (c \lor n)$

25. $\sim (p \lor \sim s) \land \sim (a \to \sim b)$

$\sim w \to (s \land a)$

_____

w

26. $(a \lor g) \to l$

$\sim l$

$g \lor (p \lor q)$

_____

$\sim p \to q$

27. $(a \lor g) \to l$

a

$(\sim m \lor w) \to \sim l$

_____

$m \to g$

28. $(a \lor s) \to n$

$a \land \sim b$

$\sim n \lor \sim p$

_____

$\sim (p \lor b)$

29. $\sim (s \to \sim q)$

$\sim q \lor p$

$(p \lor d) \to c$

_____

c

30. $\sim s \to q$
    $a \vee \sim b$
    $(\sim p \vee q) \to b$
    _____
    $\sim a \to s$

31. $(a \vee d) \to w$
    $(\sim e \vee b) \to p$
    $a \vee \sim e$
    _____
    w

32. a
    $\sim r \vee w$
    $(a \vee b) \to r$
    $(m \vee n) \to \sim w$
    _____
    $\sim m$

33. $r \vee (s \wedge p)$
    $\sim a \vee b$
    $(\sim r \to s) \to a$
    $(\sim w \vee q) \to \sim b$
    _____
    w

34. $w \to \sim a$
    $(\sim w \wedge b) \to c$
    $\sim c \vee p$
    $\sim (a \wedge \sim b)$
    _____
    $a \to p$

35. $\sim (p \vee \sim r)$
    $(r \wedge w) \to k$
    $k \to (a \vee q)$
    $a \to s$
    _____
    $w \to q$

For questions 36–37, each set of premises is followed by four possible conclusions. Select the conclusion(s) that makes the argument valid.

36. If the program is run, the computer prints headings and reads data. The withholding routine is performed if there is tax due. If data is read, then the computer performs the calculation routine. The program is run and there is not tax due. Therefore,
    a. the calculation routine is performed.
    b. the withholding routine is performed.
    c. the withholding routine is not performed.
    d. headings are printed.

37. If the stock price is high Frank will sell his stock but if the stock price is low then he buys stock. Frank gets a dividend if the stock price is high or if it is low. If Frank buys stock then the stock exchange is open. The stock price is high but the stock exchange is not open. Therefore,
    a. Frank sells his stock.   b. Frank buys stock.
    c. Frank gets a dividend.   d. the stock price is low.

Sample Exam: Chapter 3

In questions 1–8, state the illustrated equivalence or valid argument form.

1. $\sim(a \rightarrow d) \vee (r \wedge q)$

$$\dfrac{a \rightarrow d}{r \wedge q}$$

2. $(\sim n \vee c) \rightarrow (p \rightarrow \sim q)$

$$\dfrac{\sim(p \rightarrow \sim q)}{\sim(\sim n \vee c)}$$

3. $\dfrac{(l \vee \sim m) \wedge k}{l \vee \sim m}$

4. $\dfrac{\sim(g \rightarrow \sim h)}{g \wedge h}$

5. $\dfrac{\sim s \vee \sim p}{s \rightarrow \sim p}$

6. $\dfrac{a \rightarrow \sim b}{(a \rightarrow \sim b) \vee c}$

7. $\dfrac{\sim r \vee w}{\sim(r \wedge \sim w)}$

8. $\sim(a \vee m) \rightarrow (d \rightarrow j)$

$$\dfrac{\sim(a \vee m)}{d \rightarrow j}$$

In questions 9–14, each set of premises is followed by four possible conclusions. Which of the four conclusions makes the argument valid?

9. $\sim(\sim p \vee q)$
$(\sim w \vee r) \rightarrow s$
$$\dfrac{w \rightarrow \sim p}{}$$
a. r b. s
c. w d. $\sim p$

10. $(\sim m \vee c) \rightarrow l$
$l \rightarrow (p \rightarrow \sim r)$
$$\dfrac{\sim m \wedge p}{}$$
a. r b. $\sim l$
c. m d. $\sim r$

11. $\sim(p \rightarrow \sim q)$
$(q \vee r) \rightarrow \sim s$
$$\dfrac{p \rightarrow (w \vee s)}{}$$
a. $p \rightarrow \sim q$

b. $\sim s \wedge \sim w$

c. $w \vee a$

d. $\sim q \vee s$

12. $\sim w \rightarrow \sim p$
$p \wedge \sim k$
$$\dfrac{\sim w \vee n}{}$$
a. $p \rightarrow \sim w$

b. $w \rightarrow \sim p$

c. $k \vee \sim n$

d. $n \wedge \sim k$

13. $\sim p \lor q$

$\sim (\sim q \to \sim p) \lor r$

$r \to (s \lor b)$

---

a. s b. b

c. $s \lor b$ d. $\sim r$

14. $(a \lor b) \land c$

$c \to \sim a$

$b \to d$

---

a. $\sim d$ b. $c \land a$

c. d d. $\sim a \to \sim d$

In question 15, supply the reason that justifies each statement.

15. $\sim w \to \sim a$

$e \to (d \to g)$

$\sim (e \to g)$

$d \lor (c \lor a)$

$c \lor w$

Statement	Reason
1. $\sim (e \to g)$	1.
2. $e \land \sim g$	2.
3. e	3.
4. $\sim g$	4.
5. $e \to (d \to g)$	5.
6. $d \to g$	6.
7. $\sim d$	7.
8. $d \lor (c \lor a)$	8.
9. $c \lor a$	9.
10. $\sim (c \lor w)$	10.
11. $\sim c \land \sim w$	11.
12. $\sim c$	12.
13. $\sim w$	13.
14. $\sim w \to \sim a$	14.
15. $\sim a$	15.
16. c	16.
17. $c \lor w$	17.

In questions 16–17, construct a formal proof for each valid argument.

16. $(\sim s \vee r) \rightarrow \sim d$
d
$\dfrac{r \vee \sim b}{\sim b \vee n}$

17. $\sim s \rightarrow \sim a$
$d \rightarrow (a \wedge \sim c)$
$\dfrac{(s \vee q) \rightarrow r}{d \rightarrow r}$

In questions 18–21, test the argument for validity. If valid, construct a formal proof. If not, produce a counterexample chart.

18. $(b \vee \sim c) \rightarrow \sim p$
$\sim a \rightarrow (k \rightarrow b)$
$\dfrac{\sim a \wedge p}{\sim k}$

19. $g \rightarrow k$
$(p \vee a) \rightarrow c$
$\dfrac{(\sim g \vee k) \rightarrow p}{c \wedge p}$

20. $(w \rightarrow r) \vee n$
$b \rightarrow s$
$\dfrac{w \rightarrow (a \wedge b)}{w \rightarrow n}$

21. $a \vee (p \wedge \sim s)$
$a \rightarrow k$
$\dfrac{(s \vee b) \rightarrow \sim (a \vee p)}{k}$

In questions 22 and 23, symbolize each argument. If the argument is valid, construct a formal proof. If not, produce a counterexample chart.

22. If you want to eat a cannoli then you should also order cappuccino. If you do not want to eat a cannoli then you should order gelato. If you order espresso then you should not order cappuccino. You ordered espresso. Therefore, you should order gelato.

23. If you go to college you can opt to be part of Greek life. If you opt to be part of Greek life then you will either join a fraternity or a sorority. If you join a fraternity then you will go to many parties and meet a lot of men. Therefore, if you go to college you will meet a lot of men.

Valid Argument Forms

1. Modus Ponens

Given: An implication $p \rightarrow q$

The LHS of the implication p

Conclusion: The RHS of the implication q

2. Modus Tollens

Given: An implication $p \rightarrow q$

The negation of the RHS of the implication $\sim q$

Conclusion: The negation of the LHS of the implication $\sim p$

3. Disjunctive Syllogism

Given: A disjunction

The negation of one side of the disjunction $p \vee q \qquad p \vee q$

Conclusion: The other side of the disjunction $\dfrac{\sim p}{q} \qquad \dfrac{\sim q}{p}$

4. Disjunctive Addition

Given: A statement p

Conclusion: The disjunction of the given statement with any statement $p \vee z$

5. Conjunctive Addition

Given: A statement p

Another statement q

Conclusion: The conjunction of the two given statements $p \wedge q$

6. Conjunctive Simplification

Given: A conjunction $p \wedge q \qquad p \wedge q$

Conclusion: Either side of the conjunction $p \qquad\qquad q$

CHAPTER FOUR

APPLICATIONS OF LOGIC

Mathematical logic is built on verifying the truth value of *TF* statements. Logical arguments start with premises that are known to be true. The truth values of facts are easily verifiable, however in mathematics oftentimes generalizations are made. It is difficult to verify the truth value of certain generalizations using the constructs outlined in this text thus far. In Chapter Four, we will examine a construct that allows us to verify the truth values of generalizations. We will also examine an application of logic that can be used to rewrite and reduce electronic circuits.

Objectives

After completing Chapter Four, the student should be able to:
- Identify predicate statements.
- Translate quantified statements using logic symbols.
- Determine the truth value of a predicate statement given specific instances of the subject.
- Identify quantifier words.
- Determine the truth value of quantified predicate statements.
- Negate quantified statements.
- Test the validity of quantified statements using valid argument forms and/or logical equvalences.
- Determine if a given circuit has current flowing through it.
- Determine if two circuits are equivalent.
- Simplify a given circuit by finding an equivalent circuit.

4.1 Introduction to Predicates and Quantifiers

Verifying *TF* Statements

Recall that a *TF* statement is a declarative sentence to which we can assign a truth value of either true or false, but not both at the same time.

Example 1

Determine if the following *TF* statements are true or false:
  a. The Toronto Maple Leafs are a professional football team.
  b. Algebra is a topic taught in mathematics.
  c. Norway is a European country.
  d. The Nile is a river found in South America.

Solution

  a. False. The Toronto Maple Leafs are a professional hockey team.
  b. True. Algebra is a topic taught in mathematics.
  c. True. Norway is a country in Northern Europe.
  d. False. The Nile is found in Africa.

  ■

Predicates

A *predicate* is a statement that can be true or false depending on the values of input variables. The predicate statement consists of a subject or object of the statement and a condition which is applied to the subject. The subject is the variable of the sentence.

Example 2

The statement "x has four legs" is a predicate. The condition is "having four legs." This statement is true or false depending on the value of x. For $x =$ "A horse," the predicate would become "A horse has four legs." This is true.

For $x =$ "A pigeon," the predicate would be become "A pigeon has four legs." This is false.

  ■

We define the *universal set*, or simply, the universe, as a set of elements from which we construct all other sets in a particular discussion. In other words, the universe can be thought of as the set of all elements or objects under consideration. Each predicate statement has such a universal set that goes along with it. Notice that for the above example, the condition of "having four legs" is applied to objects having legs. Some examples are animals, people, chairs, and insects. Notice it would not make sense for x to equal "The color red." Therefore each predicate statement has an implied universal set or, ideally, a defined universal set.

When we give a value for x, or the subject of a predicate statement, we are defining a particular instance of the condition. Sometimes, we want to make general statements about groups of objects.

Example 3

Use the given predicate together with the given universal set to determine if the predicate statement is true or false for the given value of x.

Predicate: "x is often eaten for lunch"

$U = $ {items found in a supermarket}

   a. $x = $ "Pizza"
   b. $x = $ "Toothpaste"
   c. $x = $ "Hamburger"
   d. $x = $ "Turkey breast"
   e. $x = $ "A pack of AAA batteries"

Solution

   a. True. Replacing the variable x by "Pizza" results in the statement "Pizza is often eaten for lunch," which is certainly true.
   b. False. Replacing the variable x by "Toothpaste" results in the statement "Toothpaste is often eaten for lunch," which is certainly false.
   c. True. Replacing the variable x by "Hamburger" results in the statement "Hamburger is often eaten for lunch," which is certainly true.
   d. True. Replacing the variable x by "Turkey breast" results in the statement "Turkey breast is often eaten for lunch," which is certainly true.
   e. False. Replacing the variable x by "A pack of AAA batteries" results in the statement "A pack of AAA batteries is often eaten for lunch," which is certainly false. ■

Quantifiers

Quantifiers are words or phrases that express how many elements from the universal set should be applied to the predicate. There are two logic quantifiers. The *universal quantifier* indicates that every element in the universal set must to be applied to the predicate. The *existential quantifier* indicates that there exists at least one element in the universal set that can be applied to the predicate. To prove a universally quantified statement false, one just needs to provide a counterexample. To prove an existentially quantified statement true, one just needs to provide an instance or specific element from the universal set.

Example 4

Determine whether each quantified statement is true or false. When possible, prove your answer using an appropriate instance or counterexample:

   a. For all real numbers $x, -x$ is a negative number.
   b. Some parallelograms have four right angles.
   c. At least one U.S. college admits only women.
   d. For every integer $n, 2n + 1$ is an odd integer.

Solution

   a. False. If x is $-4, -x$ will be 4, a positive number. This one counterexample is sufficient to prove that the given universally quantified statement is false.

   b. True. Squares are parallelograms with four right angles. This one instance is sufficient to prove that the given existentially quantified statement is true.

   c. True. Mount Holyoke College in Massachusetts admits only women. This one instance is sufficient to prove that the given existentially quantified statement is true.

   d. True, however proving that this universally quantified statement is true requires more than simply replacing n by an appropriate integer. It must be proved true for all integers. Proving that this is true (and similarly proving that an existentially quantified statement is false) is much more involved and is beyond the scope of this book. ∎

Negating Quantified Statements

Recall that by negating a *TF* statement, we create a new statement whose truth value is opposite that of the original statement. To negate a universally quantified statement, we change the statement to an existentially quantified statement and negate the condition. To negate an existentially quantified statement, we change the statement to a universally quantified statement and negate the condition.

Example 5

Negate each of the following quantified statements.

a. Some Atlanta Braves fans leave the game early.

b. Every musician has a source of inspiration.

c. All professors cannot grade 150 finals in one night.

d. There are some birds that cannot fly.

Solution

a. All Atlanta Braves fans do not leave the game early.

b. Some musicians do not have a source of inspiration.

c. There are some professors that can grade 150 finals in one night.

d. All birds can fly.   ■

Negation Quantifiers

Sometimes we express ourselves using negative sentences. Two negative sentences are "No one ever listens to me!" and "Not all carefully planned events are successful." How would we verify the truth value of each sentence?

A student named Ellen might say "No one ever listens to me!" No matter how upset Ellen is and no matter how much she believes this to be true, her statement is most likely false. Why? For this to be true, Ellen would have to verify that every person that she has ever come into contact with has not listened to her. The phrase "No one" is a universal quantifier; one that quantifies a negative condition. The statement "No one (condition)" translates to "All...do not (condition)."

Now consider the statement "Not all carefully planned events are successful." This statement is probably true. If one thinks of all the events that have been carefully planned (i.e. a vacation, a surprise party, a holiday dinner, etc.) it is almost certain there was at least one time when something went wrong (i.e. a flat tire on the way to the airport, a chatty party guest ruining the surprise, a burnt roast, etc.). The phrase "Not all" is an existential quantifier; it quantifies a negative condition. "Not all (condition)" translates to "Some...do not (condition)."

Example 6

Rewrite each statement without the negation quantifier.

a. Not one pair of shoes I own matches that dress.

b. Not all hotels give their guests a free breakfast.

Solution

a. All the shoes I own do not match that dress.

b. Some hotels do not give their guests a free breakfast.

■

In-Class Exercises and Problems for Section 4.1

In-Class Exercises

For Exercises 1–5, use the given predicate along with the given universal set to determine if the predicate statement is true or false for the given value of x.

Predicate: "x needs electricity to operate"

$U = \{$the set of household items$\}$

1. $x =$ "My gas powered lawnmower"

2. $x =$ "My television"

3. $x =$ "My wine bottle opener"

4. $x =$ "My refrigerator"

5. $x =$ "My kitchen faucet"

For Exercises 6–10, use the given predicate along with the given universal set to determine if the predicate statement is true or false for the given value of x.

Predicate: "$x^2 > x$" $U = \{...,-2,-1,0,1,2,...\}$

6. $x = -2$

7. $x = 2$

8. $x = 1$

9. $x = 0$

10. $x = 5$

For Exercises 11–14, decide whether the following quantified statements are true or false. Provide a counterexample or instance where applicable.

11. Some mammals are nocturnal.

12. At least one state in the United States has no capital city.

13. There exists a real number solution to $x^2 + 1 = 0$.

14. For all positive integers x, $\frac{1}{x} < x$.

For Exercises 15–17, write the negation of each statement.

15. Some calculus texts are hard to read.

16. All subway stations are not clean.

17. Every person at the DMV is extremely helpful.

For Exercises 18–20, rewrite each statement without the negation quantifier.

18. Not every chef has a signature dish.

19. No one at the gym uses the upright bicycles.

20. No bank customer pays for checks.

Problems

For Problems 21–24, use the given predicate along with the given universal set to determine if the predicate statement is true or false for the given value of x.

Predicate: "x needs batteries to operate"
$U = \{$the set of household items$\}$

21. $x = $ "My cellphone"

22. $x = $ "My toaster oven"

23. $x = $ "My flashlight"

24. $x = $ "My microwave oven"

For Problems 25–28, use the given predicate along with the given universal set to determine if the predicate statement is true or false for the given value of x.

Predicate: "$x^3 \leq x^2$" $U = \{..., -2, -1, 0, 1, 2, ...\}$

25. $x = -3$

26. $x = 1$

27. $x = -1$

28. $x = 2$

For Problems 29–32, decide whether the following quantified statements are true or false. Provide a counterexample or instance where applicable.

29. All US citizens 18 years or older are eligible to vote.

30. Every item in your local produce aisle was grown in the United States.

31. All quadratic functions have an output value when the input value is equal to zero.

32. There is some real number, x, that satisfies the equation $2x - 3 = 11 - x$.

For Problems 33–35, write the negation of each statement.

33. There are a few roads in the city that have potholes.

34. Each time I go pick up my Chinese food, they forget the soy sauce.

35. Some gas stations offer full service.

For Problems 36–38, rewrite each statement without the negation quantifier.

36. Not all logic students have to study.

37. Not every ATM charges a transaction fee.

38. Not all squared numbers are positive.

4.2 Logic and Quantified Statements

TF statements are used in logical arguments. An argument is a collection of *TF* statements called premises, which are assumed to be true and a final statement called a conclusion. A valid argument is an argument where the conclusion is shown to be true based on the true premises. In order to construct logical arguments, we need to be able to translate sentences into symbols.

Translating Universally Quantified Statements

Consider the statement "All students do their homework." How can this be verified? We need to search a set of objects that could be students, verify that the object is a student and for every object that is a student, verify it does homework. Therefore, the universal set for this quantified statement is $U = \{x \mid x$ is a person$\}$. This statement has two conditions, being a student and doing homework. Therefore we need to symbolize each. We use function notation to represent predicates since the truth value of the predicate depends upon the value of the input variable. Therefore, we have:

$S(x) = $ "x is a student" and $H(x) = $ "x does homework."

Note that we only need to verify "doing homework" for those objects that are students. Therefore, "*if* a person is a student *then* the person does homework" conveys the meaning "All students do homework."

Let's review the truth table for the conditional statement to see why a conditional statement best conveys the meaning of a universal quantifier.

S	H	$S \rightarrow H$	Comments
T	T	T	Any object from the universal set found to be a student is also found to do homework. This conveys the meaning of "All students do homework."
T	F	F	Any object from the universal set found to be a student is found not to do homework. This does not convey the meaning of "All students do homework."
F	T	T	Any object from the universal set found not to be a student is found to do homework. The implication gives a rule about objects found to be students. Non-students may or may not do homework. Therefore, non-students found doing homework does not violate the rule of students doing homework.
F	F	T	Any object found not to be a student is also found not to do homework. The universal statement only comments about students doing homework. We don't know what non-students do; they may do homework, they may not do homework. Therefore, non-students found not doing homework does not violate the rule of students doing homework.

> The logic symbol for a universal quantifier is $\forall x$, read "for all x." Therefore, a universally quantified statement is symbolized as $\forall x \; P(x) \rightarrow Q(x)$. Other words associated with a universal quantifier are "every" and "each".

Example 1

Translate each of the following sentences using logic symbols. Define your predicate functions.

    a.   All pet owners love animals.

    b.   Every fishing boat does not need to register with the Coast Guard.

Solution

    a. Let $P(x) =$ "x is a pet owner," and let $L(x) =$ "x loves animals." Then, $\forall x \; P(x) \rightarrow L(x)$.

    b. Let $B(x) =$ "x is a fishing boat," and let $R(x) =$ "x needs to register with the Coast Guard." Then $\forall x \; B(x) \rightarrow \sim R(x)$.

■

Translating Existentially Quantified Statements

Consider the statement "Some shades of blue are unappealing." How can this be verified? We need to search a set of objects, such as a color palette, that contain "shades of blue" and verify that at least one of the objects that qualify as a "shade of blue" is unappealing. Therefore, the universal set for this particular quantified statement is $U = \{x \mid x$ is a color shade$\}$. The statement has two conditions, being a shade of blue and being unappealing. Therefore we need to symbolize each. Again, we use function notation to represent predicates since the truth value of the predicate depends upon the value of the input variable. Therefore, we have:

$B(x) =$ "x is a shade of blue " and $P(x) =$ "x is unappealing."

Note that we only need to verify a single object as being a shade of blue and unappealing to satisfy the statement "Some shades of blue are unappealing." Therefore "a color that is a shade of blue *and* unappealing" conveys the meaning that "some shades of blue are unappealing."

Let's review the truth table for the conjunction to see why a conjunction best conveys the meaning of an existential quantifier.

B	P	B∧P	Comments
T	T	T	Any object from the universal set found to be a shade of blue and unappealing would be the instance we need to show the statement "Some shades of blue are unappealing" is true.
T	F	F	Any object from the universal set found to be a shade of blue and appealing would not support the notion of "Some shades of blue are unappealing."
F	T	F	Any object from the universal set found not to be a shade of blue yet is unappealing does not support the statement "Some shades of blue are unappealing." This statement does not make a comment about shades other than blue.
F	F	F	Any object found not to be a shade of blue and appealing does not support the statement "Some shades of blue are unappealing." Again, this statement is not commenting on shades other than blue.

> The logic symbol for an existential quantifier is $\exists x$, read "there exists an x." Therefore, an existentially quantified statement is symbolized as $\exists x \, P(x) \wedge Q(x)$. Expressions associated with the existential quantifier include "some," "there exists one," "at least one," and "one."

Example 2

Translate each of the following sentences using logic symbols. Be sure to define your predicate functions.

a. Some companies give their employees a holiday bonus.

b. At least one route to New Jersey does not involve crossing a bridge.

Solution

a. $C(x) =$ "x is a company," $G(x) =$ "x gives their employees a holiday bonus." Therefore, $\exists x \, C(x) \wedge G(x)$.

b. $R(x) =$ "x is a route to New Jersey," $B(x) =$ "x involves crossing a bridge." Therefore, $\exists x \, R(x) \wedge \sim B(x)$.

■

Translating Negatively Quantified Statements

We can arrive at logical translations of negatively quantified statements by using the meaning of the negative quantifiers discussed in Section 4.1 and the logical translations of the universal and existential quantifiers discussed in this section.

Recall the statement from Example 6 in section 4.1, "Not one pair of shoes I own matches that dress." To verify "Not one," we have to go through the entire set of shoes and verify that all shoes in the set do not match the dress. Therefore, "Not one" is a universally quantified statement and can be translated as follows:

$$\sim (\exists x\ S(x) \wedge M(x)) \Leftrightarrow \forall x\ S(x) \rightarrow \sim M(x)$$

where $S(x)$ = "x are shoes I own" and $M(x)$ = "x matches that dress."

Similarly, recall from Example 6 in section 4.1, "Not all hotels give their guests a free breakfast." To verify "Not all," all we have to show is that there is at least one hotel that does not offer its guests a free breakfast." Therefore, "Not all" is an existentially quantified statement and can be translated as follows:

$$\sim (\forall x\ H(x) \rightarrow B(x)) \Leftrightarrow \exists x\ H(x) \wedge \sim B(x)$$

where $H(x)$ = "x is a hotel" and $B(x)$ = "x gives their guests a free breakfast."

Example 3

Translate each of the following sentences using logic symbols. First, define your predicate functions.

    a. Not all types of fats are bad for you.
    b. No one wants to be the bearer of bad news.

Solution

    a. $F(x)$ = "x is a type of fat," $B(x)$ = "x is bad for you." Therefore,

$$\sim (\forall x\ F(x) \rightarrow B(x)) \Leftrightarrow \exists x\ F(x) \wedge \sim B(x).$$

In English this means that there are some fats that not bad for you.

    b. $W(x)$ = "x wants," $B(x)$ = "x is a bearer of bad news." Therefore,

$$\sim (\exists x\ W(x) \wedge B(x)) \Leftrightarrow \forall x\ W(x) \rightarrow \sim B(x).$$

In English this means that no one wants to be the bearer of bad news.

■

The following table shows how we generalize finding negations of quantified statements.

Statement	Negation
All, Every, Each	Not all, Some do not
Some	None, All do not

In-Class Exercises and Problems for Section 4.2

In-Class Exercises

In Exercises 1–10, define $P(x), Q(x)$, and if necessary, $R(x)$ for each sentence. Then translate each sentence using logic symbols.

1. All country songs are sad.

2. Some cars are not hybrids.

3. Some berries found in the woods are not edible.

4. All carnival rides have a height requirement.

5. Some restaurants offer a take-out menu.

6. All vigorous exercises require you to stretch first.

7. All movies theatres have surround sound and previews.

8. Some diners offer soup or salad as a first course.

9. Not all roads close during a snowstorm.

10. No good citizen breaks the law.

Problems

In Problems 11–20, define $P(x), Q(x)$, and if necessary, $R(x)$ for each sentence. Then translate each sentence using logic symbols.

11. Some commodities are imported.

12. All actions have an equal and opposite reaction.

13. All prime numbers have only two factors.

14. Some languages have a masculine form of certain words.

15. All quadratic equations have at most two solutions.

16. Some rocks are formed by extreme pressure.

17. Some countries have suffered civil wars.

18. All fiction takes imagination to write.

19. No mammal can survive without water or food.

20. Not all poems rhyme.

4.3 Formal Proofs and Quantified Statements

Formal logical proof is a procedure that starts off with a set of true statements, called premises and by applying valid argument forms and/or logical equivalences, arrives at a final statement called a conclusion. In order to apply valid argument forms and/or logical equivalences to a statement, statements need to be translated into expressions using logical connectives. In Section 4.2, we showed how quantified statements can be translated into logical connectives. These translations are summarized in the following table:

All $P(x)$ are $Q(x)$	$\forall x \; P(x) \rightarrow Q(x)$
Some $P(x)$ are $Q(x)$	$\exists x \; P(x) \wedge Q(x)$
No $P(x)$ are $Q(x)$	$\forall x \; P(x) \rightarrow \sim Q(x)$
Not all $P(x)$ are $Q(x)$	$\exists x \; P(x) \wedge \sim Q(x)$

Since quantified statements involve conditionals and conjunctions, you may wish to first review the valid argument forms using these connectives before you go on. A summary of these forms is found on page 228.

Proofs involving quantified statements need one additional set of rules. To translate a quantified statement into a pure logic statement, i.e. without the symbols $\exists x$ or $\forall x$, we use four new rules discussed below.

Universal Instantiation

When every element of the universal set has a particular property, it follows that any particular element of the universal set has this property. This means that $\forall x \; P(x) \rightarrow Q(x)$ implies $P(c) \rightarrow Q(c)$ for any particular c that is an element of the universal set. This property is called *universal instantiation*. In proofs, we will denote this rule by UI.

Existential Instantiation

When there exists an element of the universal set that has a particular property, we write $\exists x \; P(x) \wedge Q(x)$. It follows that using logic notation, we can write this as $P(c) \wedge Q(c)$ where c is an element of the universal set with this property. This property is called *existential instantiation*. In proofs, we will denote this rule by EI.

Universal Generalization

If $P(c) \rightarrow Q(c)$ is true for an arbitrary c then it must follow that it is true for all c. Therefore we can write $\forall x \; P(x) \rightarrow Q(x)$. This rule is called *universal generalization*. In a sense, this is the "opposite" of universal instantiation. In proofs, we will denote this rule by UG.

Existential Generalization

If $P(c) \wedge Q(c)$ is true for any particular value c, then we can write $\exists x \; P(x) \wedge Q(x)$ for some x in the set. This rule is called *existential generalization*. In a sense, this is the "opposite" of existential instantiation. In proofs, we will denote this rule by EG.

Example 1

Use a formal proof to show the given argument is valid.

All elementary school students own a lunch box. Some elementary school students have a thermos. Therefore, some lunch boxes have a thermos.

Solution

First, translate each quantified statement. Let $E(x) = $ "x is an elementary school student," $L(X) = $ "x owns a lunch box," and $T(x) = $ "x has a thermos." Now we can write the argument as

$$\forall x \; E(x) \rightarrow L(x)$$
$$\underline{\exists x \; E(x) \wedge T(x)}$$
$$\exists x \; L(x) \wedge T(x)$$

Letting c stand for an instance of the objects that are elementary school students who have a thermos, we construct a proof using the same valid argument forms used in Chapter 3.

Statement	Reason
1. $\exists x \; E(x) \wedge T(x)$	Premise
2. $E(c) \wedge T(c)$	EI(1)
3. $\forall x \; E(x) \rightarrow L(x)$	Premise
4. $E(c) \rightarrow L(c)$	UI(3)
5. $E(c)$	CS(2)
6. $L(c)$	MP(4)(5)
7. $T(c)$	CS(2)
8. $L(c) \wedge T(c)$	CA(6)(7)
9. $\exists x \; L(x) \wedge T(x)$	EG(8)

Note that in step 4, since the universally quantified statement is true for *all* objects, it's obviously true for the same objects that are true for the existentially quantified statement. This is because both statements are being selected from the same universe. For this reason, *we always instantiate the existentially quantified statement first.* ∎

Example 2

Use a formal proof to show the argument given below is valid.

Not all movies are rated PG. Every nature show is rated PG. Therefore, some movies are not nature shows.

Solution

First, translate each quantified statement. Let $M(x) =$ " x is a movie," $P(x) =$ "x is rated PG," and $N(X) =$ "x is a nature show." Symbolizing the argument, we have

$$\exists x \ M(x) \wedge \sim P(x)$$
$$\underline{\forall x \ N(x) \rightarrow P(x)}$$
$$\exists x \ M(x) \wedge \sim N(x)$$

Letting c be an instance of the objects that are movies not rated PG, we have

Statement	Reason
1. $\exists x \ M(x) \wedge \sim P(x)$	Premise
2. $M(c) \wedge \sim P(c)$	EI(1)
3. $\forall x \ N(x) \rightarrow P(x)$	Premise
4. $N(c) \rightarrow P(c)$	UI(3)
5. $\sim P(c)$	CS(2)
6. $\sim N(c)$	MT(4)(5)
7. $M(c)$	CS(2)
8. $M(c) \wedge \sim N(c)$	CA(6)(7)
9. $\exists x \ M(x) \wedge \sim M(x)$	EG(8)

Again note that in step 4, since the universally quantified statement is true for *all* objects, it's obviously true for the same objects that are true for the existentially quantified statement. This is because both statements are being selected from the same universe. For this reason, we again instantiate the existentially quantified statement first. ■

In-Class Exercises and Problems for Section 4.3

In-Class Exercises

For Exercises 1–8, translate each argument and provide a formal proof.

1. People who wish to fish in some states must obtain a fishing license. All states have counties. Therefore, people who wish to fish in some counties must obtain a fishing license.

2. All people who stop at red lights obey the law. Some professional athletes do not obey the law. Therefore, some professional athletes do not stop at red lights.

3. All that glitters is not gold. Some evening gowns glitter. Therefore, some evening gowns are not gold.

4. All passports require a birth certificate. All bank transactions require ID. Some bank transactions do not require a birth certificate. Therefore, some transactions requiring ID do not need a passport.

5. Everything that is worth working for requires dedication. Each new job initially comes with long hours. Some new jobs are worth working for. Therefore, sometimes that which initially comes with long hours requires dedication.

6. No one ignorant of geometry may enter here. Some children are ignorant of geometry. Therefore, some children may not enter here.

7. Not all vacations are fun. Every trip to Grandma's is fun. Therefore, some vacations are not trips to Grandma's.

8. Not one item on the menu is appealing to me. Not all appealing items are cooked in a restaurant. Therefore, there is an item not on the menu that is not cooked in a restaurant.

Problems

For Problems 9–16, translate each argument and provide a formal proof.

9. All deadlines are firm dates that cannot be changed. Some deadlines are too short. Therefore, some firm dates that cannot be changed are too short.

10. All post office workers are off on Sunday. Some government employees are post office workers. All workers off on Sunday tend to family obligations. Therefore, some government employees tend to family obligations.

11. All baseball fans like statistics. All people who like statistics are good in math. All people who are good in math balance their checkbooks. Some people who don't balance their checkbooks can't buy tickets to the game. Therefore some people who can't buy tickets to the game are not baseball fans.

12. Not all days in April are rainy. All days when the basement floods are raining days. Therefore there are some days in April when the basement does not flood.

13. All cave crickets are insects that hibernate during the winter. Some insects in Mr. Miller's basement do not hibernate during the winter. All insects Mrs. Miller can trap are cave crickets. Therefore, some insects in Mr. Miller's basement cannot be trapped by Mrs. Miller.

14. All squares are rectangles. All squares are parallelograms. Some squares are rotated to look like diamonds. Therefore, some rectangles are parallelograms.

15. No one person can do it all. Some people are strong-willed. Therefore some person who is strong-willed cannot do it all.

16. All schools are institutions that receive state aid. Some schools are private institutions. Therefore, all institutions that do not receive state aid ask for donations.

4.4 Electronic Circuit Reduction

The ideas learned in logic can be used to simplify complicated *electronic circuits*. An electronic circuit consists of a wire, a source of electricity S, and a switch or switches that can be turned on and off. When a switch is on (i.e., the circuit is closed), current may flow through it, beginning at S, and terminating at a point T. If the switch is off (i.e., the circuit is open), current does not flow through it.

Consider the circuit below that begins at S, ends at T, and has two switches, A and B.

If both A and B are on, current will flow from S to T. If either or both switches are off, current will not flow from S to T. When both switches in a circuit must be on in order for current to flow, the switches are said to be arranged in *series*. We say A is in series with B, or there is a *series circuit* connecting A with B.

If we construct a table that shows us the conditions under which current flows in a series circuit containing two switches, A and B, it looks satisfyingly familiar.

A	B	Current flows from S to T
On	On	Yes
On	Off	No
Off	On	No
Off	Off	No

The table is analogous to the conjunction truth table in logic, with the "On" condition being equivalent to "true" and the "Off" condition being equivalent to "false". Therefore, we can symbolically represent the arrangement of two switches in series as $A \wedge B$. Light strings for Christmas trees are often wired in series. That's why when one bulb burns out, the entire string does not work.

If current can travel from S to T over two branches of a circuit as shown below, the circuit is said to be a *parallel circuit*, and switches A and B are said to be in *parallel*.

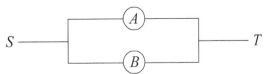

Notice that if a circuit is arranged in parallel, current will flow from *S* to *T* if either switch is on or if both switches are on.

If we construct a table that shows us the conditions under which current flows in a parallel circuit containing two switches, *A* and *B*, it too looks familiar.

A	*B*	Current flows from *S* to *T*
On	On	Yes
On	Off	Yes
Off	On	Yes
Off	Off	No

The table is analogous to the disjunction truth table in logic. Thus, we can symbolically represent the arrangement of two switches in parallel as $A \vee B$. Most lights in your home are wired in parallel. That's why when one bulb burns out, the others remain on.

If two switches are arranged such that whenever one switch is on the other is off, we say that the switches are *complementary* switches. If *A* represents a switch, then its complementary switch is denoted as $\sim A$. Often, these kinds of switches are found at the bottom and top of a staircase. This is what allows you to open and close the light over a staircase whether you are at the bottom or the top of the staircase.

If complementary switches are in series, current will not flow from *S* to *T*. In logic notation, this is expressed as $A \wedge \sim A = \text{false}$. If a switch and its complement are connected in series, current *never* flows from *S* to *T*.

On the other hand, if complementary switches are in parallel, current will flow from *S* to *T*. In logic notation, this is expressed as $A \vee \sim A = \text{true}$. That is, if a switch and its complement are connected in parallel, current *always* flows from *S* to *T*.

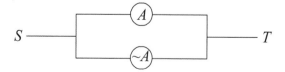

In most applications, we encounter more than two switches in a circuit.

Example 1

Consider the following circuit.

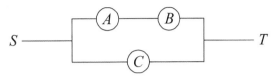

Express this circuit symbolically.

Solution

First, we notice that A and B are in series. We express this symbolically as $A \wedge B$. Switch C is in parallel with $A \wedge B$. Therefore, we can express the circuit as $(A \wedge B) \vee C$. ∎

Example 2

For the circuit shown in Example 1, under what conditions will current flow from S to T?

Solution

We answer this question by examining the equivalent logic question, i.e., under what conditions is $(A \wedge B) \vee C$ on. To do this, we construct a truth table.

A	B	C	$A \wedge B$	$(A \wedge B) \vee C$
On	On	On	On	On
On	On	Off	On	On
On	Off	On	Off	On
On	Off	Off	Off	Off
Off	On	On	Off	On
Off	On	Off	Off	Off
Off	Off	On	Off	On
Off	Off	Off	Off	Off

We see that the only times current will not flow from S to T is when either both A and C are off (the sixth row), when B and C are off (the fourth row), or when all three switches are off (the eighth row). ∎

We often use logical equivalences to reduce complex circuits into simpler ones. The distributive equivalence (see page 73) is often useful. There is another helpful logic equivalence called *absorption,* which can be expressed as $p \vee (p \wedge q) \Leftrightarrow p$ or as $p \wedge (p \vee q) \Leftrightarrow p$.

Example 3

Reduce the given circuit to an equivalent circuit with fewer switches.

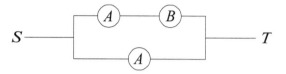

Solution

Our first step is to express the circuit in logic notation. On the upper branch of the circuit, A is in series with B. We represent this branch as $A \wedge B$. The lower branch only contains switch A. It is in parallel with the upper branch. Therefore, we can represent this circuit as $A \vee (A \wedge B)$. But, the absorption equivalence guarantees that $A \vee (A \wedge B)$ is equivalent to A. Hence, the original circuit that contained three switches can be equivalently represented as a circuit that contains only one switch, A.

This means that whether or not current flows is only dependent on switch A. If A is on, current flows, otherwise, it does not.

∎

Example 4

Reduce the given circuit to an equivalent circuit with fewer switches.

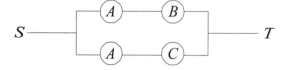

Solution

Both the upper and lower branches have two switches in series. The branches are in parallel with each other. Therefore, we can represent the above circuit as $(A \wedge B) \vee (A \wedge C)$. However, the distributive equivalence guarantees that $(A \wedge B) \vee (A \wedge C)$ and $A \wedge (B \vee C)$ are the same. This means that the circuit can be redrawn with switch A in series with a parallel branch consisting of switches B and C. The resulting reduced circuit is shown below. Notice that it has only three switches.

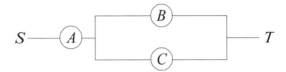

■

Example 5

Reduce the given circuit to an equivalent circuit with fewer switches.

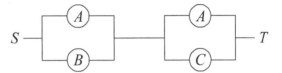

Solution

We have two parallel circuits in series with one another. The first and second parallel circuits may be expressed as $A \vee B$ and $A \vee C$ respectively. Since these two circuits are in series with each other, the entire circuit may be expressed as $(A \vee B) \wedge (A \vee C)$. Now we can use the distributive equivalence to express this as $A \vee (B \wedge C)$. The equivalent reduced circuit is shown below.

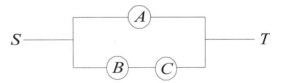

■

In the previous examples, we used either the absorption equivalence alone or the distributive equivalence alone to reduce a circuit. Sometimes, we can apply both laws to the same circuit as we try to reduce it.

Example 6

Reduce the given circuit to an equivalent circuit with fewer switches.

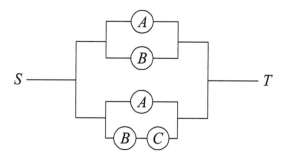

Solution

The upper branch of this parallel circuit can be expressed as the parallel sub-circuit $A \vee B$. The lower branch consists of a parallel sub-circuit, one of whose branches consists of a series circuit. Therefore, the lower branch can be expressed as $A \vee (B \wedge C)$. Hence, the entire circuit can be expressed as $(A \vee B) \vee [A \vee (B \wedge C)]$. If we apply the distributive equivanence to the lower branch, $A \vee (B \wedge C)$, we obtain $(A \vee B) \wedge (A \vee C)$. We can now express the entire circuit as $(A \vee B) \vee [(A \vee B) \wedge (A \vee C)]$. However, the absorption equivalence allows us to express this last statement as $A \vee B$. Our reduced circuit is shown below.

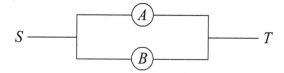

In-Class Exercises and Problems for Section 4.4

In-Class Exercises

For Exercises 1–8, use the circuit below to determine whether or not current will flow from S to T.

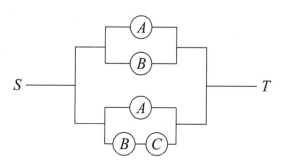

A	B	C
1. On	On	On
2. On	On	Off
3. On	Off	On
4. On	Off	Off
5. Off	On	On
6. Off	On	Off
7. Off	Off	On
8. Off	Off	Off

For Exercises 9–13, determine if each of the following pairs of circuits are equivalent.

9.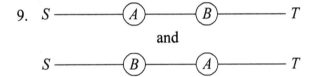

10.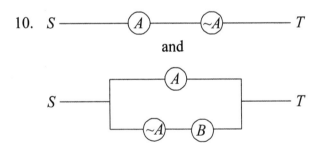

11.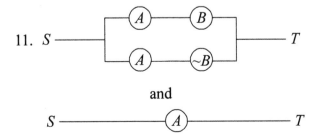

12. S
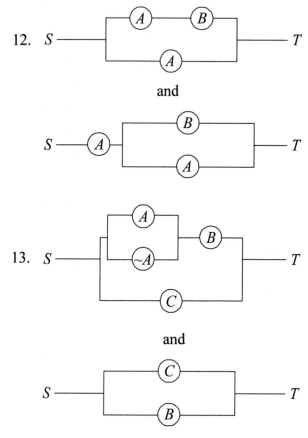

and

13. S

and

14. Design the simplest single circuit that has current flowing from S to T when either of the three conditions apply.
   · A, B and C are all on.
   · A and B are on but C is off.
   · A and C are off but B is on.

Problems

For Problems 15–22, use the circuit below to determine whether or not current will flow from S to T.

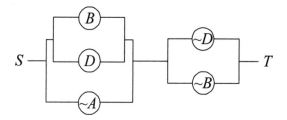

	A	B	D
15.	On	On	On
16.	On	On	Off
17.	On	Off	On
18.	On	Off	Off
19.	Off	On	On
20.	Off	On	Off
21.	Off	Off	On
22.	Off	Off	Off

For Problems 23–32, represent each circuit using logic notation. Then redesign each circuit into one which has fewer switches. Finally, express your answer in logic notation.

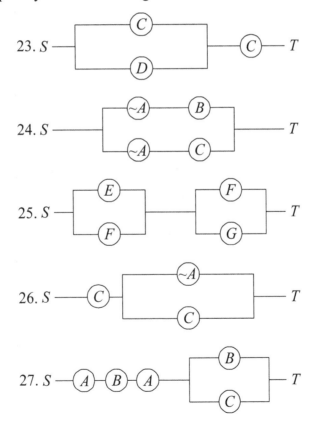

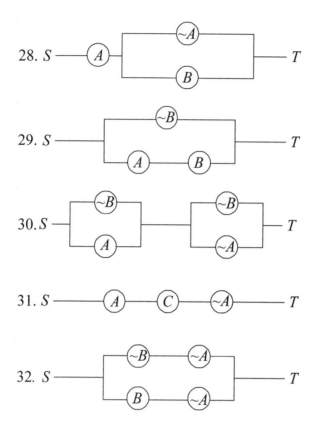

28. S ——(A)——(~A)(B)—— T

29. S ——(~B)(A)—(B)—— T

30. S ——(~B)(A)——(~B)(~A)—— T

31. S ——(A)——(C)——(~A)—— T

32. S ——(~B)—(~A)(B)—(~A)—— T

33. Think of a voting machine as having a concealed keypad for each voter. A motion is proposed and each voter votes either yes or no. Each yes vote closes a switch in the circuit and each no vote leaves the switch open. If current can get through the circuit after all the votes are cast, a motion passes and a light goes on at T. Suppose that there are three people on a committee and in order for a motion to pass, there must be a unanimous vote. That means persons A, B and C must all vote yes for the motion to pass. The set notation for this is $A \wedge B \wedge C$. The circuit resulting from this notation looks like the one shown below.

Notice that in this circuit, if anyone votes no, current does not flow through the system. Also notice that this circuit cannot be reduced, since all its components are unique. Suppose three people, *A*, *B*, and *C* vote on a motion, and the motion will pass if a majority of the three voters are in favor of it.

   a. Write the four logic expressions for the condition of majority rules.
   b. Since any of the four conditions will cause the proposal to pass, which logic connective should be used to join the four expressions?
   c. Use the distributive equivalence to simplify the expression obtained in part (b), and then draw the circuit.
   d. Suppose the motion will pass if *A* votes yes or both *B* and *C* vote yes. Write the logic expression for this condition.

Chapter 4 Review

1. Decide whether the following is true or false. Provide a counterexample or instance if necessary.

   " All zoos have a tank for a blue whale."

2. Write the negation of the statement: " Some baseball players make the Hall of Fame."

3. Write without the negation quantifier: " Not all math classes are boring."

4. Translate into symbols: "Some automobile drivers are rude and discourteous."

5. Provide a formal proof for following argument.

   All people make mistakes. Some people are professors. Therefore, some professors make mistakes.

For questions 6–13, use the circuit below to determine whether or not current will flow from S to T.

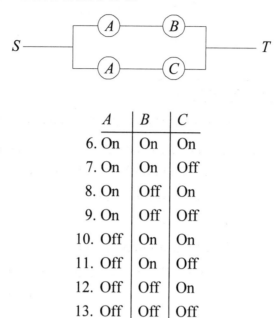

	A	B	C
6.	On	On	On
7.	On	On	Off
8.	On	Off	On
9.	On	Off	Off
10.	Off	On	On
11.	Off	On	Off
12.	Off	Off	On
13.	Off	Off	Off

For questions 14–16, represent each given circuit using logic notation. Then redesign each circuit into one that has fewer switches.

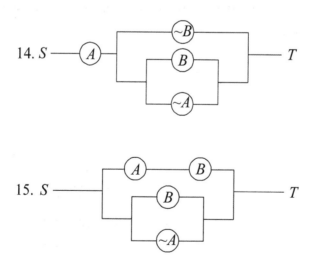

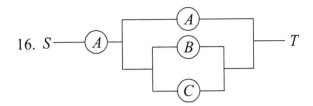

16. S

Sample Exam: Chapter 4

1. Answer true or false: All odd numbers are prime.

For questions 2–5 define $P(x), Q(x)$, and if necessary, $R(x)$ for each sentence. Then translate each sentence using logic symbols.

2. All rap music is loud.

3. No car fly.

4. All mushrooms are not edible.

5. Some NBA players are of Asian descent

6. Translate and create a formal proof for the following argument:

All guests must sign in at the desk. Not all people who enter the lobby are signed in at the desk. Therefore, some people who enter the lobby are not guests.

For questions 7–14 use the circuit below to determine whether or not current will flow from S to T.

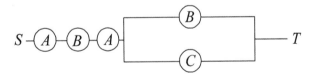

	A	B	C
7.	On	On	On
8.	On	On	Off
9.	On	Off	On
10.	On	Off	Off
11.	Off	On	On
12.	Off	On	Off
13.	Off	Off	On
14.	Off	Off	Off

15. Represent the circuit below using logic notation. Then redesign it into a circuit that has fewer switches.

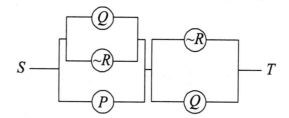

CHAPTER FIVE

SET THEORY

A concept that we are all familiar with is that of a set. We speak of a set of dishes, a set of CDs, or a tool set. The branch of mathematics that is concerned with the general study of sets is set theory. Set theory is important in the study of mathematics, and has many important applications in other areas as well.

Objectives

After completing Chapter Five, the student should be able to:

- Define a set, recognize its elements, and construct its subsets.

- Find the complement of a given set.

- Find the elements in the union, the intersection and the difference of two sets.

- Construct a Venn diagram to represent the relationships between sets.

- Solve survey problems using Venn Diagrams.

- Identify the Laws of Equal Sets.

- Use the Laws of Equal Sets to simplify compound statements about sets.

5.1 Sets and Notation

Introduction

Without getting too technical, we will think of a *set* as any collection. The objects that are in the set, called *elements,* or *members* of the set, may be numbers, letters, objects, or any other entities. Once we describe a particular set, we must be able to determine whether or not a given entity is or is not an element of that set.

As a matter of notation, braces such as { }, are used to enclose the elements of a set.

Example 1

The set consisting of the first three months of the year may be denoted as {January, February, March}.

∎

Just as we allow lowercase letters to represent simple statements in logic, we will use uppercase letters to represent sets. For example, we could say that the set consisting of the first three months of the year is called *A*. We then would write $A = $ {January, February, March}.

The phrase "is an element of" is expressed by the symbol $\in$, while the phrase "is not an element" is expressed symbolically as $\notin$.

Example 2

If $A = $ {January, February, March}, then March $\in A$, but June $\notin A$.

∎

In general, sets can be described in one of three ways. First, we could simply list the elements of a set within set braces, as we did in Example 1. This method is satisfactory as long as the number of elements is sufficiently small. If, however, the set has a large number of elements, listing them all becomes tedious or impossible. Rather, we can often just write a description of the set within set braces.

For instance, if we wanted to list the set of all whole numbers from one to one hundred, we could write {all whole numbers from one to one hundred}, or we could write {1,2,3,...,100}, where the three dots mean "and continuing in this fashion, up to." The third way a set can be described is using *set builder* notation. This notation is expressed symbolically as $S = \{x \mid x$ has a given property$\}$. The letter x denotes any element in the set, and the vertical line $\mid$ is read as "such that."

Example 3

If $S = $ {1,3,5,7,..., 99}, then S is the set of odd numbers from one to ninety-nine. Therefore, $5 \in S$, $97 \in S$, $36 \notin S$, and $103 \notin S$.

∎

Example 4

If $B = \{$whole numbers larger than five but less than 1000$\}$, then we know $7.25 \notin B$ and $976 \in B$.

■

Example 5

If $S = \{x | x$ is a sports car$\}$, the elements in set S would consist of all sports cars.

■

The Universal Set

We define the *universal set,* or simply, the *universe,* as the set of elements from which we construct all other sets in a particular discussion. We express the universe symbolically by U. If we define $U = \{1, 2, 3, 4, 5\}$, then the numbers one through five are the only numbers that exist for this particular discussion.

Example 6

Let $U = \{1, 2, 3, 4, 5\}$. Could the set $A = \{1, 3, 5, 7\}$ be constructed from this universe?

Solution

Since the element seven is not in the universe, this set cannot be constructed from the given universe.

■

The Null Set

The *null set,* is a set that contains no elements. This set is also referred to as the *empty set.* The null set is expressed symbolically as either $\varnothing$ or $\{\ \}$. At first, you may not be aware of the need to be able to talk about an empty set, but if you think for a moment, you will realize that such sets are ubiquitous. The set of all female presidents of the United States as of 2012 is an empty set, as is the set of all positive numbers less than zero.

Example 7

Find the set of all numbers that are multiples of 100 and odd.

Solution

Since all multiples of 100 are even, there are no multiples of 100 that are odd. Therefore, the set of all such numbers is empty, i.e. $\varnothing$.

■

Equal Sets

Two sets, A and B, are *equal sets* if every element of A is in B, and conversely, if every element in B is in A. Using some of our logic connectives, we can define the equality of two sets, A and B as $(A = B) \Leftrightarrow [[(x \in A) \rightarrow (x \in B)] \wedge [(x \in B) \rightarrow (x \in A)]]$.

The biconditional equivalence allows us to define the equality of two sets in a more compact way: $(A = B) \Leftrightarrow [(x \in A) \leftrightarrow (x \in B)]$.

Example 8

If $A = \{1,2,3\}$ and $B = \{3,2,1\}$, is $A = B$?

Solution

Since every element in set A is in set B and every element in set B is in set A, $A = B$.

■

Example 9

Let $A = \{$all whole numbers greater than 1 and smaller than 20$\}$. Let $B = \{2,3,4,5,6,7,8,9,10,11,12,13,14,15,16,17,18,19,20\}$. Is $A = B$?

Solution

Every element in set A is in set B. However, there is an element in set B that is not in set A, namely the number 20. Therefore, set A is not equal to set B. This is expressed symbolically as $A \neq B$.

■

Subsets

Set A is a *subset* of set B, expressed symbolically as $A \subseteq B$, if and only if every element in set A is an element of set B. This definition can be stated symbolically as $(A \subseteq B) \Leftrightarrow [(x \in A) \rightarrow (x \in B)]$. Notice that as a consequence of this definition, every set is a subset of itself. Furthermore, if $A = B$ then $A \subseteq B$ and $B \subseteq A$.

Example 10

If $A = \{1,3,5,6,7,8,9\}$ and $B = \{1,5,6\}$, is A a subset of B? Is B a subset of A?

Solution

Since there is at least one element in A that is not an element of B, A is not a subset of B. However, every element of B is an element of A. Therefore, $B \subseteq A$.

■

Example 11

Let $A = \{1,2,4,5\}$ and $B = \{2,5,7,9\}$. Is A a subset of B? Is B a subset of A?

Solution

There are elements in A that are not in B. There are also elements in B that are not in A. Hence, neither set is a subset of the other.

■

One of the most counterintuitive notions in set theory is that the null set is a subset of any set, A. To see why this is true, we need to recall that if this were not the case, there would be a least one element in $\varnothing$ that could not be found in A. This is not possible, since the empty set contains no elements.

Example 12

List all the subsets of $A = \{5, 6\}$.

Solution

The subsets of A are $\{5, 6\}, \{5\}, \{6\}$, and $\varnothing$. ∎

Example 13

Find all the subsets of $B = \{7, 8, 9\}$.

Solution

The subsets of B are $\{7, 8, 9\}, \{7\}, \{8\}, \{9\}, \{7, 8\}, \{7, 9\}, \{8, 9\}$, and $\varnothing$. ∎

Proper Subsets

Set A is called a *proper subset* of set B, expressed symbolically as $A \subset B$, if and only if set A is a subset of set B, but set A does not equal set B. This definition can be stated symbolically as $(A \subset B) \Leftrightarrow [[(x \in A) \rightarrow (x \in B)] \wedge (A \neq B)]$.

As a consequence of this definition, it should be noted that if A is proper subset of B, then there is at least one element in B that cannot be found in A.

Example 14

List all the proper subsets of $A = \{5, 6\}$.

Solution

The proper subsets of A are $\{5\}, \{6\}$, and $\varnothing$. ∎

Since every set is a subset of itself, the empty set (which has no elements) has one subset, which is the set itself. If a set has one element, it must have two subsets, itself and the empty set. In Examples 12 and 13, we saw that if a set has two elements, it has four subsets and if a set has three elements, it has eight subsets.

The following table summarizes these results.

Number of elements in A	Number of subsets of A	Number of proper subsets of A
0	1	0
1	2	1
2	4	3
3	8	7
n	2^n	$2^n - 1$

Two facts emerge from the table. It appears that as we add an element to a set, the number of subsets doubles from the previous set. Furthermore, the number of proper subsets is always one less than the total number of subsets. These results can be stated with equations that express the number of subsets and proper subsets as functions of the number of elements. Thus, if a set A has n elements, then A has 2^n subsets and $2^n - 1$ *proper* subsets.

As equations, we have:

if n is the number of elements in set A,

$$\text{the number of subsets of } A = 2^n,$$
$$\text{and}$$
$$\text{the number of proper subsets of } A = 2^n - 1.$$

Example 15

Can a set have exactly 67 subsets?

Solution

The number of subsets of a given set must be a power of two. Since 67 is not a power of two, there is no such set.

∎

The Power Set

The *power set* is the set of all subsets of a given set. If $A = \{1, 2, 3\}$, the power set of A, denoted as $\mathscr{P}(A)$, contains $2^3 = 8$ elements, and $\mathscr{P}(A) = \{\{1\}, \{2\}, \{3\} \{1, 2\}, \{1, 3\}, \{2, 3\}, \{1, 2, 3\}, \varnothing\}$. Notice that the elements of a power set are sets themselves.

Example 16

If $B = \{1, 2, 3, 4, 5\}$, how many elements are there in $\mathscr{P}(B)$, the power set of B?

Solution

The power set consists of all subsets of B. Since B has 5 elements, its power set has $2^5 = 32$ elements.

∎

In-Class Exercises and Problems for Section 5.1

In-Class Exercises

In Exercises 1–5, represent each set by listing its elements.

1. The set of whole numbers less than 20 that are divisible by 4.

2. $\{x | x$ is a multiple of 5 between 18 and 42$\}$.

3. The set of all two or three letter words that can be formed by using the letters "e", "n" or "o".

4. The set of all even numbers that can be written using the digits 1, 3, or 5.

5. The set of days of the week that do not contain the letter "s".

In Exercises 6–8, describe each set with an English sentence.

    6. $\{0, 2, 4, 6, 8, 10\}$

    7. $\{31, 33, 35, 37, 39\}$

    8. $\{a, b, c, d\}$

For Exercises 9–19, suppose that $U = \{0, 1, 2, 3, 4, 5, 6, 7, 8, 9, 10\}$, $A = \{1, 3, 5, 7, 9\}$, $B = \{0, 2, 4, 6, 8, 10\}$, $C = \{1, 2, 5, 7, 9\}$, $D = \{1, 3, 7, 9\}$, $E = \{8, 2, 10\}$, and $F = \{9, 7, 1, 3\}$.

    9. Determine if each statement is true or false.

a. $A = C$	e. $D = F$	i. $\emptyset \subset C$
b. $B \subseteq U$	f. $E \subseteq B$	j. $F \subset A$
c. $A \subset D$	g. $\emptyset \in D$	k. $F \subset D$
d. $7 \in B$	h. $F \subseteq D$	l. $D \subset A$

    10. Find the number of elements of C.

    11. Find the number of subsets of C.

    12. Find the number of proper subsets of C.

    13. Find the number of elements in $\mathscr{P}(C)$.

    14. Find the number of subsets of B.

    15. Find the number of proper subsets of B.

    16. Find the number of elements in $\mathscr{P}(D)$.

    17. Find the number of proper subsets of D.

    18. Find the number of elements in $\mathscr{P}(A)$.

    19. List $\mathscr{P}(E)$.

Problems

For Problems 20–25, represent each given set by listing its elements.

    20. The set of all Presidents of the United States before Thomas Jefferson.

    21. The set of whole numbers greater than 3 and less than 10.

    22. The set of months that have an "r" in their spelling.

    23. The set of all three-letter English words that can be formed using the letters "d", "g", "o", and "c", without repetition.

    24. The set of states in the Union that begin with the word "New".

    25. The set of negative numbers larger than 3.

For Problems 26–31, describe each set in English.

26. $\{1,3,5,7\}$

27. $\{1,3,5,7,...,21\}$

28. $\{3,6,9,12,...,36\}$

29. $\{1,4,9,16,25,36\}$

30. $\{\{1\},\{2\},\{1,2\},\{\ \}\}$

31. $S = \{x \mid x \text{ is a day of the week beginning with the letter } z\}$

For Problems 32–41, suppose $A = \{1,3,5,6,7\}$, $B = \{1,5,6\}$, $C = \{1,3,5,7,9,11\}$, and $D = \{1,3,7,9\}$. Determine if the given statement is true or false.

32. $A \subseteq B$

33. $B \subseteq A$

34. $D \subset A$

35. $A \subseteq C$

36. $1 \in B$

37. $\{1,3\} \subset D$

38. $\{1,3\} \in D$

39. $\varnothing \subseteq B$

40. $\varnothing \in B$

41. $B \subseteq D$

For Problems 42–46, suppose $A = \{1,3,5,6\}$, $B = \{1,5,6\}$, and $C = \{1,3,5,7,9,11\}$. Find:

42. The number of elements in B.

43. The number of subsets of B.

44. The number of elements in $\mathscr{P}(A)$.

45. The number of proper subsets of C.

46. The number of subsets of C less the number of proper subsets of C.

47. Is the following a correct definition for $A = B$? Explain your reasoning.

$$(A = B) \Leftrightarrow [[(x \in A) \to (x \in B)] \wedge [(x \notin A) \to (x \notin B)]]$$

The number of elements in a set is called its *cardinality*. Thus, the cardinality of set $A = \{1, 5, 6\}$ is 3, and the cardinality of set $B = \{a, b, c, ..., z\}$ is 26. Sets that have the same cardinality are said to be *equivalent*. For Problems 48-52, answer true or false for each question.

48. If two sets are equal they are equivalent.

49. If two sets are equivalent they are equal.

50. If A is a subset of B, the cardinality of A must be less than the cardinality of B.

51. The set $\{\varnothing\}$ has cardinality zero.

52. The cardinality of $A = \{4, 5, \{1, 2\}, 6\}$ is four.

JUST FOR FUN

In difficult economic times, at lunch time, Robin had taken to selling hard-boiled eggs to job seekers on Wall Street. At her first stop, she sold half her eggs and half an egg. At her second stop, she sold half her eggs and half an egg. At her third stop, she sold half her eggs and half an egg. At her fourth stop, she sold half her eggs and half an egg. At her fifth stop, she sold half her eggs and half an egg, and she had no more eggs left. How many eggs did she begin with?

Check Your Understanding

1. Represent the set of prime numbers between 16 and 24 by listing its elements.

   _____

2. Use an English sentence to describe the set $\{w, x, y, z\}$.

   _____

For questions 3–8, let $U = \{1,2,3,4,5,6,7,8,9,10\}$, $A = \{1,2,4,6,8,10\}$, $B = \{1,4,6,10\}$, $C = \{6,4,1,10\}$, and $D = \{6,4,1\}$. Determine if each statement is true or false.

3. $D \subseteq A$ _____

4. $B \subset C$ _____

5. $\varnothing \in B$ _____

6. $C \subseteq B$ _____

7. $\{1,4,6\} \subset D$ _____

8. $\{8,10\} \in A$ _____

9. If set $R = \{\text{red, orange, yellow, green, blue, indigo, violet}\}$, then what is the number of proper subsets of R?   _____

10. Set $S = \{\text{NY, NJ, CT}\}$. List $\mathcal{P}(S)$.   _____

   _____

5.2 Connectives

Introduction

In logic, we quickly discovered the need to unite *TF* statements. In this section, we will learn to combine sets. The similarities between negations, conjunctions, disjunctions, implications, and biconditionals and their set theory counterparts should become apparent.

Complementation

The operation of *complementation* in set theory is analogous to the operation of negation in logic. If set A is drawn from some universe, then the complement of A, denoted A', is the set of all elements of the universe that are not in A. This symbolization is sometimes read as "A prime." We can express this definition symbolically as $(x \in A') \Leftrightarrow [(x \notin A) \wedge (x \in U)]$.

Example 1

If $U = \{1,3,5,7,8,9\}$ and $A = \{1,5,7,9\}$ then $A' = \{3,8\}$. Notice that the numbers 2, 4, and 6 are not in the universe, so they cannot appear in A'.

∎

Example 2

If $U = \{1,3,5,7,8,9\}$ and $A = \{1,3,5,7,8,9\}$ then $A' = \varnothing$.

∎

Example 2 points out that, in general, the complement of the universal set is the empty set. Conversely, the complement of the empty set is the universal set.

Membership Tables

A *membership table* in set theory is analogous to a truth table in logic. It provides a way to show that an element is or is not a member of a set. For example, the membership table for complementation is shown in the following table.

A	A'
$\in$	$\notin$
$\notin$	$\in$

The table shows us that if an element is in set A, then it is not in A', and if an element is not in A then it must be in A'. The $\in$ symbol behaves just like the truth value of "true" and $\notin$ behaves just like the truth value of "false."

Notice that this membership table looks very much like the negation truth table in logic. We will construct membership tables for each of the connectives we encounter.

Intersection

If two sets, A and B, are drawn from some universe, U, then the *intersection* of A and B, denoted $A \cap B$, is the set of all elements that are in both A and B. We state this definition symbolically as $[x \in (A \cap B)] \Leftrightarrow [(x \in A) \land (x \in B)]$.

If two sets do not share any elements, they are said to be *disjoint*. Since disjoint sets do not share any elements, their intersection is the null set.

Example 3
Suppose $A = \{34, 56, 57, 66, 94\}$ and $B = \{35, 56, 66, 95\}$. Find $A \cap B$.
Solution
The elements that are in both sets are 56 and 66. Therefore, $A \cap B = \{56, 66\}$.   ■

Let's construct a membership table for the intersection of two sets, A and B. If an element is in both sets, it belongs in the intersection, otherwise, it does not. This is illustrated in the following membership table.

A	B	$A \cap B$
$\in$	$\in$	$\in$
$\in$	$\notin$	$\notin$
$\notin$	$\in$	$\notin$
$\notin$	$\notin$	$\notin$

Notice that this membership table is analogous to the truth table for the conjunction connective. Even the symbol we use for intersection looks quite similar to the symbol for conjunction. This should help you remember how the intersection connective behaves.

Example 4
Let $U = \{1, ..., 10\}$, $A = \{1, ..., 6\}$, $B = \{7, 8\}$, and $C = \{6, 7, 8\}$. Find $A \cap B$ and $A' \cap C$.

Solution

Set A and set B are disjoint. Therefore, $A \cap B = \varnothing$. Since A' is the set of all elements in the universe that are not in A, $A' = \{7, 8, 9, 10\}$. Therefore, $A' \cap C = \{7, 8\}$.   ∎

Union

If two sets, A and B, are drawn from some universe, U, then the *union* of A and B, denoted $A \cup B$, is the set of all elements that are either in A or B. We can state this definition symbolically as $[x \in (A \cup B)] \Leftrightarrow [(x \in A) \vee (x \in B)]$.

Example 5

Let $A = \{1, 2, 3\}$ and $B = \{2, 3, 4\}$. Find $A \cup B$.

Solution

The set of all elements that are in either A or B is $A \cup B = \{1, 2, 3, 4\}$.   ∎

Let's construct a membership table for the union of two sets, A and B. If an element is in either set, it belongs in the union, otherwise, it does not. This is illustrated in the following membership table.

A	B	$A \cup B$
$\in$	$\in$	$\in$
$\in$	$\notin$	$\in$
$\notin$	$\in$	$\in$
$\notin$	$\notin$	$\notin$

Again, notice that this membership table is analogous to the truth table for disjunction. Even the symbol we use for union looks quite similar to the symbol for disjunction. This should help you remember how the union connective behaves.

Example 6

If $U = \{1, 2, 3, 4, 5, 6, 7, 8, 9, 10\}$, $A = \{1, 2, 4, 5, 7, 8\}$, $B = \{1, 7, 8, 9, 10\}$, and $C = \{4, 8, 9, 10\}$, find $(A' \cup B) \cap C$.

Solution

We proceed by first resolving $A' \cup B$. We know that $A' = \{3, 6, 9, 10\}$, so $A' \cup B = \{1, 3, 6, 7, 8, 9, 10\}$. Hence, $(A' \cup B) \cap C = \{8, 9, 10\}$.   ∎

Set Difference

If two sets, A and B are drawn from some universe, then the *set difference*, $A - B$ is defined to be those elements in A but not in B.

Stating the definition of set difference symbolically, we have $[x \in (A-B)] \Leftrightarrow [(x \in A) \land (x \notin B)]$. Notice that the definition states that x is not in B. This means that x must be in B'. Therefore, we can also write the definition as $[x \in (A-B)] \Leftrightarrow [(x \in A) \land (x \in B')]$.

The membership table for set difference is constructed by recognizing that if an element is in the set difference $A-B$, it first must be in set A, but *cannot* be in set B.

A	B	$A-B$
$\in$	$\in$	$\notin$
$\in$	$\notin$	$\in$
$\notin$	$\in$	$\notin$
$\notin$	$\notin$	$\notin$

Example 7

Let $U = \{1,2,3,4,5\}$, $A = \{1,4,5\}$, and $B = \{1,3,4\}$. Find $A-B$.
Solution

We remove from A all those elements that are found in set B. The elements that remain in A comprise $A-B$. Therefore, $A-B = \{5\}$. ∎

Example 8

Let $U = \{1,2,3,4,5\}$, $A = \{1,4,5\}$, and $B = \{1,3,4\}$. Find $B-A$.
Solution

We remove from B all those elements that are found in set A. The elements that remain in B comprise the set $B-A$. Therefore, $B-A = \{3\}$. ∎

As Examples 7 and 8 point out, unlike the operations of union and intersection, which are commutative, set difference is not. In general, $A-B \neq B-A$.

Example 9

If $U = \{1,2,...,7\}$, $A = \{1,2,3,7\}$, $B = \{1,2,3,5,6\}$, and $C = \{1,2,3,7\}$, find $(A \cap B)-C$.
Solution

First, we resolve the intersection inside the parentheses. The set $A \cap B = \{1,2,3\}$. Now, we remove from this intersection any elements that are found in C. Therefore, $(A \cap B)-C = \varnothing$. ∎

Example 10

If $U = \{1,2,...,7\}$, $A = \{1,2,3,7\}$, $B = \{1,2,3,5,6\}$, and $C = \{2,3,7\}$, find $C-(A \cap B)$.

Solution

Again, $A \cap B = \{1, 2, 3\}$. This time, we remove all elements in this intersection from C. Therefore, $C - (A \cap B) = \{7\}$. ∎

Example 11

If $U = \{$whole numbers from 1 to 10$\}$, $A = \{2, 4, 6, 8, 10\}$, $B = \{1, 9\}$, and $C = \{1, 4, 5, 8, 9\}$, find $[(A \cap C)' - B]'$.

Solution

The solution to this more complicated problem is resolved in small steps, beginning with the contents of the inner parentheses. First, we need to find all elements common to sets A and C. We find that $A \cap C = \{4, 8\}$. Second, we find the complement of $A \cap C$. This is the set $(A \cap C)' = \{1, 2, 3, 5, 6, 7, 9, 10\}$. Next, remove from $(A \cap C)'$ any elements found in B, so $(A \cap C)' - B = \{2, 3, 5, 6, 7, 10\}$. Finally, we obtain the complement of $(A \cap C)' - B$, that is, all elements not in $(A \cap C)' - B$, to obtain $[(A \cap C)' - B]' = \{1, 4, 8, 9\}$. ∎

A Summary of Set Theory Notation

SYMBOL	HOW TO READ IT	FORMAL DEFINITION	INFORMAL DEFINITION
$x \in A$	x is an element of A x is a member of A x belongs to A		x is in set A
$x \subseteq A$	x is a subset of A	$(x \in A) \rightarrow (x \in B)$	Every member of A is a member of B
$A = B$	A equals B	$(x \in A \rightarrow x \in B)$ and $(x \in B \rightarrow x \in A)$	Both sets have the same members
$A \subset B$	A is a proper subset of B	$(x \in A \rightarrow x \in B) \wedge (A \neq B)$	A is a subset of B but $A \neq B$
$\varnothing$ or $\{\ \}$	The empty set The null set		The set with no members
U	The universal set		The set containing all the elements being used
A'	The complement of A A prime	$\{x \mid (x \in U) \wedge (x \notin A)\}$	The set of elements that are in the universal set, but not in A
$A \cap B$	A intersect B The intersection of A with B	$\{x \mid (x \in A) \wedge (x \in B)\}$	The set of elements that A and B have in common
$A \cup B$	A union B The union of A with B	$\{x \mid (x \in A) \vee (x \in B)\}$	The set of elements that are in A or B or both
$A - B$	A minus B A take away B	$\{x \mid (x \in A) \wedge (x \notin B)\}$	The elements that are in A, but not in B.

___

In-Class Exercises and Problems for Section 5.2

In-Class Exercises

In Exercises 1–20, assume $U = \{1, 2, 3, 4, 5, 6, 7\}$, $A = \{1, 3, 5, 7\}$, $B = \{2, 4, 6\}$, $C = \{1, 2, 5, 6\}$, and $D = \{2, 3, 4\}$. Find the elements in each of the given sets.

1. $C \cup D$
2. $C \cap D$
3. $C - D$
4. $D - C$
5. $B \cup D'$
6. $D - A'$
7. $(A \cup B)'$
8. $(A \cap B) \cup B'$
9. $(D \cap A)' - B$
10. $(C' \cup A)' - D$
11. $(A \cap D') \cup C'$
12. $(A \cap C) - (A \cap D)$
13. $(B \cap D) \cup (B - D)$
14. $[(A \cup C)' \cap (C - D)]'$
15. $(A - B) \cup (C - D')$
16. $[(A' \cap D) - C']'$
17. $(C \cup B') - (A' \cap D)$
18. $C' - [(B \cup D) \cap (D - A')]$
19. $[(A' - C)' \cap (B \cup C')] - (A \cup D)$
20. $[(A - C) \cap (B - D)'] \cup [(C - A) \cap (D - B)']$

For questions 21–24, assume that U = {the days of the week}, W = $\{x \mid x$ is a weekend}, S = $\{x \mid x$ contains 6 letters in its spelling}, E = $\{x \mid x$ contains 8 letters in its spelling} and M = {Monday, Wednesday, Friday}. Find the elements in each of the following sets.

21. $W' \cap S$

22. $E - W$

23. $(M \cup W) \cap E$

24. $(W \cup M)' - E$

Problems

For Problems 25–34, assume $U = \{1,2,3,...,10\}$, $A = \{1,3,4,7,10\}$, $B = \{7,8,10\}$, $C = \{3,5,7\}$, and $D = \{2,3,8,9\}$. Find the elements of each set.

25. $A \cap C$

26. $A \cup B'$

27. $D' - B$

28. $A \cap D'$

29. $A' \cap B'$

30. $A \cap B \cap C \cap D$

31. $(A \cap B') \cup (C' \cap D)$

32. $(A - C) - D$

33. $[(A' \cap C)' - (B' \cup D)]'$

34. $(D - C') \cap (B - A')$

For Problems 35–44, suppose that U = {the months of the year}. If set $A = \{x \mid x$ has 30 days}, set $B = \{x \mid x$ has an r in its spelling}, set C = {April, May, June}, and set D = {May,...,October}, find

35. $A \cap B$

36. $A \cup B'$

37. $D - B$

38. $A \cap D$

39. $A' \cap B'$

40. $A \cap B \cap C \cap D$

41. $C \cap D$

42. $(A - C) - D$

43. $[(A' \cap C)' - (B' \cup D)]'$

44. $U - A - B - C - D$

Check Your Understanding

For questions 1–5, assume that $U = \{1, 2, 3, 4, 5, 6, 7, 8, 9, 10\}$, $A = \{2, 4, 6, 8, 10\}$, $B = \{1, 3, 5, 7, 9\}$, $C = \{1, 2, 3\}$, $D = \{4, 7, 8, 9\}$ and $E = \{7, 9\}$. Find the elements in each of the following sets.

1. $(E - B)'$ _____

2. $(A \cap C') - D$ _____

3. $(B \cup D) \cap A$ _____

4. $(A' \cap D') \cup C$ _____

5. $B - [(D - A)' \cap C]$ _____

For questions 6–11, let Z represent an unknown set. Answer true or false for each question.

6. $Z \cap \emptyset = \emptyset$

7. $Z \cup U = Z$

8. $Z \cap Z' = \emptyset$

9. $Z \cup Z' = U$

10. $Z - Z' = \emptyset$

11. $\emptyset \subset Z$

5.3 Venn Diagrams

A *Venn diagram* is a pictorial way of representing sets and the relationships that exist among them. A Venn diagram consists of a rectangle, the interior of which represents the universe. Within the rectangle, circular regions represent the sets under discussion.

Example 1

Suppose in a given universe, set B is a proper subset of set A. Draw a possible Venn diagram for this relationship.

Solution

Inside a rectangle, we draw two circles, one entirely within the other.

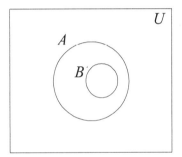

Example 2

Assume that in some universe, sets A and B share some, but not all of their elements. Draw a Venn diagram that depicts this condition.

Solution

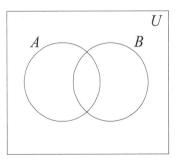

Example 3

Assume that in some universe, sets A and B share some, but not all of their elements, and that C is a proper subset of $A \cap B$. Draw a Venn diagram that depicts this condition.

Solution

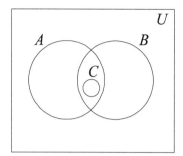

■

Example 4

Suppose sets A and B are disjoint, and C is a proper subset of A. Draw a Venn diagram that depicts this condition.

Solution

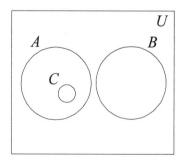

■

Example 5

Let $U = \{1, 2, 3, 4, 5, 6\}$, $A = \{1, 2, 3\}$, $B = \{2, 3, 4\}$. Place the elements into their proper regions in a Venn diagram.

Solution

Since sets A and B share elements, we will draw two intersecting circles and place the elements 2 and 3 within the intersection. The element 1 is not in B but is in A. Similarly, the element 4 is not in A but is in B. The elements 5 and 6 are in neither set.

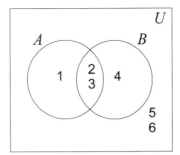

■

Example 6

Suppose that $U = \{1, 2, 3, ..., 10\}$, $A = \{1, 3, 4, 7, 8\}$, $B = \{3, 4, 9, 10\}$, and $C = \{1, 2, 4, 10\}$. Place the elements into their proper regions in a Venn diagram.

Solution

Since there are three sets, we must have three circles. Each circle shares some of its elements with each of the other two circles. Therefore, each circle must intersect the other two circles. The element 4 is in all three sets, so it must be placed in their common intersection. The element 3 is in sets A and B, but it is not in set C. The element 1 is in sets A and C but not in set B. The element 10 is in sets B and C, but not in set A. The elements 5 and 6 are not in any of the three sets. Finally, the elements 7 and 8 are only in A, the element 9 is only in B and the element 2 is only in C. These results are shown in the following Venn diagram.

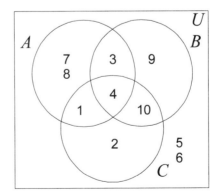

■

The previous example illustrates the need to be able to identify and name the various regions in any Venn diagram. We will begin by returning to the situation in which we have only two sets in some universe.

In the Venn diagram below, we have four distinct regions. We have labeled these regions with Roman numerals so that we may refer to the regions in a concise manner. The order in which we label the regions is immaterial, however if the regions are labeled as shown below, the rows of a corresponding membership table for the Venn diagram will correspond to the numbered regions.

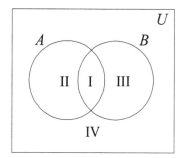

Notice that the region $A \cap B$ is denoted by the Roman numeral I, while $A \cup B$ consists of the *three* regions I, II, and III. The region B' consists of regions II and IV.

Example 7

Name the region or regions in the above Venn diagram that represent $A \cap B'$.

Solution

In order to be in $A \cap B'$, an element must be in set A and outside set B. This is region II.

■

Example 8

Name the region or regions in the above Venn diagram that represent $(B - A)'$.

Solution

First, we need to find $B - A$. This set consists of all elements in B that are not in A. This corresponds to region III. Then we take the complement of this region, so our answer consists of regions I, II, and IV.

■

With three intersecting circles, there are eight distinct regions. Again, the order in which we assign the numbering is immaterial, but if the regions are labeled as shown below, each of the rows of a corresponding membership table for the Venn diagram will correspond to the numbered regions.

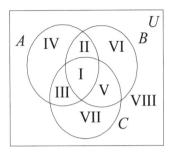

Note that the intersection of all three sets, i.e., $A \cap B \cap C$, is represented by region I. This is the region that is common to all three sets. Note also that the intersection of *any* two of the three sets consists of two regions. For example, $A \cap B$ is represented by regions I and II. It is also important to realize that each of the three sets A, B, and C consists of four regions. For example, set A is represented by regions I, II, III and IV.

If we wish to identify an element that is *only* in set A, we must look in region IV, while an element that is only in set B is found in region VI. Region VII consists of elements only in set C.

Example 9
Name the numbered region in terms of sets A, B, and C.

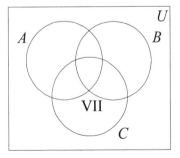

Solution
We begin by noting that region VII is in set C, but C consists of more than just region VII. The regions that C shares with both A and B have been removed from C. Therefore one way to express region VII is $C - (A \cup B)$. Another way to name region VII is $(C - A) - B$. ∎

Example 10

Use a membership table to decide which region or regions in the Venn diagram below belong to $(A \cup B)' - C$.

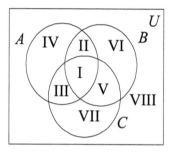

Solution

Since there are three sets under discussion, our membership table will have eight rows. We will have columns for each set, for $A \cup B$, $(A \cup B)'$, and finally for $(A \cup B)' - C$. If an element is in either A or B, it belongs to $A \cup B$, and therefore does not belong to $(A \cup B)'$. Due to the nature of the set difference operation, we are interested only in those elements that are in $(A \cup B)'$, but are not in C. This occurs in row eight of the following membership table.

Region	A	B	C	$A \cup B$	$(A \cup B)'$	$(A \cup B)' - C$
I	$\in$	$\in$	$\in$	$\in$	$\notin$	$\notin$
II	$\in$	$\in$	$\notin$	$\in$	$\notin$	$\notin$
III	$\in$	$\notin$	$\in$	$\in$	$\notin$	$\notin$
IV	$\in$	$\notin$	$\notin$	$\in$	$\notin$	$\notin$
V	$\notin$	$\in$	$\in$	$\in$	$\notin$	$\notin$
VI	$\notin$	$\in$	$\notin$	$\in$	$\notin$	$\notin$
VII	$\notin$	$\notin$	$\in$	$\notin$	$\in$	$\notin$
VIII	$\notin$	$\notin$	$\notin$	$\notin$	$\in$	$\in$

Therefore, the only elements that are in $(A \cup B)' - C$ are those elements that are not in A nor B nor C. Hence, the only region that describes $(A \cup B)' - C$ in the previous Venn diagram is region VIII. Notice that the membership table's eighth row shows the elements that are in neither A, nor B, nor C.

In-Class Exercises and Problems for Section 5.3

In-Class Exercises

For Exercises 1–5, construct a Venn diagram for each situation. Be sure to place the elements in their correct regions.

1. $U = \{1,3,5,7,9,11\}$, $A = \{3,5,7\}$, $B = \{3\}$

2. $U = \{1,4,9,16,25,36\}$, $C = \{1,4,16,36\}$, $D = \{4,16\}$, $E = \{9,36\}$

3. $U = \{2,4,6,8,10,12,14\}$, $F = \{2,4,6\}$, $G = \{8,12,14\}$, $H = \{12\}$

4. $U = \{1,2,3,4,5,6,7,8,9,10\}$, $A = \{1,3,4,7,8\}$, $B = \{3,4,6,9\}$
   $C = \{1,3,9,10\}$

5. $U = \{2,3,5,7,11,13,17,19\}$, $W = \{2,5,7,13\}$, $X = \{3,5,11,17\}$
   $Y = \{2,7\}$, $Z = \{11,17\}$

For Exercises 6–9, construct a Venn diagram that satisfies the conditions stated.

6. Sets D, E, and F are proper subsets of U. $D \cap E \neq \varnothing$, and neither is a subset of the other. Also, $D \cap F = \varnothing$, $E \cap F = \varnothing$.

7. Sets A, B, C and D are proper subsets of U. $C \subset A$, $D \subset B$, and $A \cap B = \varnothing$.

8. Sets A, B, and C are proper subsets of U. $B \subset A$, $C \subset A$, and $B \cap C = \varnothing$.

9. Sets A, B, C and D are proper subsets of U. $A \cap B \neq \varnothing$, $C \subset A$, $D \subset (A \cap B)$, and $B \cap C = \varnothing$. Neither A nor B is a subset of the other.

Use the Venn diagram below for Exercises 10–19. Name the region or regions described.

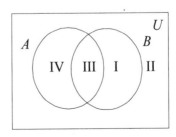

10. A 11. B 12. A' 13. B' 14. $A \cap B$ 15. $A - B$

16. $(A \cap B)'$ 17. $A' \cup B$ 18. $(A \cap B') \cup (B - A)$

19. $(A - B')' - (A' \cap B)$

Use the Venn diagram below for Exercises 20–34. Name the region or regions described.

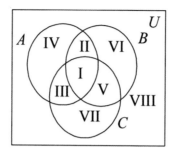

20. B

21. C'

22. $B \cap C$

23. $A' \cup C$

24. $B - C$

25. $(A - B) \cap C$

26. $(A \cap B \cap C)' - B'$

27. $(A \cup C') - (A \cap B)$

28. $(A' \cup B') \cap (B \cup C)$

29. $[B' - (A \cap C')]'$

30. $[C - (A - B)]' \cap (A' \cup B')$

31. Only set A

32. Set B but not set A

33. Set B and set C but not set A

34. Set A or set C but not in both.

For each Venn diagram in Exercises 35–38, name the designated region or regions in terms of sets A, B and C. There are many correct answers for each Venn diagram.

35.
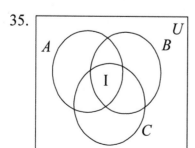

36.
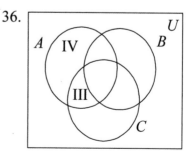

37.

38.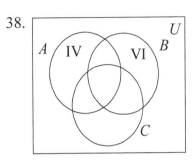

Problems

For Problems 39–44, construct a Venn diagram for each situation. Be sure to place the elements in their correct regions.

39. $U = \{10,11,12,13,14\}$, $A = \{11,12\}$, $B = \{10,11,14\}$.

40. $U = \{1,2,...,6\}$, $A = \{1,2,3\}$, $B = \{4,5,6\}$.

41. $U = \{2,4,6,8,10\}$, $A = \{2,4,6,10\}$, $B = \{4,10\}$, $C = \{8\}$.

42. $U = \{1,3,5,...,11\}$, $A = \{1,3,5\}$, $B = \{3,5,7\}$, $C = \{1,3,5,7,9\}$.

43. $U = \{1,2,3,5,8,13,21\}$, $A = \{1,2,3,5\}$, $B = \{2,3,8\}$, $C = \{3\}$.

44. $U = \{1,2,3,...,10\}$, $A = \{2,3,7,10\}$, $B = \{2,3\}$, $C = \{3,5,7\}$.

For Problems 45–49, construct a Venn diagram that satisfies the conditions stated.

45. Sets A, B, and C are proper subsets of U. $A \subset B$, $B \subset C$.

46. Sets A, B, and C are proper subsets of U. $A \subset B$, $B \cap C \neq \emptyset$, neither B nor C is a subset of the other, $A \cap C = \emptyset$.

47. Sets A, B, and C are proper subsets of U. $A \cap B = \emptyset$, $C \subset B$.

48. Let A, B, and C be three proper subsets of U with the following conditions: $A \cap B \neq \emptyset$, neither A nor B is a subset of the other, $C \subset (A \cap B)$.

49. Sets A, B, and C are proper subsets of U. $A \cap B \neq \emptyset$, $B \cap C \neq \emptyset$, $A \cap C = \emptyset$. No set is a subset of any other set.

For Problems 50–59, use the Venn diagram below. Name the region or regions described.

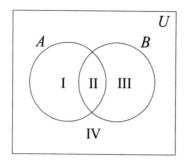

50. A

51. $A \cap B$

52. $A \cup B$

53. $(A \cup B)'$

54. $A' \cap B'$

55. $A - B'$

56. $(A' \cap B')'$

57. $(A \cup B) - (A \cap B)$

58. $(A' \cap B')' \cap (A' \cap B)$

59. $[(A \cap B') \cap (B \cap A')]'$

For Problems 60–79, use the Venn diagram below. Name the region or regions described.

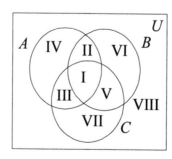

60. A

61. $A \cap B$

62. $A \cup B$

63. $(A \cup B)'$

64. $A' \cap B'$

65. $A - B'$

66. $(A' - B')'$

67. $(A \cup B) \cap C$

68. $(A - C) \cap B'$

69. $(A \cup C) - B$

70. $A' \cap B' \cap C'$

71. $(A - B) - C$

72. $A - (B - C)$

73. $[(A \cap B) \cap C]'$

74. $(A - B) \cup (A - C) \cup (B - C)$

75. $(A - B) \cap (A - C) \cap (B - C)$

76. Sets A and B, but not set C

77. Only set C

78. Sets B or C but not set A

79. Set B but not set A nor set C

For Problems 80–83, use a membership table to decide which region or regions in the Venn diagram below belong to the given set.

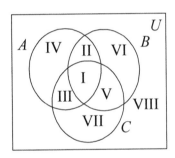

80. $(A \cap B')' \cap C$

82. $(A \cap C') - B$

81. $(A' - B) \cap C'$

83. $(B - C)' \cap (A \cup B)$

For Problems 84–91, name the designated region or regions in terms of sets A, B and C. There may be more than one correct way to name the region.

84.

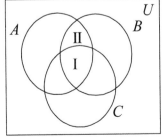

85.

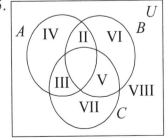

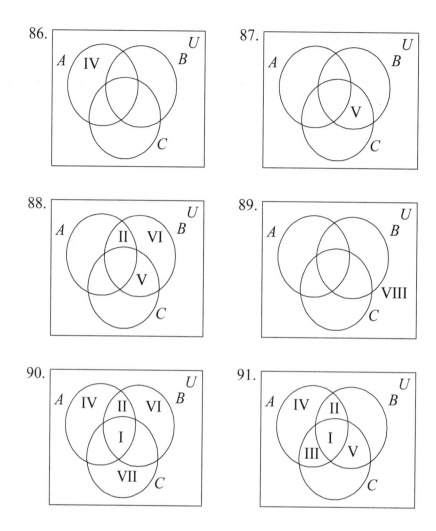

DID YOU KNOW?

John Venn (1834–1923) was an English mathematician and logician who invented the Venn diagram. He was elected as a Fellow of the Royal Society in 1883. His diagrams visually point out the similarities and differences for two or three different sets as well as indicate their unions and/or intersections. Venn's diagrams were first published in 1880, in "Philosophical Magazine and Journal of Science."

To extend Venn diagrams to four, five or six sets, a Karnaugh map is often used. The Karnaugh map was invented in 1952 by Edward W. Veitch. It was further developed in 1953 by Maurice Karnaugh (hence the name), a telecommunications engineer at Bell Labs, to help simplify digital electronic circuits. Karnaugh maps are useful for expressions of perhaps up to six variables. When there are more than six sets, even these maps become unwieldy and too complex to use.

Check Your Understanding

For questions 1–3, construct a Venn diagram for each situation. Be sure to place the elements in their correct regions.

1. $U = \{1, 3, 5, 7, 9\}$, $A = \{1, 3, 5, 9\}$, $B = \{1, 9\}$, $C = \{5\}$

2. $U = \{3, 6, 9, 12, 15, 18, 21, 24, 27, 30\}$, $D = \{3, 6, 9, 12, 18, 21, 30\}$,
   $E = \{3, 6, 12, 30\}$, $F = \{9, 12, 18, 30\}$, $G = \{15, 24\}$

3. Sets R, S, and T are proper subsets of U. Neither R nor S is a subset of each other. $R \cap S \neq \varnothing$, $S \cap T = \varnothing$, and $T \subset R$.

For questions 4–7, use the Venn diagram below. Name the region or regions described.

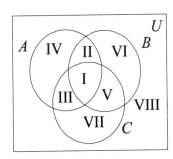

4. $(B - A) \cap C'$ _____

5. $(A' - C') \cap (A \cup B)'$ _____

6. $[(A - B) \cup (A - C)] \cap B'$ _____

7. Not set A but set B and set C _____

5.4 Venn Diagrams and Survey Problems

One of the most useful applications of set theory is solving what have come to be known as *survey problems*. Typically, such problems involve splitting the results of a survey into disjoint sets in order to quickly arrive at solutions to questions that are not easily answered by considering the information in the survey as it is initially presented.

Example 1

Suppose there is a meeting of the softball and basketball teams. All in attendance play either softball or basketball. There are 28 people who play softball and 33 who play basketball. How many people were in attendance at this meeting?

Solution

At first, one would guess that there were $28 + 33 = 61$ people present. This situation is represented by the Venn diagram below.

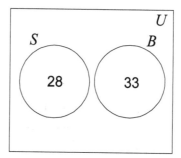

However, this solution may not be correct. Consider the possibility that all 28 softball players also play basketball. If this were the case, a total of only 33 people would be needed to satisfy the conditions given.

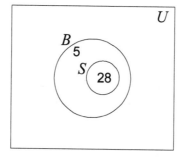

Armed only with the given information, the best we can do is say that there were between 33 and 61 people present.

■

Example 2

Suppose we are now told that there are exactly 8 people who play both softball and basketball. How can we refine our previous answer given this new information?

Solution

The new information guarantees that the two sets of athletes are not disjoint, and neither is a subset of the other. In fact, we are told that there are eight athletes in the intersection of the two sets.

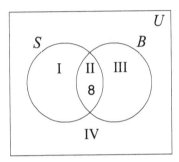

Softball players are distributed in regions I and II. Since there are 28 softball players, and eight of them have been already accounted for in region II, there must be exactly 20 in region I. Similarly, the 33 basketball players are distributed in regions II and III. Since eight of them have been already accounted for in region II, there must be exactly 25 in region III.

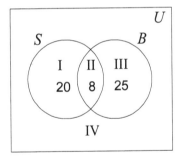

We now see that there were exactly 20 + 8 + 25 = 53 people in attendance at the meeting. ∎

Example 3

In a restaurant parking lot there were 15 black cars, 20 sedans and 5 black sedans. In all, there were 41 cars in the lot.

    a. How many cars were black, but not sedans?

    b. How many sedans were not black?

    c. How many cars were neither black nor sedans?

Solution

We construct a Venn diagram with two intersecting circles. We know there are 5 black sedans, so we place 5 cars in region II. Sedans are distributed in regions I and II. Knowing there are 20 sedans, and five of them have been already accounted for in region II, there must be exactly 15 in region I. Similarly, the 15 black cars are distributed in regions II and III. Since five of them have been already accounted for in region II, there must be exactly 10 in region III.

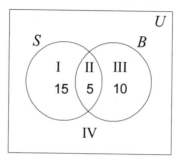

So far, we have accounted for 30 cars. The information given states that there were 41 cars in the lot, so the remaining 11 cars must be in region IV.

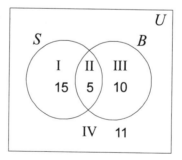

In order to answer part (a), we need to identify the elements that are in B but not in S. These are in region III. Hence, there are 10 cars that are black, but not sedans. For part (b), we need to identify the elements that are in S but not in B. These are in region I. This means that 15 cars are sedans, but not black. Finally, the cars that are neither black nor sedans are in region IV. Therefore, there were 11 cars that were neither black nor sedans.

∎

Example 4

At a sorority party attended by 46 people, it was discovered that 26 people took a mathematics course, 28 took an English course, and 30 took a psychology course. There were 15 people who took both mathematics and English, 21 who took both English and psychology, and 18 who took both mathematics and psychology. Ten people took all three courses.

  a How many students took at least one of the courses?

  b. How many students took none of the three courses?

  c. How many students took exactly one of the courses?

  d. How many students took mathematics and English, but not psychology?

Solution

We begin by drawing three mutually intersecting circles and labeling them in the usual way.

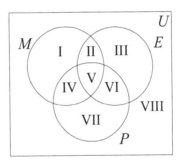

The 46 people must be divided into the eight disjoint regions. We might try to begin with the fact that 26 people took mathematics, but the mathematics set consists of four disjoint regions and at this point, it is unclear how the 26 people distribute themselves into these four regions.

However, the people who took all three courses can be assigned to exactly one region, namely region V. Next, consider those people who can be assigned to exactly two sets, that is, the people who took either mathematics and English, English and psychology, or mathematics and psychology. First, there were 15 people who took mathematics and English. The regions that make up mathematics and English are II and V, so the 15 people must be distributed between these two regions. Since there are already 10 people in region V, we put the remaining 5 people in region II. This is shown next.

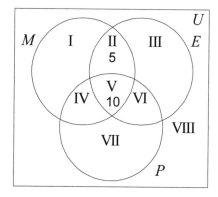

We now continue with the 21 students who took English and psychology. The regions that make up English and psychology are V and VI. Since there are already 10 people in region V, we put the remaining 11 people in region VI.

Similarly, regions IV and V describe the students who took mathematics and psychology. Again, there are already 10 people in region V, so the remaining 8 people are in region IV.

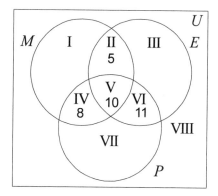

Now we are able to turn our attention to those people who took mathematics. The 26 people who took mathematics must be assigned to regions I, II, IV, and V. Since regions II, IV, and V already account for 23 people, there are exactly 3 people left for region I. In the same way regions III and VII have 2 people and 1 person, respectively. Summing all the people in the disjoint regions considered thus far (all except region VIII), accounts for 40 people. Since there were 46 people at the party, there are 6 people left for region VIII. Now we can begin to answer the questions posed.

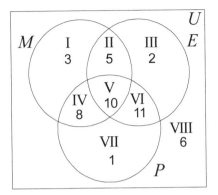

Part (a) asked how many people took at least one course. These are the people in regions I through VII. There are 40 people distributed in these regions.

The answer to part (b) is obtained by recognizing that the people who took none of the three courses are found in region VIII. There are six people who took none of the three courses.

For part (c), we are interested in only regions I, III, and VII, since all other regions within the given sets are in two or three sets. Therefore, there are 6 people who took only one of the three courses.

In part (d), we are interested in the set $M \cap E \cap P'$. This is region II. Hence, five people took mathematics and English, but not psychology. ■

Example 5

At a doctor's office, the following data was gathered over the course of one week by the office staff. There were 535 patients who saw a doctor. Of these patients, 90 were treated for heart disease, 65 were treated for diabetes, and 105 were treated for emphysema. There were 35 patients being treated for both heart disease and diabetes, 70 were treated for both heart disease and emphysema, and 40 were treated for both emphysema and diabetes. Exactly 30 people were treated for all three conditions.

    a. How many patients were treated for conditions other than heart disease, diabetes, or emphysema?

    b. How many patients were treated for only one of the three conditions mentioned?

    c. How many patients were treated for at least two of the conditions mentioned?

Solution

Begin by placing 30 in region V, which contains the elements of $H \cap D \cap E$. Now assign people to regions shared by two sets. Regions II and V make up the set $H \cap D$. We know that there are 35 people distributed in these two regions, and 30 are already in region V. Therefore, 5 people belong in region II. Reasoning in this fashion for the sets $H \cap E$ and $D \cap E$ we find that region IV has 40 people and region VI has 10 people. Since there are 90 people in set H who must be distributed over regions I, II, IV, and V, we assign 15 people to region I. Similarly, we find there are 20 people in region III, and 25 people in region VII. We have accounted for 145 people in regions I through VII. Therefore, there must be 390 people in region VIII.

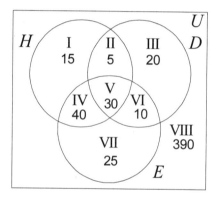

Now we can begin to answer the questions asked. For part (a) we are being asked for $(H \cup D \cup E)'$, which is region VIII. Therefore, there were 390 people who were treated for something other than heart disease, diabetes, or emphysema.

In part (b), the people who were treated for only one condition are found in regions I, III, and VII. Hence, a total of 60 people were treated for only one of the three conditions mentioned.

To answer part (c), we look to regions II, IV, V, and VI. Thus, there were 85 people who were treated for at least two of the conditions mentioned.

∎

In-Class Exercises and Problems for Section 5.4

In-Class Exercises

1. There were 25 people who went to dinner. There were two choices of vegetables: carrots and string beans. There were 3 people who ordered both, 10 had only carrots and 8 had neither string beans nor carrots. How many had only string beans?

2. There were 60 people at a party. There were 30 who were Irish, 30 were Polish, 35 were German, and 7 shared all three nationalities. Thirteen were Polish and German and not Irish. Eight were Polish and Irish and not German. Ten were Irish and German and not Polish.
   a. How many were only Irish?
   b. How many were neither Irish, nor Polish nor German?

3. In a certain high school, there are 50 students in the high school band. Twenty students can play the clarinet, 10 students can play the saxophone and six students can play the oboe. Five students indicated that they can play both the clarinet and saxophone, 2 indicated that they can play both the clarinet and oboe, while 1 student can play both the saxophone and oboe. No student can play all three instruments.

   a. How many students cannot play any of these instruments?
   b. How many students can play the clarinet or oboe?
   c. How many students cannot play the clarinet and saxophone?
   d. How many students cannot play the saxophone or oboe?

4. In a certain college, every freshman was required to take at least one of the following courses: social science, physical science or humanities. A total of 380 students took a humanities course, 420 students took a social science course and 500 students took a physical science course. One hundred forty students took both a social science course and a physical science course, while 180 students took both a social science and a humanities course. Also, 120 students took both a physical science course and a humanities course. One hundred students took all three types.
   a. How many freshman are there at this college?
   b. How many freshman are taking only humanities courses?

    c. How many freshman are taking only one of the required courses?

    d. How many freshman are taking a social and a physical science course, but not a humanities course?

    e. How many freshman are taking a social science or physical science course, but not a humanities course?

5. A survey was conducted in a school cafeteria to determine what foods students like to eat. Altogether, 400 students indicated that they liked hamburgers, 280 liked frankfurters and 340 liked pizza. Some of these students also indicated that they preferred more than one of the choices. There were 50 who said they would eat hamburgers and frankfurters but not pizza and 130 students would eat only pizza. There were 200 students would eat only hamburgers while 120 would eat both hamburgers and frankfurters. There were 700 students in the cafeteria at the time of the survey.

    a. How many liked at least one of the three foods?

    b. How many liked none of the three foods?

    c. How many liked only frankfurters?

    d. How many liked hamburgers or frankfurters, but not pizza?

    e. How many liked hamburgers and frankfurters, but not pizza?

    f. How many liked pizza or frankfurters but not both?

    g. How many liked only one of these foods?

    h. How many liked only two of these foods?

6. A survey was taken to see what kind of chocolate candy some people preferred. Snickers were liked by 12 people, 14 people liked Kit Kat and 13 people preferred Milky Way. Eight people liked both Snickers and Milky Way, 10 liked Snickers and Kit Kat but 2 liked only Kit Kat. Four people didn't like chocolate and 7 liked them all.

    a. How many people participated in the survey?

    b. How many like only Milky Way?

    c. How many liked Kit Kat and Milky Way but not Snickers?

    d. How many liked exactly two kinds of chocolate?

    e. How many like Snickers or Milky Ways but not Kit Kat?

7. A student designed a survey for her statistics course. One question on the survey was to determine the number of people who saw three old movies: Star Wars, The Godfather, and The Princess Bride. After surveying 55 students, she determined the following: 17 had seen The Godfather and 23 had seen The Princess Bride. Six had seen Star Wars and The Godfather, 10 had seen The Godfather and The Princess Bride, and 7 had seen only The Princess Bride. Two students indicated they had seen all three movies and 20 hadn't seen any of these movies.
   a. How many students had seen Star Wars but neither of the other two movies?
   b. How many students had seen exactly one of these movies?
   c. How many students had seen at least two of these movies?

Problems

For Problems 8–14, use a Venn diagram to help solve each of the survey problems.

8. This past semester, a community college offered vaccinations for measles and mumps to their students, faculty and staff, totaling 20,000 individuals. The Health Center at the college reported that 2,400 of these individuals received the vaccination for both measles and mumps, 14,000 received the vaccination for measles, and 8,200 received the vaccination for mumps.
   a. How many members of the college community did not receive either vaccination this past semester?
   b. How many members of the college community received only the measles vaccination?

9. Anne and Sterling each made a list of their favorite television shows. Anne's list had nine shows and Sterling's list had seven shows. When they compared their lists, they found that a total of 12 different shows had been mentioned.
   a. How many shows were on both of their lists?
   b. How many shows were on Anne's list but not on Sterling's list?
   c. How many shows were on Sterling's list but not on Anne's list?

10. One hundred middle school children belong to a YMCA. Kaysha volunteers her time to run an after school homework assistance center for students needing help with math, social studies, and English. Half of the students need help in their math homework, one-quarter need help in their social studies homework, and 40% need help with their English homework. Eleven students need help with both math and social studies and seven students need help with both math and English homework. While 2 students require help with all three subject areas, 25 need help only in English.
    a. How many students need only social studies help?
    b. How many students don't require Kaysha's help?

11. In the 2002 winter Olympic games, 19 countries won gold medals, 19 countries won silver medals, and 20 countries won bronze medals. Fifteen countries won both gold and silver medals, 14 countries won both gold and bronze medals, and 16 countries won both silver and bronze. Twelve countries won gold, silver, and bronze medals.
    a. How many countries won only silver medals?
    b. How many countries won gold and silver but no bronze medals?
    c. How many countries won gold or bronze but no silver medals?
    d. How many countries won at least one medal?

12. Seventy-five faculty members are in the Mathematics, Statistics, and Computer Processing Department. Each faculty member teaches at least one course. Twenty-nine faculty members teach statistics courses, 44 faculty members teach computer courses, 30 faculty members teach both mathematics and computer courses, 28 faculty members teach both mathematics and statistics courses, 3 faculty members teach computers and statistics courses, and 2 faculty members teach all three.
    a. How many faculty members teach only mathematics?
    b. How many faculty members do not teach statistics?
    c. How many faculty members teach only one subject area?

13. A computer printout for the employees of a company indicated that 300 employees grossed over $40,000 while 600 employees grossed under $50,000. Since there are only 700 employees in this company, some of them must have been counted twice.

a. How many employees fell into neither category?

b. How many grossed over $40,000 but under $50,000?

c. How many grossed $40,000 or less?

d. How many grossed $50,000 or more?

14. A marketing firm wanted to gather information about cell phones, Kindles and iPads for an advertising campaign it was handling. The firm obtained the following information: 1,850 people had a cell phone, 450 people had an iPad. No one owned just a kindle, but 2000 people owned a Kindle or a cell phone. There were 250 people who owned a cell phone and an iPad, while 550 owned a cell phone and a Kindle, but not an iPad. One hundered and fifty people owned all three items, while 20 people did not have any of the items.

a. How many people own a Kindle?

b. How many people own just an iPad?

c. How many people were surveyed?

d. How many people own a cell phone and an iPad but not a Kindle?

e. How many people own an iPad or a Kindle, but not both?

JUST FOR FUN

1. There are five houses, each of a different color and inhabited by men of different nationalities, with different pets, drinks, and cigarettes.

2. The Englishman lives in the red house.

3. The Spaniard owns the dog.

4. Coffee is drunk in the green house.

5. The Ukrainian drinks tea.

6. The green house is immediately to the right (your right) of the ivory house.

7. The Old Gold smoker owns snails.

8. Kools are smoked in the yellow house.

9. Milk is drunk in the middle house.

10. The Norwegian lives in the first house on the left.

11. The man who smokes Chesterfields lives in the house next to the man with the fox.

12. Kools are smoked in the house next to the house where the horse is kept.

13. The Lucky Strike smoker drinks orange juice.

14. The Japanese smokes Parliaments.

15. The Norwegian lives next to the blue house.

Now, who drinks the water? And who owns the zebra?

Check Your Understanding

For questions 1–3, use a Venn diagram to help solve each survey problem.

1. Twenty women were randomly selected to answer the following question. "If someone were to send you flowers, would you like getting roses or orchids?" Five women said they liked both types of flowers. Nine women said they like orchids while three women were allergic to flowers and didn't want either.

   a. How many women like roses?

   b. How many didn't like orchids?

2. The television show "Friends" was one of the most popular sitcoms of all time. Both men and women were asked, "which female lead did you like?" Six people couldn't decide and picked all three actresses. Forty-five people picked Lisa Kudrow while twenty-five people said they only liked Jennifer Aniston. Sixteen indicated they liked both Jennifer Aniston and Courtney Cox while fifteen liked Jennifer Aniston and Lisa Kudrow. Four people said they liked Courtney Cox and Lisa Kudrow but not Jennifer Aniston while five people said they never saw the show and, therefore, couldn't pick any of the actresses. If 100 people were surveyed, answer the following questions.

   a. How many people only liked Lisa Kudrow?

   b. How many people liked Courtney Cox?

   c. How many people liked Jennifer Aniston or Lisa Kudrow but not Courtney Cox?

3. Since music and cell phones play an important role in the life of college students, an advertising agency was interested in knowing whether college students used AT&T as their cell phone carrier, if they had an iPhone and if they owned an iPod. One hundred students said AT&T was their cell phone carrier, fifty students had an iPhone and seventy-five students owned an iPod. Ten students had all three and five students had none of the three. While every student who had an iPhone used AT&T as their cell phone carrier, forty-five students who owned an iPod did not use AT&T as their carrier.

   a. How many college students were surveyed?

   b. How many students didn't own an iPod?

5.5 Laws of Equal Sets

Introduction

Many of the equivalences that we studied in Chapter One have identical counterparts in set theory. In this section, we will use membership tables and Venn diagrams to verify these set equivalences. A summary of the laws of equal sets can be found on page 307.

Double Complement Law

The double negation equivalence in logic was expressed symbolically as $\sim(\sim p) \Leftrightarrow p$. In set theory, the corresponding equivalence is expressed symbolically as $(A')' = A$, and is called the *double complement law*. Let's use a membership table to show that the sets on left and right sides of the equal sign are the same.

A	A'	$(A')'$	$A = (A')'$
$\in$	$\notin$	$\in$	T
$\notin$	$\in$	$\notin$	T

We see that the first and third columns always have the same elements. Therefore, $(A')' = A$.

DeMorgan's Law

DeMorgan's equivalences state that $\sim(p \wedge q) \Leftrightarrow (\sim p \vee \sim q)$, and $\sim(p \vee q) \Leftrightarrow (\sim p \wedge \sim q)$. In set theory, these equivalences correspond to $(P \cap Q)' = (P' \cup Q')$ and $(P \cup Q)' = (P' \cap Q')$ respectively, and are called *DeMorgan's laws*.

Example 1

Use a membership table to show that $\sim(a \wedge b) \Leftrightarrow (\sim a \vee \sim b)$ is true for sets.

Solution

The corresponding form of DeMorgan's equivalence in set theory is expressed as $(A \cap B)' = A' \cup B'$. The membership table is constructed below.

A	B	A'	B'	$A \cap B$	$(A \cap B)'$	$A' \cup B'$	$(A \cap B)' = A' \cup B'$
$\in$	$\in$	$\notin$	$\notin$	$\in$	$\notin$	$\notin$	T
$\in$	$\notin$	$\notin$	$\in$	$\notin$	$\in$	$\in$	T
$\notin$	$\in$	$\in$	$\notin$	$\notin$	$\in$	$\in$	T
$\notin$	$\notin$	$\in$	$\in$	$\notin$	$\subset$	$\in$	T

Since the sixth and seventh columns are identical, we have shown that $(A \cap B)' = A' \cup B'$.

■

Distributive Law

Example 2

Use a Venn diagram to verify a distributive law for sets. In particular, verify that $A \cup (B \cap C) = (A \cup B) \cap (A \cup C)$.

Solution

The regions that correspond to $B \cap C$ are I and V. Set A consists of regions I, II, III, and IV. Therefore, $A \cup (B \cap C)$ consists of regions I, II, III, IV, and V.

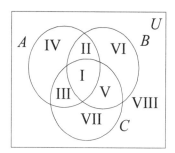

The set $A \cup B$ consists of regions I, II, III, IV, V, and VI, while set $A \cup C$ consists of regions I, II, III, IV, V, and VII. Therefore, $(A \cup B) \cap (A \cup C)$ consists of regions I, II, III, IV, and V. These are exactly the same regions we obtained for $A \cup (B \cap C)$. Hence, $A \cup (B \cap C) = (A \cup B) \cap (A \cup C)$.

■

Absorption Law

In set theory, as well as in electronic circuit theory, we often refer to the *absorption law*. One form of the absorption law is expressed symbolically as $A \cap (A \cup B) = A$.

Example 3

Use a membership table to verify the absorption law shown above.

Solution

A	B	$A \cup B$	$A \cap (A \cup B)$	$A \cap (A \cup B) = A$
$\in$	$\in$	$\in$	$\in$	T
$\in$	$\notin$	$\in$	$\in$	T
$\notin$	$\in$	$\in$	$\notin$	T
$\notin$	$\notin$	$\notin$	$\notin$	T

Since the first and fourth columns are identical, $A \cap (A \cup B) = A$ ■

There are other Laws of Equal Sets that are also worth mentioning. The *Idempotent Law* simply states that the union (or intersection) of any set with itself is exactly the set itself. The *Inverse Law* states that the intersection of a set with its complement is empty, while the union of a set with its complement is the universe. The *Complement Law* states that the complement of the empty set is the universe, while the complement of the universe is the empty set. The *Identity Law* tells us how to combine a set with the universal set and the null set using the operations of union and intersection. *Set difference* allows us to write a difference of two sets as an intersection. All eleven of the Laws of Equal Sets are summarized in the table below.

Laws of Equal Sets

1. Double Complement Law $\quad (A')' = A$

2. Commutative Law
$$A \cap B = B \cap A$$
$$A \cup B = B \cup A$$

3. Associative Law
$$A \cap (B \cap C) = (A \cap B) \cap C$$
$$A \cup (B \cup C) = (A \cup B) \cup C$$

4. Distributive Law
$$A \cap (B \cup C) = (A \cap B) \cup (A \cap C)$$
$$A \cup (B \cap C) = (A \cup B) \cap (A \cup C)$$

5. DeMorgan's Law
$$(A \cap B)' = A' \cup B'$$
$$(A \cup B)' = A' \cap B'$$

6. Absorption Law
$$A \cap (A \cup B) = A$$
$$A \cup (A \cap B) = A$$

7. Idempotent Law
$$A \cup A = A$$
$$B \cap B = B$$

8. Inverse Law
$$A \cap A' = \varnothing$$
$$A \cup A' = U$$

9. Complement Law
$$\varnothing' = U$$
$$U' = \varnothing$$

10. Identity Law
$$A \cap U = A \qquad A \cup U = U$$
$$A \cap \varnothing = \varnothing \qquad A \cup \varnothing = A$$

11. Set Difference Law $\quad A - B = A \cap B'$

Simplifying Sets

The Laws of Equal Sets can be used to simplify compound statements about sets.

Example 4

Simplify the set $A \cap (A' \cup B)$.

Solution

First, use the distributive law to obtain

$$A \cap (A' \cup B) = (A \cap A') \cup (A \cap B).$$

The inverse law allows us to express $(A \cap A')$ as $\varnothing$. Therefore,

$$A \cap (A' \cup B) = (A \cap A') \cup (A \cap B) = \varnothing \cup (A \cap B).$$

Then, using the identity law, $\varnothing \cup (A \cap B) = A \cap B$. ∎

Example 5

Use Laws of Equal Sets to verify that $(A \cup B) \cap (A' \cap B)' = A$.

Solution

First use DeMorgan's Law to obtain $(A' \cap B)' = A \cup B'$. Now we can write

$$(A \cup B) \cap (A' \cap B)' = (A \cup B) \cap (A \cup B').$$

Using the distributive law we obtain

$$(A \cup B) \cap (A \cup B') = A \cup (B \cap B').$$

Since $B \cap B' = \varnothing$ we have $A \cup (B \cap B') = A \cup \varnothing = A$. Therefore, we have $(A \cup B) \cap (A' \cap B)' = A$. ∎

Often, we choose to simplify a set using statements and reasons, much like we did in writing formal proofs.

Example 6

Write a formal proof to show the relationship in Example 5 is true.

Solution

Statement	Reason
1. $(A \cup B) \cap (A' \cap B)'$	Given
2. $(A \cup B) \cap (A \cup B')$	DeMorgan's Law
3. $A \cup (B \cap B')$	Distributive Law
4. $A \cup \varnothing$	Inverse Law
5. A	Identity Law

∎

In-Class Exercises and Problems for Section 5.5

In-Class Exercises

For Exercises 1–2, supply the correct Law of Equal Sets that justifies the reasoning from one step to the next.

1. $A - (B' \cap A)$ given
   $A \cap (B' \cap A)'$ a.
   $A \cap (B \cup A')$ b.
   $(A \cap B) \cup (A \cap A')$ c.
   $(A \cap B) \cup \varnothing$ d.
   $A \cap B$ e.

2. $[A \cup (A \cap B)] \cup (B' - A')$ given
   $A \cup (B' - A')$ a.
   $A \cup (B' \cap A)$ b.
   $(A \cup B') \cap (A \cup A)$ c.
   $(A \cup B') \cap A$ d.
   $A \cap (A \cup B')$ e.
   A f.

For Exercises 3–5, use a Venn diagram to verify that each of the following statements is true.

3. $(A \cap B)' - C = (A' \cup B') \cap C'$

4. $A \cap (C - D)' = (A - C) \cup (A - D')$

5. $(R \cup S) - (C - S) = S \cup (R - C)$

For Exercises 6–10, simplify each set.

6. $B \cap (B' \cup C)$

7. $A' \cup (B \cap B')$

8. $(H' \cup H) \cap B'$

9. $(A \cup B) \cap (A \cup B')$

10. $(A \cup B) \cup A'$

Problems

For Problems 11–14, use a membership table to verify each of the following Laws of Equal Sets.

    11. $A \cap B = A - B'$

    12. $(A \cup B)' = A' \cap B'$

    13. $A \cap (B \cup C) = (A \cap B) \cup (A \cap C)$

    14. $A \cup (A \cap B) = A$

For Problems 15–19, use the Venn diagram below with the technique shown in Example 2 to verify each of the following Laws of Equal Sets.

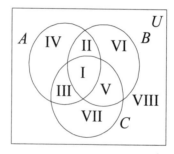

    15. $A - B = A \cap B'$

    16. $(A \cup B)' = A' \cap B'$

    17. $(A \cup B) \cup C = A \cup (B \cup C)$

    18. $A \cap (B \cup C) = (A \cap B) \cup (A \cap C)$

    19. $A \cup (A \cap B) = A$

For Problems 20–22, supply the Law of Equal Sets that justifies the reasoning from one step to the next.

    20. $(P - R)' \cup R'$ given

          $(P \cap R')' \cup R'$ *a.*

          $(P' \cup R) \cup R'$ *b.*

          $P' \cup (R \cup R')$ *c.*

          $P' \cup U$ *d.*

          U *e.*

21. $S' \cap (P \cap S')'$ given

  $S' \cap (P' \cup S)$ *a.*

  $(S' \cap P') \cup (S' \cap S)$ *b.*

  $(S' \cap P') \cup \varnothing$ *c.*

  $S' \cap P'$ *d.*

  $S' - P$ *e.*

22. $[(C \cap U) - (A' \cup \varnothing)]' \cap C$ given

  $[C - (A' \cup \varnothing)]' \cap C$ *a.*

  $(C - A')' \cap C$ *b.*

  $(C \cap A)' \cap C$ *c.*

  $(C' \cup A') \cap C$ *d.*

  $C \cap (C' \cup A')$ *e.*

  $(C \cap C') \cup (C \cap A')$ *f.*

  $\varnothing \cup (C \cap A')$ *g.*

  $C \cap A'$ *h.*

  $C - A$ *i.*

For Problems 23-32, simplify each set.

23. $A' \cup (B \cap B')$

24. $(C' \cup C) \cap D'$

25. $A \cap (A' \cup C)$

26. $(E \cup A') \cap (E \cup A)$

27. $B' \cap (B \cap A')'$

28. $G \cap (G' \cap H')$

29. $(C' \cap D) \cup (C \cup D)'$

30. $[(R \cap R')' \cap (R - S)] \cap S$

31. $P \cap (P - R)'$

32. $\{[C' \cap (C' \cup D)] \cap C\}'$

JUST FOR FUN

Ole Santa's pack held thirty toys

Made by his elfin crew;

And though none made the same amount,

Each elf made more than two.

The elf named Cher made one more toy

Than the elf who dressed in reds,

But Cher made one less Christmas toy

Than the elf who made the sleds.

Spry Johnny Elf made racing cars;

Five toys were made by Jane.

The elf who dressed in yellow suits

Made each and every train.

The elf who always dressed in green

Made one-third as many as Sue.

Cute Marcia Elf was dressed in orange,

And one elf dressed in blue.

The elf who made the spinning tops

Made the most toys of them all.

Another perky, smiling elf

Made each and every ball.

Ole Santa's pack held 30 gifts

All tagged for girls and boys.

From the clues that you've been given,

Now guess who made what toys.

Chapter 5 Review

For questions 1–5, use set notation to represent each given set by listing its elements.

1. The set of vowels found in the word "rhythm."

2. The set of all continents.

3. The set of all whole numbers between 9 and 10.

4. The set of all months with 30 days.

5. The set of all planets in our solar system.

For questions 6–10, assume that $U = \{1,2,3,4,5,6,7,8,9\}$, $R = \{2,5,8,9\}$, $S = \{1,2,7,8\}$, $T = \{3,7,9\}$ and $W = \{4,8\}$. Find the elements in each of the given sets.

6. $(R \cap S') - W$

7. $(T' \cup S) \cap (R - S)$

8. $(S \cup R')' \cap (T \cap W)$

9. $(R \cap S \cap W) - (R' \cup S')'$

10. $(R - W')' - (R \cup S \cup T \cup W)'$

For questions 11–18, assume that the universe is $U = \{1, 2, 3, ..., 10\}$. Let set $A = \{2, 4, 6, 8\}$, set $B = \{2, 4, 6\}$, set $C = \{1, 3, 5, 7\}$, and set $D = \{4, 6, 8\}$. Indicate whether each of the following statements is true or false.

11. $A - B = D - B$

12. $B - A = C \cap D$

13. $A \cup C = U$

14. $(A - D) \subset B$

15. $A' \cap C' \neq \varnothing$

16. $B \cup D = A \cap U$

17. $A - (B \cup C) = A - B$

18. $B \cap C' = A \cap D$

19. Let the universe be $U = \{1, 2, 3, 4, 5, 6, 7, 8, 9, 10\}$, $D = \{1, 5\}$, $E = \{2, 3, 7, 8, 10\}$, $F = \{1, 3, 5, 8, 9\}$, $G = \{7, 10\}$, and $H = \{8\}$. Construct a Venn diagram for this situation. Be sure to place each element in its proper region.

20. Construct a Venn diagram that satisfies the following conditions: Sets A, B, C, D and E are all proper subsets of U, $A \cap C = \emptyset$, $A \cap B \neq \emptyset$, $B \cap C \neq \emptyset$, $D \subset (A - B)$, and assume $E \subset [B - (A \cup C)]$.

For questions 21–29, let $M = \{m,a,t,h\}$ and $E = \{e,n,g,l,i,s,h\}$.

21. Find the number of elements in M.

22. Find the number of elements in $\mathscr{P}(M)$.

23. Find the number of elements in $\mathscr{P}(E)$.

24. Find the number of proper subsets of E.

25. Find $\mathscr{P}(M \cap E)$.

26. Is $a \in M$?

27. Is $\{g\} \subset E$?

28. Is $\{mat\} \subseteq M$?

29. Is $\emptyset \in (M \cup E)$?

30. Given the following Venn diagram, place an "**X**" in the region or regions that represents $(A \cap B') \cup B$.

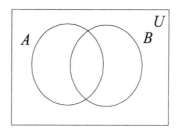

31. Given the following Venn diagram, place an "**X**" in the region or regions that represents $(A - B)' \cap (C \cup A)$.

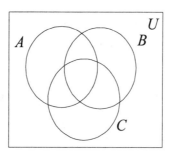

For questions 32–33, consider the Venn diagram below.

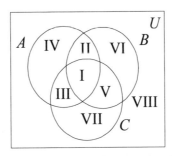

32. Name the numbered region or regions described by the following conditions.

a. $(A \cap B) - C$

b. $(A \cup B) \cap (A - C)$

c. $[(A \cap B') \cup B]'$

d. $[(B - C) \cup (C - A)']'$

33. Name the designated region or regions in terms of sets A, B and C. There may be more than one correct way to name the region.

a. Regions II and V.

b. Region VI.

c. Regions I, IV and VII.

d. Region III.

e. Regions VII and VIII.

34. Use a membership table to determine if the compound statement $[A \cup (A \cup B')'] - C = (A \cap C') \cup (B \cap C')$ is true.

35. Match each numbered statement with the appropriate Law of Equal Sets.

i. $R' \cup (R' \cap S) = R'$ a. DeMorgan's Law

ii. $R' \cup (P \cup S') = R' \cup (S' \cup P)$ b. Distributive Law

iii. $(R \cup S')' \cap P = (R' \cap S) \cap P$ c. Commutative Law

iv. $R' \cup (P \cup S') = (R' \cup P) \cup S'$ d. Absorption Law

v. $R' \cup (P \cap S') = (R' \cup P) \cap (R' \cup S')$ e. Associative Law

36. Describe all the relationships that are exhibited in the Venn diagram below.

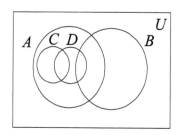

37. For each of the following compound statements, supply the Law of Equal Sets that justifies each step.

a. 1. $(R - S')' - R'$ 1. given
     2. $(R \cap S)' - R'$ 2.
     3. $(R' \cup S') - R'$ 3.
     4. $(R' \cup S') \cap R$ 4.
     5. $R \cap (R' \cup S')$ 5.
     6. $(R \cap R') \cup (R \cap S')$ 6.
     7. $\varnothing \cup (R \cap S')$ 7.
     8. $R \cap S'$ 8.
     9. $R - S$ 9.

b. 1. $[A \cup (A - B)] \cup [A - (B \cup A')]$ 1. given
     2. $[A \cup (A \cap B')] \cup [A - (B \cup A')]$ 2.
     3. $A \cup [A - (B \cup A')]$ 3.
     4. $A \cup [A \cap (B \cup A')']$ 4.
     5. $A \cup [A \cap (B' \cap A)]$ 5.
     6. $A \cup [A \cap (A \cap B')]$ 6.
     7. $A \cup [(A \cap A) \cap B')]$ 7.
     8. $A \cup (A \cap B')$ 8.
     9. A 9.

38. Use the Laws of Equal Sets to verify each of the following statements.

    a. $(A-B)\cup(A-C')=A-(B\cap C')$

    b. $[P'\cap(P'\cup S)]-(P'\cap Q)=(P\cup Q)'$

    c. $(A\cup C)'-[A'\cup(A'\cap B)]=\varnothing$

For questions 39–41, use a Venn diagram to answer each question.

39. In the census of a small town, 830 people claimed to be of European ancestry, 115 people claimed to be of Asian ancestry, 325 claimed to be of African ancestry, while 30 claimed to be of other ancestry. However, some of the people surveyed claimed dual ancestry. Ten claimed they were of European and Asian ancestry, 5 were of African and Asian ancestry while 20 were of African and European ancestry. No one claimed ancestry from all three regions.

    a. How many people in the town were surveyed?

    b. How many were of only Asian ancestry?

    c. How many were not of European and African ancestry?

    d. How many were of European or Asian ancestry, but not African?

    e. How many were not of European ancestry?

40. In a particular lab experiment, a scientist was working with thirty mice. Sixteen mice were male, 20 were trained and 18 were well-fed. Ten of the mice were trained, well-fed males. Twelve were trained males, 11 were well-fed males and 13 were trained and well-fed.

    a. How many male mice were starved?

    b. How many were females?

    c. How many were trained, starved female mice?

41. A restaurant owner wanted to know customers' fondness for three items on the dessert menu. Questioning a group of 50 people, he discovered that 20 people liked raspberry tarts, 15 people liked ice cream, and 25 people liked chocolate mousse. Three people indicated that they liked only raspberry tarts and ice cream, six people liked only raspberry tarts and chocolate

mousse, while 10 people liked just raspberry tarts. Nine people said that they didn't like any of the three choices.

   a. How many people liked all three of these desserts?

   b. How many people liked only one of these desserts?

   c. How many people liked raspberry tarts or ice cream but not chocolate mousse?

   d. How many people like ice cream and chocolate mousse but not raspberry tarts?

JUST FOR FUN

As you can see above, there are three cans. The cans contain marbles. The first can is labeled as two black marbles; the second can is labeled as one black and one white marble; and the third can is labeled as two white marbles. Unfortunately, *each of the three cans is labeled incorrectly.*

You can pick one marble at a time, record its color and put the marble back into the same can. You may do this again for the same can or for another can, as many times as you would like. Is it possible to label the cans correctly? If not, state why not. If yes, state the least number of picks it takes to correctly relabel the cans.

Sample Exam: Chapter 5

For questions 1-10, assume that $U = \{1,2,3,4,5,6,7,8,9,10\}$, $S = \{3,6,9\}$, $P = \{1,2,4,7,10\}$, $Q = \{2,3,4,6,9\}$, and $R = \{1,6,9,10\}$.

1. Find the elements in $P' \cap Q$.
2. Find the elements in $(S - Q)'$.
3. Find the elements in $(R \cup S) - (R \cap S)$.
4. Find the elements in $Q' - (R \cap P)$.
5. Find the elements in $(Q \cup S)' \cap P$.
6. How many elements are in $\mathscr{P}(Q \cap S)$?
7. Does $(P \cap S') = P \cup (P \cap R')$?
8. Is $\{369\} \in \mathscr{P}(S)$?
9. Which of the following is a true statement?
    a. $(Q - S) \subset (P \cap Q)$
    b. $(Q - S) \subseteq (P \cap Q)$
    c. $(Q - S) \cap (P \cap Q) = \varnothing$
    d. $(Q - S)' \cup (P \cap Q) = \varnothing$
    e. none of these
10. Which of the following is a true statement?
    a. $\varnothing \subset S$
    b. $\varnothing \in \mathscr{P}(R)$
    c. $P \cap S = \varnothing$
    d. all of the above
    e. none of the above

For questions 11–20, indicate whether the statement is true or false.

11. If $x \in (A \cap B)'$ then $x \in A'$ or $x \in B'$.
12. $(B \subset D) \rightarrow (B \subseteq D)$.
13. $[x \in (C - D)] \Leftrightarrow [(x \in C) \wedge (x \notin D)]$.
14. $(C \subseteq A) \Leftrightarrow (A \subseteq C)$.
15. If $P \subseteq Q$ then $P = Q$.
16. If $M = \{1,2,5,8\}$ and $N = \{8,2,5,1\}$ then $M \subset N$.
17. $(R = S) \Leftrightarrow [\,[(x \in R) \rightarrow (x \in S)] \vee [(x \in S) \rightarrow (x \in R)]\,]$.
18. If $x \in (S - W)$ then $x \in W'$.
19. $G = U - G'$.
20. If $x \in [P \cap (P \cup Q)]$ then $x \in Q$.

21. For the given Venn diagram, sets A, B, and C are proper subsets of the universal set.

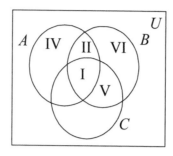

Which of the following represents the numbered regions?
  a. $(A \cup B) - C$
  b. $B \cup C'$
  c. $B \cup [A - (B \cup C)]$
  d. $(A \cup B) - (A \cap C)$
  e. none of these

22. Let $U = \{1, 2, 3, 5, 7, 11, 13, 17, 19, 23, 29\}$, $A = \{1, 2, 3, 5, 13, 17\}$, $B = \{1, 2\}$, $C = \{3, 5, 7, 11, 19, 23\}$, and $D = \{7, 11, 23\}$. Construct a Venn diagram that satisfies the given conditions. Be sure to place all of the elements in their correct regions.

23. Use a membership table to determine if the compound statement $P - (R \cap S)' = (P - R') \cup (P - S')$ is true.

24. For the given Venn diagram, sets A, B, P and W are proper subsets of the universal set.

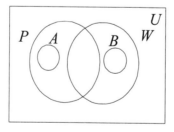

Which of the following is a true statement?
  a. $A \cap W \neq \emptyset$
  b. $P \cap W = \emptyset$
  c. $(x \in P) \rightarrow (x \in A)$
  d. $B \subset (P \cup W)$
  e. none of these

For questions 25–33, supply the Law of Equal Sets that justifies each step.

$$C \cap [(C - A)' \cup A]$$ given

$$C \cap [(C \cap A')' \cup A]$$ 25.

$$C \cap [A \cup (C \cap A')']$$ 26.

$$C \cap [A \cup (C' \cup A)]$$ 27.

$$C \cap [A \cup (A \cup C')]$$ 28.

$$C \cap [(A \cup A) \cup C']$$ 29.

$$C \cap (A \cup C')$$ 30.

$$(C \cap A) \cup (C \cap C')$$ 31.

$$(C \cap A) \cup \varnothing$$ 32.

$$C \cap A$$ 33.

34. Use the Laws of Equal Sets to verify that the compound statement $W \cap (S \cap W)' = W - S$ is true.

For questions 35–37, use a Venn diagram to help solve each problem.

35. A group of 47 co-workers planned a New Year's Eve party. Each person was surveyed about three dessert preferences: apple pie, cheesecake, and brownie sundae. The following data was compiled: 25 liked cheesecake, 24 liked brownie sundae, 4 liked only apple pie, 3 liked only brownie sundae, 6 like all three desserts, 9 like apple pie and cheesecake but not brownie sundae, and 14 like cheesecake and brownie sundae.
    a. How many people liked apple pie?
    b. How many people didn't like any of these desserts?
    c. How many people liked apple pie but not cheesecake?
    d. How many people liked cheesecake or brownie sundae but not both?
    e. How many people did not like brownie sundae?

36. A number immigrants were asked to list three presidents who they felt most changed the culture of the United States. The following information was gathered: 32 listed Abraham Lincoln, 32 listed John Kennedy, and 28 picked Barack Obama. There were 11 who listed both Lincoln and Kennedy, 10 listed only Lincoln and Obama, 12 listed only Kennedy, 5 listed all three presidents, and 7 did not like any of these presidents.

a. How many immigrants were surveyed?
b. How many immigrants did not list both Obama and Kennedy?
c. How many immigrants listed exactly two presidents?

37. Twenty percent of the population of Springhill has been to Italy at least once in the last five years and thirty percent of the population of Springhill has been to France at least once in the last five years. Therefore, half of the population of Springhill has been to Europe at least once in the last five years. The argument is faulty because it ignores the possibility that:

   a. Some of the population of Springhill has been neither to Italy nor to France in the last five years.
   b. Some of the population of Springhill may have been both to Italy and to France in the last five years.
   c. Some of the population of Springhill has been either to Italy or to France in the last five years, but not to both.

CHAPTER SIX

APPLICATIONS OF SET THEORY

One application of set theory is in the area of probability. We can use the concepts of sets to compute the probability for a particular event occurring. Set theory is also the basis of one of the most important concepts in higher mathematics, that of functions and relations. In this chapter, we will examine these applications.

Objectives

After completing Chapter Six, the student should be able to:

- Write the set theory counterparts of the premises and the conclusion of a given argument.

- Test the validity of arguments using Venn Diagrams.

- Determine the probability of an event, based on the number of ways for the event to occur and the number of possible outcomes.

- Determine the probability that event "A or B" will occur.

- Determine the probability that "A and B" will occur if A and B are independent events.

- Determine the probability of the complement of an event, if the probability of this event is known.

- Identify a relation and a function

- Identify the domain and range of a relation and a function

- Identify a one-to-one function

6.1 Testing the Validity of Arguments Using Venn Diagrams

Since the operations of negation, disjunction and conjunction in logic are analogous respectively to complementation, union and intersection in set theory, it is not surprising that there should be a way to test the validity of arguments using set theory. Venn diagrams provide a means for doing this.

A four step algorithm is used to test for validity.

1. Construct a Venn diagram with as many intersecting circles as there are variables in the argument, and label all the regions.
2. Write the set theory counterparts of the premises and the conclusion of the argument. In the case of conditional statements, first use the conditional equivalence to express the conditional statement as a disjunctive statement.
3. Identify the region(s) in the Venn diagram containing the intersection of all the premises.
4. Identify the region(s) in the Venn diagram containing the conclusion.
5. If the intersection of the premises is a subset of the conclusion, the argument is valid. If not, the argument is invalid.

Example 1

Use a Venn diagram to show that the following argument is valid.

$$p \vee q$$
$$\dfrac{\sim p}{q}$$

Solution

First, construct a Venn diagram with two intersecting circles, and label the regions.

Next, write the set theory counterpart of the given argument. We now have the following Venn diagram and argument.

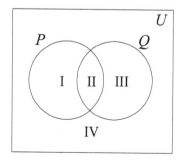

$$P \cup Q$$
$$\dfrac{P'}{Q}$$

The first premise corresponds to regions I, II, and III. The second premise corresponds to regions III and IV. The conclusion corresponds to regions II and III. The intersection of the premises is region III which is a subset of the conclusion region. Hence, the argument is valid. ■

Example 2

Use a Venn diagram to show that the following argument is valid.

$$p \to (q \vee r)$$
$$\sim q$$
$$\sim r$$
$$\overline{}$$
$$\sim p$$

Solution

Our Venn diagram will have three intersecting circles. Using the conditional equivalence, $p \to (q \vee r)$ can be written as $\sim p \vee (q \vee r)$. We now have the following Venn diagram and argument.

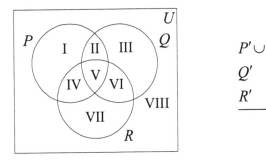

$$P' \cup (Q \cup R)$$
$$Q'$$
$$R'$$
$$\overline{}$$
$$P'$$

The first premise is represented by regions II through VIII, the second premise by regions I, IV, VII, and VIII, and the third premise by regions I, II, III, and VIII. The conclusion is represented by regions III, VI, VII, and VIII. The intersection of these premises occurs only in region VIII. Since this is a subset of the conclusion region, the argument is valid. ■

Example 3

Use a Venn diagram to show that the following argument is invalid.

$$p$$
$$p \to (q \vee r)$$
$$\overline{}$$
$$r$$

Solution

Our Venn diagram will have three intersecting circles. Using the conditional equivalence, $p \rightarrow (q \vee r)$ can be written as $\sim p \vee (q \vee r)$. Next, write the set theory counterpart of the given argument. We now have the following Venn diagram and argument.

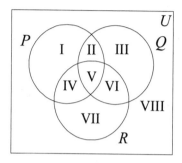

$$P$$
$$\frac{P' \cup (Q \cup R)}{R}$$

The first premise is represented by regions I, II, IV, and V, the second premise by regions II, III, IV, V, VI, VII, and VIII. The intersection of the premises is regions II, IV, V. The conclusion is represented by regions IV, V, VI, and VII. Since the intersection of the premises is not a subset of the conclusion, the argument is invalid. ∎

In-Class Exercises and Problems for Section 6.1

In-Class Exercises

For Exercises 1–8, use a Venn diagram to test the validity of each argument.

1. $a \vee \sim b$
    $\sim b \rightarrow c$
    $\underline{\sim a}$
    $\sim c$

2. $p \rightarrow q$
    p
    $\underline{\sim q \vee r}$
    r

3. $r \rightarrow s$
    $\underline{s \rightarrow w}$
    $r \rightarrow w$

4. $\sim (\sim a \vee b)$
    $\underline{a \rightarrow c}$
    $c \wedge a$

5. $\sim b$

$a \rightarrow b$

$\dfrac{c \rightarrow a}{c \wedge b}$

6. $\sim (r \rightarrow s)$

$s \rightarrow p$

$\dfrac{\sim p}{r \rightarrow p}$

7. $p \vee \sim q$

$\sim p$

$\dfrac{\sim q \rightarrow \sim r}{\sim (r \vee p)}$

8. $\sim (n \wedge \sim p)$

$m \rightarrow n$

$\dfrac{\sim p}{\sim m \vee p}$

Problems

For Problems 9–16, use a Venn diagram to test the validity of each argument.

9. $(p \rightarrow q) \vee r$

$\dfrac{\sim r}{p \wedge \sim q}$

10. $(p \vee q) \rightarrow r$

$\dfrac{\sim p \rightarrow q}{r}$

11. $\sim p \vee r$

$\dfrac{p \vee r}{r}$

12. $(p \vee q) \rightarrow r$

$\dfrac{\sim p \rightarrow \sim q}{r}$

13. $(p \rightarrow q) \rightarrow r$

$\dfrac{\sim p \rightarrow q}{r \vee p}$

14 $(p \rightarrow q) \rightarrow r$

$\dfrac{\sim p \vee q}{r \vee q}$

15. $\sim a$

$a \vee \sim b$

$\dfrac{\sim b \rightarrow c}{c \rightarrow a}$

16. w

$x \rightarrow z$

$\dfrac{\sim w \vee x}{z \rightarrow w}$

Check Your Understanding

For questions 1–3, write the set theory counterpart to each logic statement.

1. $\sim s \vee b$ _____

2. $\sim (w \wedge \sim r)$ _____

3. $\sim (\sim a \rightarrow c)$ _____

4. If the intersection of the premises of an argument is regions II, III and V and the conclusion of the argument is regions II, III and V, is the argument valid?

    _____

5. If the intersection of the premises of an argument is regions IV and V and the conclusion of the argument is region V, is the argument valid?

    _____

6. If the intersection of the premises of an argument is $\varnothing$, and the conclusion of the argument is regions II and VIII, is the argument valid?

    _____

For question 7, use a Venn diagram to test the validity of the argument.

7. $a \rightarrow b$

    $\dfrac{a \wedge c}{b \wedge c}$

6.2 Sets and Probability

Introduction

A useful application of set theory is in an area of mathematics known as probability. We shall illustrate the nature of probability by considering several examples. In a coin toss, we will employ the common usage of saying a coin "lands heads" to mean the head side of the coin is facing up.

Example 1

To determine which football team will kick off to begin the game, a coin is tossed in the air. How likely is it that the coin will land heads?

Solution

To answer the question, we try to determine what fraction of time we would expect to see a head if we tossed the coin many times. This fraction is called the probability of heads. We say that the probability of getting heads is ½. This is because, if we tossed the coin lots of times, we'd expect to get heads one-half the time. ∎

Probability Definition

In order to deal with more complicated problems, we need a more precise way to define probability. The *probability* that an event E will occur is defined as

$$P(E) = \frac{\text{the number of ways event } E \text{ may occur}}{\text{the total number of possible outcomes}}.$$

Our set theory notation gives us an easy way to evaluate the probability fraction for a particular event. We define a *sample space* to be the set of all possible outcomes that may occur when we perform an experiment. The sample space corresponds to the universe in set theory. In Example 1, the experiment is tossing a coin, and the sample space would be the set $S = \{\text{heads, tails}\}$. The event whose probability we wish to determine would be the set $E = \{\text{heads}\}$. The outcome "heads" is called a *sample point* in the sample space. Notice that in general, E must be a subset of S. Since the number of ways event E can occur is the same as the number of elements in set E, and the total number of possible outcomes is the same as the number of elements in S,

$$P(E) = \frac{\text{the number of elements in set } E}{\text{the number of elements in the entire sample space } S}.$$

In Example 1,

$$P(E) = P(\text{head}) = \frac{\text{the number of elements in set } E}{\text{the number of elements in set } S} = \frac{1}{2}.$$

Example 2

John Zakie has complained that when course registration is done in alphabetical order, the people whose last names start with "A" always get first choice. In order to make the registration process fairer, it was decided to hold an alphabet lottery to pick the letter of the last name with which to begin the registration process. Since there are 26 letters in the alphabet, 26 slips of paper, each containing a different letter of the alphabet, are placed in a jar and then one is selected. We assume that each slip of paper has an equal opportunity of being chosen. How likely is it that the letter "z" will be the first one chosen? How likely is it that the first letter chosen will be a vowel?

Solution

The sample space is the set consisting of the letters of the alphabet, i.e., $S = \{a, b, c, \ldots, z\}$, and $E = \{\text{the letter } z \text{ is chosen}\}$. We know that

$$P(E) = \frac{\text{the number of elements in set } E}{\text{the number of elements in the entire sample space}}.$$

Since there are 26 elements in the sample space, and only one way to satisfy the event "z will be chosen" we have

$$P(z \text{ will be chosen}) = \frac{1}{26}.$$

To determine how likely is it that the first letter chosen will be a vowel we count the number of elements in this event. There are five vowels in the alphabet. They form the set $E = \{a, e, i, o, u\}$. Then,

$$P(E) = \frac{\text{the number of elements in set } E}{\text{the number of elements in the entire sample space}},$$

so,

$$P(\text{a vowel is selected}) = \frac{5}{26}.$$

Example 3

A die is a small cube with dots on each of its six faces, numbered 1 through 6. Suppose a fair die is rolled. What is the probability that the side facing up is the number 5? What is the probability that it is a number less than 3? What is the probability that it is an even number? Note that a *fair* die is one such that each of its faces is equally likely to show if the die is rolled. Unless otherwise noted, all dice we discuss will be assumed fair.

Solution

For the die, the sample space is $S = \{1, 2, 3, 4, 5, 6\}$, so the total number of elements in the sample space is 6. The set representing a roll of 5 has just one point in it, the number 5 itself, so $E_1 = \{5\}$. Therefore,

$$P(5) = \frac{1}{6}.$$

The set that contains numbers less than 3 is $E_2 = \{1, 2\}$. Therefore,

$$P(\text{less than 3}) = \frac{2}{6} = \frac{1}{3}.$$

The set that contains only even numbers is $E_3 = \{2, 4, 6\}$. Using the same reasoning as before,

$$P(\text{even}) = \frac{3}{6} = \frac{1}{2}.$$

■

Example 4

An urn contains seven colored chips: four green chips, two red chips and one blue chip. One chip is selected at random. Determine $P(\text{green})$, $P(\text{red})$ and $P(\text{blue})$.

Solution

The number of elements in the sample space is the total number of chips, 7. The subset of green chips has 4 elements in it; the subset of red chips has 2 elements; and the subset of blue chips has only 1 element. This gives us

$$P(\text{green}) = \frac{4}{7}, \ P(\text{red}) = \frac{2}{7}, \text{ and } P(\text{blue}) = \frac{1}{7}.$$

■

Mutually Exclusive Events

In Example 4, there are only three possible events that can occur if we select a chip at random. The chip must be green, red, or blue. These events are all *mutually exclusive,* that is, they cannot happen at the same time. For example, a chip cannot be both red and green at the same time. Moreover, these three events represent all possible outcomes. If we add the probabilities of all three mutually exclusive events, we obtain

$$P(\text{green}) + P(\text{red}) + P(\text{blue}) = \frac{4}{7} + \frac{2}{7} + \frac{1}{7} = 1.$$

This illustrates an important result in probability: *The sum of the probabilities of all the mutually exclusive events in a sample space is always one.*

Complementary Probability

If $P(E)$ represents the probability that event E occurs, we let $P(E')$ represent the probability that event E will not occur. We call $P(E')$ the *complementary probability* of event E.

Example 5

Consider a deck of 52 ordinary playing cards. One card is selected at random. Determine the probability of not getting an ace.

Solution

The total number of elements in the sample space is 52. Notice that it is easier to find the number of cards that are aces than it is to count the cards that are not aces. Since there are 4 aces,

$$P(\text{ace}) = \frac{4}{52}.$$

The number of cards that are not aces is $52 - 4 = 48$. Therefore,

$$P(\text{not an ace}) = P(\text{ace}') = \frac{48}{52}. \qquad \blacksquare$$

We observe that the event "ace" and the event "not an ace" are mutually exclusive events. If a card is selected, either it will or will not be an ace. Thus, $P(\text{ace}) + P(\text{not an ace}) = 1$. This was illustrated in Example 5 where we saw that

$$P(\text{ace}) + P(\text{ace}') = \frac{4}{52} + \frac{48}{52} = 1.$$

The solution to Example 5 illustrates a formula that can be used to compute the complementary probability of any event E. Since $P(E) + P(E') = 1$, we can write

$$P(E') = 1 - P(E).$$

Example 6

A die is rolled. Find the probability that the number rolled is not greater than four.

Solution

The total number of elements in the sample space is 6. Two of these are numbers that are greater than four, so

$$P(\text{not greater than 4}) = 1 - P(\text{greater than 4}) = 1 - \frac{2}{6} = \frac{4}{6} = \frac{2}{3}. \quad \blacksquare$$

Probability and Percents

Sometimes we may not know the total number of elements in a sample space, but we may be given the relationship among the various events in terms of percentages.

Example 7

At a certain movie theater, 70% of the women are wearing shoes, 20% are wearing sneakers, and 10% are wearing boots. If a woman is chosen at random, find the probability that she is wearing shoes, the probability that she is wearing sneakers, and the probability that she is not wearing boots.

Solution

Since percent means "for every hundred", we can treat the sample space as if it contains 100 elements. The fact that seventy percent of the women are wearing shoes indicates that 70 of every 100 women are wearing shoes, so $P(\text{shoes}) = \dfrac{70}{100} = 0.7$. Since 20% of the women are wearing sneakers, $P(\text{sneakers}) = \dfrac{20}{100} = 0.2$. Finally, since 10% of the women are wearing boots, the probability that a woman is wearing boots is $P(\text{boots}) = 0.1$. Therefore, the probability a woman is not wearing boots is $P((\text{boots})') = 1 - P(\text{boots}) = 1 - 0.1 = 0.9$. $\quad \blacksquare$

In-Class Exercises and Problems for Section 6.2

In-Class Exercises

1. Twenty-six blocks, each containing one letter of the alphabet, are placed in a carton. One of the blocks is picked at random. Find the probability that the block is
   a. the letter "x".
   b. the letter "p".
   c. not the letter "q".
   d. not a vowel.
   e. a letter contained in the word "dog".
   f. not a letter contained in the word "frog".
   g. neither an "a" nor a "b".

2. A die is rolled. Find:
   a. $P(2)$.
   b. $P(1)+P(2)+P(3)+P(4)+P(5)+P(6)$.
   c. $P(2\ or\ 3)$.
   d. $P(\text{odd})$.
   e. $P(\text{not }4)$.
   f. $P((5)')$.
   g. $P(\text{neither 5 nor 6})$.
   h. $P(7)$.
   i. $P(\text{less than }4)$.
   j. $P(\text{not less than }3)$.
   k. $P((\text{greater than }5)')$.
   l. $P((\text{less than }9)')$.

3. The results for a television survey at 9 pm on a certain night were as follows: 20% of the homes surveyed were tuned to NBC, 10 % were tuned to ABC, 5% were tuned to CNN and 15% were tuned to FOX. The survey company believes that this reflects the viewing habits of the entire nation. If a home is called at random and one television is on in that home, find the probability that the television is tuned to
   a. CNN.
   b. either NBC or FOX.

c. one of the four networks cited in the survey.

d. none of the four networks.

4. A card is selected from an ordinary deck of fifty-two cards. Find

   a. $P(\text{club})$.

   b. $P(\text{not a club})$.

   c. $P(\text{not a black card})$.

   d. $P(\text{not a king})$.

   e. $P(\text{the six of spades})$.

   f. $P(\text{a queen or a king})$.

   g. $P(\text{neither a queen nor a king})$.

   h. $P(\text{the card is less than five})$, assuming that the ace counts as a one.

Problems

5. An urn contains seven colored chips: two red, four green and one blue. Find the probability that a chip selected at random is
   a. red.
   b. green.
   c. blue.
   d. not red.
   e. not blue.
   f. neither green nor blue.
   g. white.
   h. red or green.

6. A certain die is "loaded" to favor the numbers 1, 2, 3 and 4. That is, $P(1) = P(2) = P(3) = P(4) = \dfrac{1}{5}$, but $P(5) = P(6) = \dfrac{1}{10}$. Find

   a. $P(1) + P(2) + P(3) + P(4) + P(5) + P(6)$.

   b. $P(\text{not a 2})$.

   c. $P(\text{not a 4})$.

   d. $P(\text{not a 5})$.

7. In a particular mathematics class, 30% of the students are freshmen, 20% are sophomores, 40% are juniors and 10% are seniors. Find the probability that a student chosen at random from this class is

   a. a junior.

   b. a freshman.

   c. not a senior.

   d. a sophomore or above.

   e. neither a freshman nor a junior.

8. A deck of ordinary playing cards is shuffled thoroughly. If a single card is drawn at random, find the probability that the card is

   a. an ace of diamonds.

   b. a black ace.

   c. a king.

   d. a picture card.

   e. a number greater than 5 but less than 8.

   f. not the queen of hearts.

   g. not a jack.

   h. either an ace or a king.

   i. a red card.

   j. not a diamond.

   k. an even-numbered card.

   l. not an odd-numbered card, if the ace is considered a one.

9. A computer is programmed to generate a random whole number from 1 to 12. Find

   a. $P(7)$.

   b. $P((3)')$.

   c. $P(\text{an even number})$.

   d. $P(19)$.

   e. $P((13)')$.

   f. $P(2 \text{ or } 12)$.

   g. $P((1 \text{ or } 10)')$.

   h. $P(\text{a number less than 10})$.

   i. $P((\text{a number greater than 5})')$.

Check Your Understanding

1. Sample space S contains 24 elements. Suppose E is a subset of S and E contains 7 elements. If an element is selected at random from sample space S, what is the probability that this element will belong to set E?_____

2. If the probability that event E occurs is 2/11, find the probability that event E does not occur.   _____

3. There are 9 Republicans, 8 Democrats and 4 Independents in a room. One of these is selected at random. Find the probability that the person selected is:
   a. a Republican.   _____
   b. not an Independent.   _____

4. A survey was conducted in a college cafeteria at lunch. The data showed that 40% of the students purchased only pizza at the cafeteria, 27% purchased only hamburgers at the cafeteria, and 19% brought their lunch from outside. (There were no students that both brought lunch from outside and also purchased food from the cafeteria). Furthermore, 14% of the students were dieting, and neither brought food from outside nor purchased food at the cafeteria. Find the probability that a student selected at random:
   a. purchased a hamburger at the cafeteria.   _____
   b. did not bring lunch from outside.   _____
   c. purchased lunch at the cafeteria.   _____

JUST FOR FUN

1. In a two child family, what is the probability that one child is a boy and the other child is a girl?

2. In a two child family, one child is a girl. What is the probability that the other child is a girl? (Hint: Think carefully about the sample space.)

6.3 Venn Diagrams and Probability

Introduction

We have already seen that many of the ideas of probability can be explained using set theory notation. Using the notions of union, intersection, and Venn diagrams, we now expand our previous results to find probabilities of compound events.

We begin our discussion by considering situations that require finding the probability that either of two mutually exclusive events will occur. Events that are mutually exclusive can be represented by disjoint sets in a Venn diagram.

Example 1

Suppose a die is rolled. What is the probability that a number less than 4 or greater than 5 is rolled?

Solution

Our sample space consists of the set of all the possible numbers we can roll on the die. Thus, $S = \{1, 2, 3, 4, 5, 6\}$. Letting L and G be subsets of S, where L represents the set of numbers less than 4, and G represents the set of numbers greater than 5, we have $L = \{1, 2, 3\}$ and $G = \{6\}$. We observe that L and G have no elements in common, i.e., $L \cap G = \varnothing$. Sets L and G represent mutually exclusive events. A Venn diagram representing this situation is shown below.

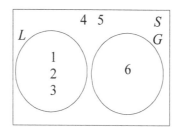

Since L has 3 elements and S has 6 elements,

$$P(\text{a number less than 4}) = P(L) = \frac{3}{6} = \frac{1}{2}.$$

Also, since G only contains one element,

$$P(\text{a number greater than 5}) = P(G) = \frac{1}{6}.$$

The union of the two sets consists of the elements that are in either L or G, that is, $L \cup G$. In this case, $L \cup G = \{1, 2, 3, 6\}$. Therefore,

$$P(\text{a number is less than 4 or greater than 5}) = P(L \cup G) = \frac{4}{6} = \frac{2}{3}. \quad \blacksquare$$

We observe that since set L and set G are disjoint, we have $P(L \cup G) = P(L) + P(G)$. This result can be generalized as follows:

Given any two mutually exclusive events A and B, that is, $A \cap B = \varnothing$, it follows that $P(A \cup B) = P(A) + P(B)$.

Example 2

Suppose we select a card at random from an ordinary deck of 52 playing cards. What is the probability that the card chosen is either an 8 or a picture card?

Solution

Letting the sample space S be the set of all possible outcomes, we find that S contains 52 elements. If we let E represent the set of eights, we find 4 elements belonging to E, namely the eight of hearts, the eight of diamonds, the eight of clubs, and the eight of spades. If T is the set of picture cards, then T contains the kings, the queens and the jacks in all four suits, so T contains 12 elements. No card is both an eight and a picture card, so events E and T are mutually exclusive.

Since $P(E) = \dfrac{4}{52}$ and $P(T) = \dfrac{12}{52}$, and $E \cap T = \varnothing$, we have

$$P(\text{an eight or a picture card}) = P(E) + P(T) = \frac{4}{52} + \frac{12}{52} = \frac{16}{52} = \frac{4}{13}. \quad \blacksquare$$

Example 3

In a certain mathematics class, the grades on a particular examination were 3 A's, 5 B's, 12 C's, 5 D's and 5 F's. Find the probability that a student selected at random from this class received either an A or a D.

Solution

The total number of possible elements in the sample space is $3 + 5 + 12 + 5 + 5 = 30$. Since events A and D are mutually exclusive,

$$P(A \text{ or } D) = P(A \cup D) = P(A) + P(D) = \frac{3}{30} + \frac{5}{30} = \frac{8}{30} = \frac{4}{15}. \quad \blacksquare$$

The Addition Theorem

If event A and event B are not mutually exclusive, then the probability of obtaining event A or event B is a little more complicated.

Example 4

Suppose a die is rolled. What is the probability that the number rolled is greater than 2 and an even number? What is the probability that the number rolled is greater than 2 or an even number?

Solution

There are six elements in the sample space, $S = \{1, 2, 3, 4, 5, 6\}$. If we let the set G represent the event "a number greater than 2 is rolled", then $G = \{3, 4, 5, 6\}$. If we let set E represent the event "an even number is rolled", then $E = \{2, 4, 6\}$. The events are not mutually exclusive, i.e., $G \cap E \neq \varnothing$. In fact, their intersection is $G \cap E = \{4, 6\}$, which are the numbers that are both greater than 2 and even. The corresponding Venn diagram is shown below.

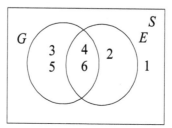

Observe that

$$P(\text{greater than 2 and an even number}) = P(G \cap E) = \frac{2}{6} = \frac{1}{3}.$$

We must now determine the probability that the roll of a die produces a number that is greater than 2 *or* even. Since the events G and E are not mutually exclusive, we cannot simply add $P(G)$ to $P(E)$ in order to find $P(G \cup E)$. If we did, the elements of $G \cap E$ would be counted twice, once as elements of G and then again as elements of E. To correct this problem, we subtract $P(G \cap E)$ from the sum of $P(G)$ and $P(E)$. Symbolically, we have

$$P(G \cup E) = P(G) + P(E) - P(G \cap E).$$

Thus,

$$P(G \cup E) = \frac{4}{6} + \frac{3}{6} - \frac{2}{6} = \frac{5}{6}.$$

We can check this result in the Venn diagram. The number of elements in the universe is 6. The elements that are either greater than 2 or even are members of $G \cup E$, and we count 5 of them. Since there are 6 elements in the universe,

$$P(G \cup E) = \frac{5}{6}.$$

■

As a result of the reasoning used in Example 4, we get the general *Addition Theorem* for the probability of event A or event B. *Given two events, A and B,*

$$P(A \text{ or } B) = P(A \cup B) = P(A) + P(B) - P(A \cap B).$$

Example 5

Suppose we select a card at random from an ordinary deck of 52 playing cards. Determine:

  a. P(the card is an ace or a heart),
  b. P(the card is a five or a queen),
  c. P(the card is a picture card and black).

Solution

  a. There are 4 aces in a deck of cards. There are 13 cards in the heart suit. These events are not mutually exclusive, because there is one card, namely the ace of hearts, that satisfies both events. If we let A represent the event "ace" and H represent the event "hearts", the addition theorem gives

$$P(A \cup H) = P(A) + P(H) - P(A \cap H)$$

$$= \frac{4}{52} + \frac{13}{52} - \frac{1}{52} = \frac{16}{52} = \frac{4}{13}.$$

  b. There are 4 fives in the deck and there are 4 queens. There is no card that is both a five and a queen, so these events are mutually exclusive. Using F for "fives" and Q for "queens", we can still apply the addition theorem, giving us

$$P(F \cup Q) = P(F) + P(Q) - P(F \cap Q) = \frac{4}{52} + \frac{4}{52} - 0 = \frac{8}{52} = \frac{2}{13}.$$

  c. If we let T be the set of picture cards and B be the set of black cards, we wish to find $P(T \cap B)$. We simply count the number of elements in the intersection $T \cap B$, which contains the jack, queen, and king of spades and the jack, queen, and king of clubs.

Since $T \cap B$ contains 6 elements, $P(T \cap B) = \frac{6}{52} = \frac{3}{26}.$ ■

Using DeMorgan's Law

We have seen how some of the rules of probability are related to the set theory concepts of union, intersection and complementation. Remember that $P(A \cap B)$ represents the probability that an event satisfies outcome A and outcome B simultaneously, while $P(A \cup B)$ represents the probability that A or B will occur. Recall that the complementary probability, the probability that event A does *not* happen, is given by

$$P(A') = 1 - P(A).$$

When complementary probability is combined with union or intersection, we can simplify the computations needed to solve the problem with the help of DeMorgan's Law. The two forms of DeMorgan's Law of Equal Sets are given by

$$(A \cap B)' = A' \cup B' \text{ and } (A \cup B)' = A' \cap B'.$$

Example 6

A card is selected from a deck of playing cards. Find the probability that the card is not a king or not a spade.

Solution

If we let K represent the set of kings, and S the set of spades, then the set of cards that are not kings or not spades is $K' \cup S'$. Thus, we are looking for $P(K' \cup S')$. DeMorgan's Law allows us to rewrite $P(K' \cup S')$ as $P((K \cap S)')$. Using the idea of complementary probability, $P((K \cap S)') = 1 - P(K \cap S)$. We know that there is only one card that is both a king and a spade, namely the king of spades.

Therefore, $P(K \cap S) = \dfrac{1}{52}$. Putting this all together, we have

$$P(\text{not a king or not a spade}) = P(K' \cup S') = P((K \cap S)')$$

$$= 1 - P(K \cap S) = 1 - \frac{1}{52} = \frac{51}{52}. \quad \blacksquare$$

Example 7

What is the probability that a randomly selected card is not a club and not a picture card?

Solution

Using C for the set of clubs and T for the set of picture cards, we obtain

$$P(\text{not a club and not a picture card}) = P(C' \cap T') = P((C \cup T)')$$

$$= 1 - P(C \cup T).$$

From the addition theorem,

$$P(C \cup T) = P(C) + P(T) - P(C \cap T)$$
$$= \frac{13}{52} + \frac{12}{52} - \frac{3}{52} = \frac{22}{52}.$$

Therefore,

$$P(C' \cap T') = P((C \cup T)') = 1 - P(C \cup T)$$
$$= 1 - \frac{22}{52} = \frac{52}{52} - \frac{22}{52} = \frac{30}{52} = \frac{15}{26}.$$

■

In-Class Exercises and Problems for Section 6.3

In-Class Exercises

For Exercises 1–5, assume a card is selected at random from an ordinary deck of 52 playing cards. Determine if the two given events are mutually exclusive.

1. The card has on odd number on it; the card is a picture card.
2. The card is a queen; the card is a picture card.
3. The card is a club; the card is a spade.
4. The card is a diamond; the card is red.
5. The card is a 5; the card is a king.

Use the following information to answer Exercises 6-7. There are freshman students, honors students and teachers in a library study room. In all, there are ten people. Let F represent the set of freshman students, H represent the set of honors students, and T represent the set of teachers.

6. Draw a Venn diagram to illustrate the relationship that exists among these three sets if Chris, Jan and Mike are freshmen honors students; Lou, Val and Nan are freshmen non-honors students; Rachel and Will are non-freshmen honors students; and Kate and Pat are teachers.

7. Suppose one person is selected at random from the study room. Use your Venn diagram to determine the probability that the person selected is:

    a. a freshman.

    b. an honors student.

    c. a freshman and an honors student.

    d. a freshman or an honors student.

    e. a teacher.

    f. a teacher and a freshman.

    g. a teacher or a freshman.

    h. not an honors student.

    i. neither an honors student nor a freshman.

    j. not an honors student and not a teacher.

For Exercises 8–17, assume E, F, and G are three events from a sample space of all possible outcomes. Let $P(E) = \dfrac{3}{17}$, $P(F) = \dfrac{10}{17}$, $P(G) = \dfrac{7}{17}$, $P(E \cap F) = \dfrac{1}{17}$, $P(F \cap G) = \dfrac{4}{17}$, and $P(E \cap G) = 0$. Apply the rules of probability to find the following probabilities. Venn diagrams are not necessary.

    8. $P(E \cup F)$ 13. $P(G')$

    9. $P(F \cup G)$ 14. $P((E \cap F)')$

    10. $P(E \cup G)$ 15. $P(E' \cup F')$

    11. $P(E')$ 16. $P(E' \cap F')$

    12. $P(F')$ 17. $P(E' \cap G')$

For Exercises 18–29, consider a bingo machine containing seventy-five balls numbered 1 through 75. Determine the probability that the number on the first ball picked is:

    18. 30.

    19. less than 30.

    20. greater than 30.

21. less than 30 or greater than 30.

22. even.

23. odd.

24. less than 10.

25. odd and less than 10.

26. odd or less than 10.

27. not odd and not less than 10.

28. not even and does not end in 3.

29. not even and not greater than 60.

For Exercises 30–39, assume one card is chosen from a half-deck of playing cards that only contains the red cards: the diamonds and the hearts. Assuming that the ace card acts like the number one, find the probability that the card is:

30. a king or a queen.

31. a king or a picture card.

32. a king and a picture card.

33. a diamond or a heart.

34. a diamond and a heart.

35. an even number less than 9.

36. an even number or a number less than 9.

37. a diamond and a picture card.

38. not a diamond or not a picture card.

39. not a heart and not a picture card.

Problems

Use your understanding of mutually exclusive events to answer Problems 40–41.

40. A die is rolled. Which events are mutually exclusive?

a. The number is even; the number is odd.

b. The number is a 3; the number is odd.

c. The number is greater than 4; the number is less than 5.

d. The number is less than 3; the number is greater than 4.

41. A man and a woman are selected from a room full of people. Which events are mutually exclusive?

    a. The woman has brown hair; the man has brown hair.

    b. The woman has brown hair; the woman has blue eyes.

    c. The woman has blue eyes; the woman has brown eyes.

    d. The man has a blonde hair; the man has black hair.

Draw Venn diagrams to help you complete Problems 42–43.

42. A card is selected at random from the 13 hearts in a deck of cards: A, 2, 3, 4,…,10, J, Q, and K. In the Venn diagram, let T represent the elements that are picture cards and E represent the elements that are even-numbered cards. Find:

    a. P(a picture card).

    b. P(an even-numbered card).

    c. P(a picture card or an even-numbered card).

    d. P((a picture card or an even-numbered card)′).

    e. P(not a picture card and not an even-numbered card).

43. Ten chips are placed into a container. Four are colored red and six are blue. The four red chips are numbered 1 through 4, while the six blue chips are numbered 5 through 10. In the Venn diagram, let B be the subset of elements that are blue and let O be the subset of elements that are odd. Find:

    a. $P(B)$.

    b. $P(O)$.

    c. $P(B \cap O)$.

    d. $P(B \cup O)$.

    e. $P((B \cup O)')$.

    f. $P((B \cap O)')$.

    g. $P(B' \cup O')$.

    h. $P(B' \cap O')$.

Do Problems 44–47 by applying the rules of probability. It is not necessary to draw Venn diagrams.

44. A and B represent two events that are mutually exclusive. Event A can be expected to occur 30% of the time and event B is expected 40% of the time. Find each of the following probabilities.

   a. $P(A')$

   b. $P(B')$

   c. $P(A \cup B)$

   d. $P(A \cap B)$

   e. $P((A \cap B)')$

   f. $P((A \cup B)')$

   g. $P(A' \cup B')$

   h. $P(A' \cap B')$

45. Suppose $P(Q) = \dfrac{5}{31}$, $P(R) = \dfrac{7}{31}$, and $P(Q \cap R) = \dfrac{3}{31}$. Find the value of each of the following:

   a. $P(Q')$.

   b. $P(R')$.

   c. $P(Q \cup R)$.

   d. $P((Q \cup R)')$.

   e. $P((Q \cap R)')$.

   f. $P(Q' \cup R')$.

   g. $P(Q' \cap R')$.

46. Suppose that one hundred slips of paper numbered from 1 to 100 are placed into a bag. Find the probability that the number on a slip of paper picked at random is:

    a. greater than 50.

    b. greater than 50 or less than 21.

    c. not greater than 50 and not less than 21.

    d. greater than 50 or a multiple of 10.

    e. not greater than 50 and not a multiple of 10.

    f. an odd number or a number ending in 4.

    g. not an odd number and not a number ending in 4.

    h. an even number or a multiple of 5.

    i. not an even number and not a multiple of 5.

47. From an ordinary deck of fifty-two playing cards one card is chosen at random. Find the probability that the card is:

    a. black or a heart.

    b. black or a picture card.

    c. a queen or a spade.

    d. black and not a 2.

    e. an ace or not a diamond.

    f. not a jack or not a diamond.

    g. not an ace and not a club.

    h. not a spade and not a 2, 3, 4 or 5.

    i. neither red nor a 2.

    j. not a queen or not a picture card.

    k. not an ace or not a picture card.

    l. a picture card but not a jack.

Check Your Understanding
---

1. Use DeMorgan's Law and the Rule of Complementary Probability to rewrite $P(A' \cap B')$.   _____

For questions 2–6, assume all 52 cards in a deck of ordinary playing cards are placed face down on a table. One card is turned over. Let P: the card is a picture card, S: the card has a number on it that is less than 6.

2. Are events P and S mutually exclusive?   _____

3. What is the probability that event P occurs?   _____

4. What is the probability that event S occurs?   _____

5. What is the probability that the card is a picture card or a number less than 6?

    _____

6. What is the probability that the card is not a picture card and not a number less than 6?   _____

DID YOU KNOW?

Blaise Pascal (1623–1662) was a French mathematician, physicist, and religious philosopher. He was a child prodigy who was educated by his father. To make his father's work as a tax official easier, Pascal invented a calculating machine for addition and subtraction and oversaw its construction. Pascal was a great mathematician. In 1653, he wrote his *Traité du Triangle Arithmétique* ("Treatise on the Arithmetical Triangle") in which he described a convenient tabular presentation for binomial coefficients, now called Pascal's triangle.

Pascal's development of probability theory was his most influential contribution to mathematics. Originally applied to gambling, today probability is extremely important in economics, especially in actuarial science. Another significant application of probability theory in everyday life is reliability. Many consumer products, such as automobiles and electronics, use reliability theory in order to reduce the probability of product failure. The probability of failure is, of course, closely associated with the product's warranty.

6.4 Independent Events

Introduction

The calculation of the probability of two events, A and B, denoted $P(A \cap B)$, can be complicated. Thus far, we have determined $P(A \cap B)$ by listing all of the elements in the sample space. This calculation can be done more easily if the two events are independent.

Two events, E_1 and E_2, are said to be *independent* if the occurrence of one of the events does not alter the probability that the other event will occur. As an example, if a coin is tossed into the air and at the same time, a die is rolled, the result of the coin toss does not affect the outcome of the die roll.

The Counting Principle

Given two activities like tossing a coin and rolling a die, it is often important to determine how many outcomes are possible when *both* activities are performed. The *Counting Principle* provides a way to to do this. The counting principle states that if one activity can occur in m ways, and a second activity can occur in n ways, then there are $m \times n$ ways for both activities to occur.

Example 1

A coin is tossed and a die is rolled. How many outcomes are possible?

Solution

The coin can land two ways (heads and tails) and the die can have one of six outcomes. The combined event can happen in $2 \times 6 = 12$ ways.

■

Example 2

A coin is tossed and a die is rolled. What is the probability that the coin will land heads and the die will show a 5?

Solution

The outcome of the coin toss and the outcome of the die roll have no effect on one another, that is, the two events are independent. We can list the elements of the sample space as ordered pairs of events, using the notation (coin toss, die roll). If we use H to represent heads and T to represent tails, then our sample space is the set

$$S = \begin{Bmatrix} (H,1),(H,2),(H,3),(H,4),(H,5),(H,6), \\ (T,1),(T,2),(T,3),(T,4),(T,5),(T,6) \end{Bmatrix}.$$

There are exactly 12 outcomes for this coin toss and die roll. This is exactly what the counting principle guaranteed in Example 1. The combined event that concerns us, heads on the coin paired with 5 on the die, is represented by just one element in S, namely, $(H, 5)$. Therefore,

$$P(H \cap 5) = P(H, 5) = \frac{\text{the number of ways the event can occur}}{\text{the number of elements in the sample space}}$$
$$= \frac{1}{12}.$$

■

Independent Probability Formula

There is another way to evaluate this type of "and" probability for independent events. If we consider just the coin in Example 2, $P(H) = \dfrac{1}{2}$, and if we consider just the die, $P(5) = \dfrac{1}{6}$. Observe that the product of $P(H)$ and $P(5)$ is $P(H) \times P(5) = \dfrac{1}{2} \times \dfrac{1}{6} = \dfrac{1}{12}$. This is the same result we obtained when we found $P((H, 5))$ above. This suggests a generalized *independent probability formula. Given two independent events A and B,*

$$P(A \text{ and } B) = P(A \cap B) = P(A) \times P(B).$$

Example 3

Four chips, numbered 1, 2, 3 and 4, and four playing cards, a jack, a queen, a king and an ace, are all placed in a hat. A chip and a card are selected at random. What is the probability that an even-numbered chip and a picture card are selected?

Solution

We observe that the two events are independent. We first solve the problem by listing all of the members of the sample space, using J for jack, Q for queen, K for king and A for ace. Since there are 4 chips and 4 cards, there are $4 \times 4 = 16$ ways for this compound event to occur. Therefore, S will have 16 elements, as shown below.

$$S = \left\{ \begin{matrix} (1, J), (1, Q), (1, K), (1, A), (2, J), (2, Q), (2, K), (2, A), \\ (3, J), (3, Q), (3, K), (3, A), (4, J), (4, Q), (4, K), (4, A) \end{matrix} \right\}.$$

The set E containing ordered pairs of the form (even number, picture card) is the set

$$E = \{(2, J), (2, Q), (2, K), (4, J), (4, Q), (4, K)\}.$$

Then,

$$P(\text{even} \cap \text{picture card}) = \frac{\text{number of ways the event can occur}}{\text{number of elements in the sample space}}$$

$$= \frac{6}{16} = \frac{3}{8}.$$

If we solve the problem using the independent probability formula, we obtain

$$P(\text{even} \cap \text{picture card}) = P(\text{even}) \times P(\text{picture card}) = \frac{2}{4} \times \frac{3}{4} = \frac{6}{16} = \frac{3}{8}.$$

■

Example 4

A card is selected from an ordinary deck of 52 playing cards and a die is rolled. Find the probability that a diamond card is selected and an odd number is rolled.

Solution

The sample space for this problem is large. In fact, there are $52 \times 6 = 312$ outcomes. We will solve the problem using the independent probability formula. Since the events are independent,

$$P(\text{diamond} \cap \text{odd number}) = P(\text{diamond}) \times P(\text{odd number})$$

$$= \frac{13}{52} \times \frac{3}{6} = \frac{39}{312} = \frac{1}{8}.$$

■

The Probability of Several Events

Both the addition theorem for mutually exclusive events and the formula for independent probability can be extended to cases involving more than two events.

Suppose that $E_1, E_2, E_3, ..., E_n$ represent n events. If these events are mutually exclusive, then the probability of E_1 or E_2 or E_3 or...or E_n is found by adding the probabilities of each event. Symbolically, we write

$$P(E_1 \text{ or } E_2 \text{ or } E_3 \text{ or...or } E_n) = P(E_1 \cup E_2 \cup E_3 \cup ... \cup E_n)$$

$$= P(E_1) + P(E_2) + P(E_3) + ... + P(E_n).$$

Similarly, if these n events are *all* independent events, then the probability of E_1 and E_2 and E_3 and...and E_n is found by multiplying the probabilities of each event. Symbolically, we write

$$P(E_1 \text{ and } E_2 \text{ and } E_3 \text{ and...and } E_n) = P(E_1 \cap E_2 \cap E_3 \cap...\cap E_n)$$

$$= P(E_1) \times P(E_2) \times P(E_3) \times...\times P(E_n).$$

Example 5

The Lucky-Three Lottery requires the bettor to choose a three-digit number from 000 to 999. The winning number is picked at random. Determine

    a. P(the winning number is 111 or 222 or 333 or 444 or 555 or 666 or 777 or 888 or 999),

    b. P(the same three-digit number occurs five days in a row).

Solution

To answer part (a), we note that there are 1,000 possible elements in the sample space, one for each possible number. The probability that a particular number is the winning number is $\dfrac{1}{1,000}$. The events 111 or 222 or 333 or ... or 999 are all mutually exclusive, and each has a probability of $\dfrac{1}{1,000}$ of occurring. Therefore,

$$P(111 \text{ or } 222 \text{ or...or } 999) = \frac{1}{1,000} + \frac{1}{1,000} +...+ \frac{1}{1,000} = \frac{9}{1,000}.$$

To answer part (b), we note that each day's winning number is independent of the previous day's number. For the same three-digit number to occur five days in a row, it does not matter which number wins on the first day. We need only calculate the probability that the same number reoccurs on the next four days. If we let E represent the event "the number that won on the first day wins again," the probability that the same number occurs on the next four days in a row is

$$P(E \text{ and } E \text{ and } E \text{ and } E) = P(E) \times P(E) \times P(E) \times P(E)$$

$$= \frac{1}{1,000} \times \frac{1}{1,000} \times \frac{1}{1,000} \times \frac{1}{1,000}$$

$$= \frac{1}{1,000,000,000,000}.$$ ■

Example 6

Two dice are rolled. Determine

    a. P(the sum of the two dice is 11),

    b. P(the sum of the two dice is 5).

Solution

There are two ways that the dice can add up to 11. The first die can be 6 and the second 5, or the first die can be 5 and the second 6. Since the events are mutually exclusive,

$$P(5 \text{ and } 6) \text{ or } P(6 \text{ and } 5) = P(5 \text{ and } 6) + P(6 \text{ and } 5).$$

Also, since each die roll is independent of the other die roll, we have

$$P(5 \text{ and } 6) = P(5) \times P(6) = \frac{1}{6} \times \frac{1}{6} = \frac{1}{36}. \text{ In a similar manner, we find}$$

that $P(6 \text{ and } 5) = P(6) \times P(5) = \dfrac{1}{36}.$ Putting this all together, we obtain

$$P(\text{the sum is } 11) = P(5 \text{ and } 6) + P(6 \text{ and } 5)$$

$$= \frac{1}{36} + \frac{1}{36} = \frac{2}{36} = \frac{1}{18}.$$

We answer part (b) in a similar fashion. There are four different ways to produce the number 5 as a sum. We have

$$P(\text{the sum is } 5) = P(1 \text{ and } 4) + P(2 \text{ and } 3) + P(3 \text{ and } 2) + P(4 \text{ and } 1)$$

$$= \frac{1}{36} + \frac{1}{36} + \frac{1}{36} + \frac{1}{36}$$

$$= \frac{4}{36} = \frac{1}{9}.$$

■

Example 7

Alicia has three friends, Brittany, Carla and Dara. Find

    a. P(none of Alicia's friends is born in the same month as Alicia)

    b. P(at least one of Alicia's friends is born in a different month than Alicia)

Solution

The month in which Alicia was born does not matter. We let B represent the event "Brittany was born in Alicia's birth month," C

represent the event "Carla was born in Alicia's birth month," and D represent the event "Dara was born in Alicia's birth month." Then,

$$P(B) = P(C) = P(D) = \frac{1}{12}, \text{ and } P(B') = P(C') = P(D') = \frac{11}{12}.$$

In part (a), we want to find the probability that all three friends were born in months different from Alicia's month. Therefore we want $P(B' \text{ and } C' \text{ and } D')$. Since the three events are independent,

$$P(B' \cap C' \cap D') = P(B') \times P(C') \times P(D')$$

$$= \frac{11}{12} \times \frac{11}{12} \times \frac{11}{12} = \frac{1331}{1728}.$$

For part (b), the probability that at least one of Alicia's friends is born in a different month than hers is $P(B' \cup C' \cup D')$. These events are not mutually exclusive. It is possible, for instance, to have both Brittany and Carla born in months different from Alicia's. Thus, we cannot use the addition theorem for mutually exclusive events. But, we can use an extended form of DeMorgan's Law and the concept of complementary probability. Using our set theory notation

$$P(B' \cup C' \cup D') = P((B \cap C \cap D)')$$

$$= 1 - P(B \cap C \cap D)$$

$$= 1 - [P(B) \times P(C) \times P(D)]$$

$$= 1 - \left[\frac{1}{12} \times \frac{1}{12} \times \frac{1}{12}\right] = \frac{1727}{1728}.$$

■

When we calculate the probability of event A and event B, it is critical to know whether the two events are independent. If they are independent, we can apply the independent probability formula. If not, we must list the members of the sample space in order to calculate the probability. This distinction is illustrated in the following two examples.

Example 8

Shira has three tiles lettered d, g, and o in a bag. She will use the tiles to form a word, using the first tile picked as the first letter, the second as the second letter and the remaining tile as the third letter. She does not replace the tiles in the bag. What is the probability that Shira will spell the word "dog"?

Solution

We cannot say that Shira's attempts are independent. Whichever letter she picks first will affect the choice of letters for the second and third attempts, so we cannot use the formula for independent events. One way to answer the question is to list the sample space of all possible outcomes. $S = \{$dgo, dog, gdo, god, odg, ogd$\}$. If we let D represent the event "the first letter is d," O represent the event "the second letter is o," and G represent the event "the third letter is g," then

$P(\text{dog}) = P(D \cap O \cap G) = \dfrac{1}{6}$. The Venn diagram for the solution is

shown in the following illustration.

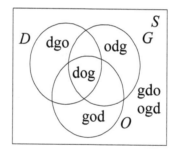

■

Example 9

Suppose Maxine took the bag containing the letters d, g and o from Shira. She picked one tile from the bag but she wrote down the letter. Then she placed that tile back in the bag and picked again. She then wrote down the second letter, replaced the tile, and picked a third time. What is the probability that she will spell the word "dog"?

Solution

Since Maxine replaced each tile after looking at the letter, each of her attempts to pick a tile is independent of every other attempt. For

each attempt, $P(D) = P(O) = P(G) = \dfrac{1}{3}$. Then,

$$P(\text{dog}) = P(D \cap O \cap G) = P(D) \cdot P(O) \cdot P(G) = \frac{1}{3} \cdot \frac{1}{3} \cdot \frac{1}{3} = \frac{1}{27}.$$

The Venn diagram for the solution is shown below.

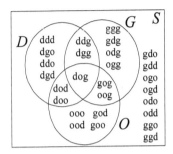

■

In-Class Exercises and Problems for Section 6.4

In-Class Exercises

For each pair of events in Exercises 1–2, answer the question "Are the two events independent?"

1. A coin is flipped and a die is rolled.

   a. heads on the coin; 5 on the die.

   b. heads on the coin; tails on the coin.

   c. 4 on the die; 2 on the die.

2. A slot machine consists of three spinning wheels with pictures of different objects (cherries, lemons, bells,...) on them. The gambler on the machine pulls a lever or pushes a button to activate the machine.

   a. The first wheel shows cherries; the second wheel shows cherries.

   b. The first wheel shows any object; the second wheel shows the same object as is shown on the first wheel.

   c. The third wheel shows a fruit; the third wheel shows a lemon.

   d. The first wheel shows a bell; the first wheel shows a fruit.

For Exercises 3–5, apply the rules of probability to answer each question.

3. Suppose H, J and K are three independent events with $P(H) = \dfrac{1}{3}$, $P(J) = \dfrac{1}{2}$, and $P(K) = \dfrac{3}{4}$. Find:

   a. $P(H \cap J)$.

   b. $P(H \cap K)$.

   c. $P(H \cap J \cap K)$.

   d. $P((J \cap K)')$.

   e. $P(J' \cup K')$.

   f. $P(J' \cap K')$.

   g. $P((H \cup J \cup K)')$.

   h. $P(H' \cup J' \cup K')$.

4. Ruven and Debbie are playing a game in which each calls out a random number from 1 to 5 simultaneously. Find the probability that:

   a. both call out 5.

   b. both call out 3.

   c. Ruven calls out 3 and Debbie calls out 5.

   d. either one of them calls out 3 and the other calls out 5.

   e. Debbie calls out 5 and Ruven a number less than 3.

   f. they call out the same number.

   g. each calls out an even number.

   h. neither calls out 2.

   i. either one of them calls out 4 while the other does not.

5. Two fair dice are rolled. Determine:

   a. P(the sum of the dice is 2).

   b. P(the sum of the dice is 12).

   c. P(the sum of the dice is 3).

   d. P(the sum of the dice is 8).

   e. P(the sum of the dice is 7).

Problems

6. Joshua's wardrobe consists of ten pairs of pants (five black, three green and two brown), eight shirts (four white, two green, one brown and one blue) and twelve pairs of socks (six black, four white and two brown). He selects one of each type of garment, but because of a power failure, he must select his clothes in the dark. Find the probability that Joshua selects:

   a. a black pair of pants and a white shirt.

   b. a green pair of pants and a blue shirt.

   c. a green shirt but not a black pair of pants.

   d. neither a brown pair of pants nor a brown shirt.

   e. brown pants and a brown shirt.

   f. green pants and a green shirt.

   g. the same color pants and shirt.

   h. the same color pants, shirt and socks.

   i. a white shirt and white socks.

   j. a white shirt or white socks.

   k. a white shirt or a brown shirt or a blue shirt.

   l. not a white shirt and not a brown shirt and not a blue shirt.

7. Four pills and three capsules are in a box. One of the pills is poisonous while one of the capsules is its antidote. Find the probability that a person who swallows a pill and a capsule will be poisoned and not saved.

8. Thirty percent of the women in a large investment firm earn over $100,000 while forty percent of the men earn over $100,000. A woman and a man are both randomly chosen to work together. Determine the probability that:

   a. both people earn over $100,000.

   b. at least one of them earns over $100,000.

   c. neither of them earns over $100,000.

   d. only one of them earns over $100,000.

9. A card is selected at random from an ordinary deck of playing cards. It is replaced in the deck, the deck is shuffled, and another card is selected. Determine the probability that:

   a. both cards are aces.

   b. the first card is an ace and the second card is a king.

   c. the first card is a picture card and the second is a club.

   d. the cards are a jack of diamonds and a queen of spades, in any order.

   e. the cards are an ace and a king, in any order.

   f. both cards are exactly the same.

10. A pair of dice has been altered so that for each die, $P(1) = P(2) = P(3) = \dfrac{1}{4}$ and $P(4) = P(5) = P(6) = \dfrac{1}{12}$. The two dice are rolled. Find:

    a. P(a 1 on the first die and a 1 on the second).

    b. P(a 2 on the first die or a 2 on the second).

    c. P(the sum of the two dice is 12).

    d. P(the sum of the two dice is 11).

    e. P(the sum of the two dice is 8).

    f. P(the sum of the two dice is 7).

    g. P(both dice show the same number).

Check Your Understanding

1. Suppose that one activity can produce m different outcomes, and a second activity can produce n different outcomes. How many outcomes are possible if both activities are performed? _____

2. Suppose A and B are two independent events. Write the formula for $P(A \cap B)$.

_____

For questions 3–8 assume that a wheel of fortune has eight equally likely outcomes, numbered one through eight. The wheel is spun twice and each time the winning number is recorded. Let events F and S be defined as follows:

F: The first spin produces an odd number.
S: The second spin produces an odd number.

3. How many different outcomes can be observed if the wheel is spun twice?_____

4. Are events F and S independent? _____

5. What is the probability that event F occurs? _____

6. What is the probability that both spins produce odd numbers? _____

7. Are events F and S mutually exclusive? _____

8. What is the probability that neither of the spins produces an odd number?_____

6.5 Relations and Functions

In Chapter 5, we explored the basic notions of sets. Now let us consider a particular type of set: one whose elements are *ordered pairs* of the form (x, y). Consider two sets: set $A = \{2, 3, 5, 7\}$ and set $B = \{4, 7, 8\}$. We can construct many ordered pairs in which x is any element of set A and y is any element of set B. For example, one of these ordered pairs would be $(3, 8)$ since 3 belongs to set A and 8 belongs to set B. Another ordered pair would be $(2, 7)$ since $2 \in A$ and $7 \in B$, and still another would be $(2, 8)$. If we were to consider *all* possible ordered pairs for which $x \in A$ and $y \in B$, we would have

$$(2,4), (2,7), (2,8), (3,4), (3,7), (3,8),$$
$$(5,4), (5,7), (5,8), (7,4), (7,7), (7,8).$$

Now let's create a set whose elements are the ordered pairs listed above:

$$\{(2,4), (2,7), (2,8), (3,4), (3,7), (3,8),$$
$$(5,4), (5,7), (5,8), (7,4), (7,7), (7,8)\}.$$

The set of *all* such ordered pairs is called the *product set* (or *Cartesian product*) of sets A and B, and is denoted by the symbol $A \times B$. Mathematically, we write $A \times B = \{(x, y) \mid x \in A, y \in B\}$.

Example 1

Let $M = \{a, b, c\}$ and let $N = \{r, s, t, u\}$. Find the product set, $M \times N$.

Solution

$$M \times N = \{(a,r), (a,s), (a,t), (a,u), (b,r), (b,s),$$
$$(b,t), (b,u), (c,r), (c,s), (c,t), (c,u)\}$$ ∎

Example 2

Find any non-empty subset of $M \times N$ in Example 1, above.

Solution

Since there are 12 elements in $M \times N$ there are $2^{12} - 1 = 4095$ non-empty subsets. One of these is $\{(a,r), (b,s), (c,s), (c,t)\}$. Of course there are 4094 other correct answers! ∎

Given any two sets, A and B, any non-empty subset $A \times B$ is called a *relation* from A to B.

Example 3

Using sets $A = \{2,3,5,7\}$ and $B = \{4,7,8\}$, find any two relations from A to B.

Solution

Since a relation is any non-empty subset of $A \times B$, two of the 127 possible subsets are

$$\{(2,7),(2,8),(7,7)\} \text{ and } \{(3,7)\}.$$ ∎

Notice that any set of ordered pairs (x, y) in which $x \in A$ and $y \in B$ is a relation from A to B. The set of all first coordinates of a relation is called the *domain* of the relation. The set of all second coordinates of a relation is called the *range* of the relation. Notice that the domain of a relation from A to B is a subset of set A, and the range of a relation from A to B is a subset of set B.

Example 4

Let set $A = \{3,5,7,9,11\}$, and set $B = \{2,3,5,6\}$. Is set S consisting of the ordered pairs $S = \{(3,2),(3,5),(7,5),(11,6)\}$ a relation from A to B? If so, state its domain and range.

Solution

Since a relation is simply a set of ordered pairs, S is certainly a relation. The domain of set S is $\{3,7,11\}$ and the range of set S is $\{2,5,6\}$. ∎

Example 5

Let set $A = \{2,3,5,7\}$. Find a relation R from set A to set A *i.e.,* $A \times A$.

Solution

One of many possible relations is $R = \{(3,3),(3,7),(5,2)\}$. ∎

When a relation is from a set A to itself, we say the relation is simply a *relation on set A*. Notice that a relation on set A is a subset of the product set $A \times A$, also called A^2.

The *inverse of a relation* is a new relation that is formed by interchanging the first and second coordinates in each ordered pair of the original relation. The inverse of a relation R is denoted by R^{-1}. Mathematically, we write

$$R^{-1} = \{(b,a) \mid (a,b) \in R\}.$$

Example 6

Construct the inverse of relation S given in Example 4.

Solution

Since $S = \{(3,2),(3,5),(7,5),(11,6)\}$, to find S^{-1} we interchange the x and y components of each of the ordered pairs. Therefore, $S^{-1} = \{(2,3),(5,3),(5,7),(6,11)\}$.

■

Arrow Diagrams

There are other methods by which relations may be represented. One useful representation is called an *arrow diagram*. To illustrate, consider the sets $A = \{2,3,5,7\}$ and $B = \{4,7,8\}$. Then one relation from A to B would be $R = \{(3,4),(3,8),(5,8)\}$. An arrow diagram representing relation R is shown below.

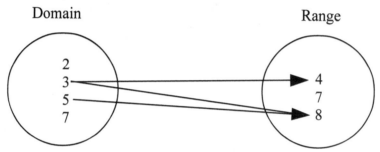

Example 7

Let $Q = \{1,2,3,4,5,6,7\}$ and $M = \{2,4,6,8,10,12,14\}$. Suppose the relation R is defined by $R = \{(1,6),(2,6),(3,2)\}$. Construct an arrow diagram that represents this relation.

Solution

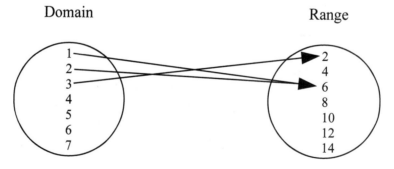

■

Functions

A *function* is a relation that has two special properties. First, each member of the domain must be used. Second, each member of this domain is paired with a unique (i.e. exactly one) member of the range. We define a *function*, or *mapping* from set A to set B to be a relation in which each member of set A is paired with exactly one member of set B. The notation

$$f : A \rightarrow B$$

is read "f is a function, or mapping, from set A to set B." Notice that for a function from A to B:

- The domain is the entire set A.
- The range is set B or any subset of set B. That is, the range may not contain every element of set B.
- Each member of set A must be paired with a unique (just one!) element of set B.
- It's okay for two different members of the domain to be paired with the same member of the range.

Example 8

Let $A = \{2, 4, 6, 8\}$ and $B = \{3, 5, 7, 9\}$. Is $\{(2,7), (4,3), (6,5), (8,5)\}$ a function from A to B?

Solution

Every element of set A is paired with exactly one element from set B. Therefore, the set $\{(2,7), (4,3), (6,5), (8,5)\}$ is a function from A to B. Notice that the numbers 6 and 8 from set A are paired with 5 from set B. This is perfectly acceptable since the definition of a function states that each member of set A must be paired with a unique element of set B. ∎

Example 9

Let $A = \{2, 4, 6, 8\}$ and $B = \{3, 5, 7, 8\}$. Is $\{(2,7), (2,8), (4,3), (8,5)\}$ a function from A to B?

Solution

There are two reasons that this relation is not a function from A into B. First, the number 2 in the domain is paired with more than one member of the range. It is paired with both 7 and 8. This violates the requirement that each member of set A must be paired with a unique element of set B. Second, it is also the case that not every element in set A was used in the relation. The arrow diagram clearly illustrates this idea.

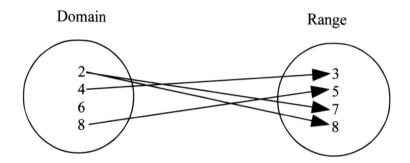

A *one-to-one function* is a function in which no two elements of the domain are ever paired with the same element of the range.

Example 10

Let $A = \{2, 4, 6, 8\}$ and $B = \{a, b, c\}$. Is $\{(2, b), (4, a), (6, b), (8, c)\}$ a one-to-one function from A into B?

Solution

This relation is a function from A into B, since each element of set A is paired with a unique element of set B. However, some members of the domain are paired with the same member of the range (2 and 6 are both paired with b). Thus, it is not a one-to-one function from A into B.

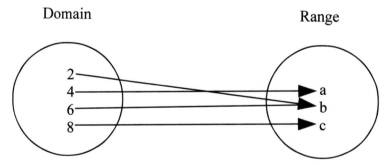

Example 11

Let $A = \{2,3,5,7\}$ and $B = \{a,b,c,d\}$. Suppose we let the function

$f = \{(2,c),(3,a),(5,b),(7,d)\}$. Is f a one-to-one function from A to B?

Solution

The relation f is certainly a function from A to B, since each member of set A is paired with a unique member of set B. Also, every element of B is paired with a unique member A. Hence, f is a one-to-one function from set A to set B.

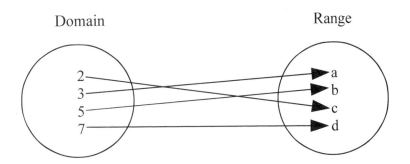

■

In-Class Exercises and Problems for Section 6.5

In-Class Exercises

For Exercises 1–8, let $P = \{a,b\}$ and $Q = \{1,3,5\}$.

1. Find the members of the product set $P \times Q$.

2. Find the members of the product set $P \times P$.

3. Let $R = \{(a,1),(b,3)\}$ be a relation from P to Q.
   a. What is the domain of R?
   b. What is the range of R?
   c. Is R a function from P to Q?

4. Let $S = \{(a,3),(a,5),(b,1)\}$ be a relation from P to Q.
   a. What is the domain of S?
   b. What is the range of S?
   c. Is S a function from P to Q?

5. Let $T = \{(a,5),(b,3),(b,5)\}$ be a relation from P to Q.
   a. What is the domain of T?
   b. What is the range of T?
   c. Is T a function from P to Q?

6. Let $U = \{(a,5),(b,5)\}$ be a relation from P to Q.
   a. What is the domain of U?
   b. What is the range of U?
   c. Is U a function from P to Q?

7. Let $V = \{(a,1),(a,3),(a,5)\}$ be a relation from P to Q.
   a. What is the domain of V?
   b. What is the range of V?
   c. Is V a function from P to Q?

8. Let $W = \{(b,3)\}$ be a relation from P to Q.
   a. What is the domain of W?
   b. What is the range of W?
   c. Is W a function from P to Q?

For Exercises 9–12, consider the three sets A, B, and C, where $A = \{1,2,3,4,5\}$, $B = \{6,7,8,9\}$, and $C = \{10,11,12,13,14\}$. In each case, determine if the given function is one-to-one.

9. The given function from set A to set B is defined by $f = \{(1,6),(2,7),(3,7),(4,6),(5,8)\}$.

10. The function from B to C is $g = \{(6,11),(7,12),(8,10),(9,14)\}$

11. The given function from set A to set B is defined by $h = \{(1,6),(2,7),(3,6),(4,9),(5,8)\}$.

12. The function $k = \{(1,14),(2,13),(3,12),(4,11),(5,10)\}$ which is from A to C.

Problems

For Problems 13–14, find the members of the given product set.

13. $C \times D$ if $C = \{5,7\}$ and $D = \{1,2,3,4,5\}$.

14. $E \times G$ if $E = \{a,b,c,d\}$ and $G = \{p,q,r\}$.

For Problems 15–20, let $A = \{1, 2, 3, 4, 5\}$ and $B = \{6, 7, 8, 9, 10, 11\}$. For each problem, state

    a. the domain of the given relation.

    b the range of the given relation.

    c. whether the relation is a function from A to B.

15. $R = \{(1, 7), (2, 7), (3, 8), (3, 10), (4, 7)\}$.

16. $S = \{(1, 11), (2, 9), (3, 8), (4, 8), (5, 10)\}$.

17. $T = \{(1, 8), (1, 10), (2, 6), (3, 9), (4, 7), (5, 11)\}$.

18. $V = \{(2, 7), (3, 8), (4, 9), (5, 11)\}$.

19. $W = \{(1, 11), (2, 6), (3, 8), (4, 9), (5, 9)\}$.

20. $Z = \{(1, 10), (2, 9), (3, 6), (3, 7), (4, 10), (5, 11)\}$.

For Problems 21–28, let $L = \{1, 2, 3, 4, 5, 6, 7\}$, $M = \{q, r, s, t, u\}$, and $N = \{10, 20, 30, 40, 50, 60, 70\}$. For each problem, determine if the given function is one-to-one.

21. Let $f = \{(1, q), (2, r), (3, s), (4, t), (5, u), (6, q), (7, r)\}$, where f is a function from L to M.

22. Let $g = \{(q, 10), (r, 30), (s, 40), (t, 10), (u, 70)\}$, where g is a function from M to N.

23. $h = \{(1, 10), (2, 20), (3, 30), (4, 40), (5, 50), (6, 60), (7, 70)\}$, and h is a function from L to N.

24. $i = \{(q, 1), (r, 2), (s, 3), (t, 5), (u, 6)$, where i is a function from M to L.

25. $j = \{(q, 70), (r, 60), (s, 50), (t, 40), (u, 30)$, where j is a function from M to N.

26. $k = \{(10, r), (20, s), (30, t), (40, q), (50, r), (60, u), (70, s)\}$, and k is a function from N to M.

27. $y = \{(q, 1), (r, 7), (s, 6), (t, 6), (u, t)\}$, and y is a function from M to L.

28. Let $z = \{(10, 7), (20, 2), (30, 5), (40, 3), (50, 1), (60, 6), (70, 4)\}$, and z is a function from N to L.

Chapter 6 Review

For questions 1–2, use a Venn diagram to test the validity of each argument.

<table>
<tr><td>

1. $a \rightarrow (b \vee c)$

$\dfrac{\sim (b \wedge a)}{c \rightarrow \sim a}$

</td><td>

2. $\sim m \rightarrow \sim n$

$n \wedge w$

$\dfrac{m \vee \sim w}{m \rightarrow (n \rightarrow w)}$

</td></tr>
</table>

For questions 3–7, assume 100 M&M's are randomly chosen from a bag of M&M's with the following results: nine red, eleven blue, eleven yellow, fifteen orange, sixteen green, and thirty-eight brown. If Carmine randomly selects 3 M&M's, one at a time, from the 100 M&M's and replaces each one before making his next selection, find the probability that he selects:

3. a brown followed by a red followed by a green.
4. three green or three blue.
5. a yellow followed by two browns.
6. a green or an orange followed by a red and then a yellow.
7. no red ones.

For questions 8–11, assume that ninety percent of the cars manufactured by Speedy Motors have automatic transmissions, while 10% have standard transmissions. Ten percent have four-cylinder engines, 50% have six-cylinder engines and 40% have eight-cylinder engines. Assume any combination of transmission and engine is possible. If one of Speedy Motors' cars is selected at random, find the probability of it having:

8. an automatic transmission and a four-cylinder.
9. a standard transmission or a eight-cylinder.
10. not an eight-cylinder.
11. neither a four-cylinder nor a standard transmission.

For Questions 12–15, let $M = \{w, x, y\}$ and let $N = \{3, 5, 7, 9\}$. Is the given relation a function?

12. The relation $\{(w, 9), (x, 5), (y, 9)\}$ from M to N.
13. The relation $\{(9, w), (5, x), (7, w), (3, y)\}$ from M to N.
14. The relation $\{(9, w), (5, x), (7, w), (3, y)\}$ from N to M.

15. The relation $\{(w,9,)(w,3),(w,7)\}$ from N to M.

16. Let $R=\{5,7,12\}$. Is $A=\{(5,7),(5,12),(7,5)\}$ a relation on $R \times R$?

17. Let the set $Q=\{3,5,7\}$ and the set $R=\{45,50,55,60\}$. Is $f=\{(3,45)(5,60)(7,50)\}$ a one-to-one function from Q to R?

Sample Exam: Chapter 6

For questions 1–2, use a Venn diagram to test the validity of each argument.

$$1. \sim(r \rightarrow \sim s)$$
$$a \vee \sim s$$
$$\underline{r \qquad\qquad}$$
$$a \wedge r$$

$$2. \sim(p \wedge \sim r)$$
$$\sim(r \vee s)$$
$$\underline{s \rightarrow p}$$
$$p$$

For questions 3–6, consider a deck of playing cards and a group of 26 checkers, sixteen of which are red and ten of which are black. Two selections are made. Either two cards are selected with replacement, two checkers are selected with replacement, or a card and a checker may be selected. Determine the probability that:

3. if a card is selected and then a checker is selected, the card is black and the checker is black.

4. two red items are chosen.

5. if a card is selected and then a checker is selected, both are the same color.

6. a three or a black card and a red checker is selected.

7. Let $R=\{\{a,b\},\{1,2,3\}\}$. Find the product set $R \times R$.

8. Let $M=\{2,3\}$ and $N=\{a,b,c\}$. Is $T=\{(2,a),(2,b),(2,c)\}$ a relation from M to N?

9. Let $R=\{(0,0),(1,3)\}$ and $S=\{5,8,9\}$. Notice that the elements of R are themselves, ordered pairs. Is the function $f=\{((0,0),5),((1,3),9)\}$ a one-to-one function from R to S?

10. Let $A=\{@,!,?\}$ and let $B=\{\#,\$\}$.

    a. Is it possible to define a one-to-one function from A to B? Explain your answer.

    b. Is it possible to define a one-to-one function from B to A? Explain your answer.

Glossary

Absorption equivalence: Expressed symbolically as either $p \vee (p \wedge q) \Leftrightarrow p$ or $p \wedge (p \vee q) \Leftrightarrow p$.

Absorption law: Expressed symbolically as either $A \cap (A \cup B) = A$ or $A \cup (A \cap B) = A$. It is analogous to the Absorption equivalence in logic.

Antecedent: The left hand side (LHS) of a conditional statement.

Argument: A collection of *TF* statements, called premises, which are always assumed to be true, followed by a final statement, called the conclusion.

Associative equivalence: Expressed symbolically as $[(p \wedge q) \wedge r] \Leftrightarrow [(p \wedge (q \wedge r)]$ or $[(p \vee q) \vee r] \Leftrightarrow [(p \vee (q \vee r)]$. Note that the order in which the variables appear is unchanged, but the order in which the operations are performed is altered. This equivalence pertains only to the conjunction and disjunction connectives.

Associative law: Expressed symbolically as $(A \cap B) \cap C = A \cap (B \cap C)$ or as $(A \cup B) \cup C = A \cup (B \cup C)$. Note that the order in which the sets appear is unchanged, but the order in which the operations are performed is altered. This equivalence pertains only to the intersection and union connectives. It is analogous to the Associative equivalence in logic.

Biconditional equivalence: Indicates that a biconditional statement is equivalent to the conjunction of a conditional statement with its converse. It is expressed symbolically as $(p \leftrightarrow q) \Leftrightarrow [(p \rightarrow q) \wedge (q \rightarrow p)]$.

Biconditional statement: A compound statement of the form "*p* if and only if *q*", denoted $p \leftrightarrow q$. Note that $(p \leftrightarrow q) \Leftrightarrow [(p \rightarrow q) \wedge (q \rightarrow p)]$.

Biconditional negation equivalence: Allows us to express the negation of a biconditional as the disjunction of two specified conjunctions. It is expressed symbolically as $\sim (p \leftrightarrow q) \Leftrightarrow [(p \wedge \sim q) \vee (q \wedge \sim p)]$.

Braces: Symbols used to enclose the elements of a set: { }.

Cardinality: The number of elements in a set.

Cartesian product: See "Product set".

Commutative equivalence: Expressed symbolically as $(p \vee q) \Leftrightarrow (q \vee p)$ or as $(p \wedge q) \Leftrightarrow (q \wedge p)$; states that the order in which the LHS and RHS occur in a conjunction or in a disjunction is immaterial.

Commutative law: Expressed symbolically as $A \cup B = B \cup A$ or $A \cap B = B \cap A$. Note that the order in which the sets occur in an intersection or a union is immaterial. It is analogous to the Commutative equivalence in logic.

Complement: The complement of a set A, denoted by A', is the set of all elements in the universe that are not in set A; the operation of complementation in set theory is analogous to the operation of negation in logic.

Complement law: The complement of the empty set is the universal set while the complement of the universal set is the empty set.

Complementary probability: The probability that an event will not occur.

Complementary switches: Two switches arranged in such a way that when one is turned on the other is turned off.

Compound statement: Two or more simple statements joined by one or more connectives.

Conclusion: A statement inferred from the premises of an argument, often preceded by the word "therefore."

Conditional equivalence: Expressed symbolically as $(p \to q) \Leftrightarrow (\sim p \vee q)$; allows us to express a conditional statement as a disjunction to which it is logically equivalent.

Conditional negation equivalence: Allows us to express the negation of an implication as the conjunction of its LHS with the negation of its RHS. It is expressed symbolically as $\sim (p \to q) \Leftrightarrow (p \wedge \sim q)$.

Conditional proof: A formal proof that assumes the LHS of a conditional conclusion is true. If its RHS can be shown to be true under this assumption, the argument is valid.

Conditional statement: A compound statement of the form "If p then q" or "p implies q", denoted as $p \to q$.

Conjunction: A compound statement in which two *TF* statements are joined by the word *and*; the symbol for a conjunction is $\wedge$; the statement $p \wedge q$ is read "p and q" or "p conjunction q".

Conjunctive addition: A valid argument form which states that if two statements are known to be true, they can be joined with the conjunction connective and the resulting conjunction will be true.

Conjunctive simplification: A valid argument form which states that if a conjunction is true, each of its components must be true.

Connective: Indicates an operation on two *TF* statements; the four basic connectives are denoted symbolically by $\wedge$ (and), $\vee$ (or), $\rightarrow$ (implies), and $\leftrightarrow$ (if and only if).

Consequent: The right hand side (RHS) of a conditional statement.

Contrapositive: The contrapositive of a given conditional statement $p \rightarrow q$ is another conditional statement of the form $\sim q \rightarrow \sim p$. Note that the LHS and RHS of the original conditional statement have been interchanged *and* negated.

Contrapositive equivalence: Expressed symbolically as $(p \rightarrow q) \Leftrightarrow (\sim q \rightarrow \sim p)$; states that a conditional statement is logically equivalent to its contrapositive (see contrapositive).

Converse: The converse of a given conditional statement $p \rightarrow q$ is another conditional statement of the form $q \rightarrow p$. Note that the LHS and RHS of the given conditional have been interchanged.

Counterexample: An illustration that assigns truth values to the variables such that all the premises are true yet the conclusion is false.

DeMorgan's equivalence: Expressed symbolically as $\sim (p \vee q) \Leftrightarrow (\sim p \wedge \sim q)$ or as $\sim (p \wedge q) \Leftrightarrow (\sim p \vee \sim q)$. It allows us to express the negation of a conjunction as a disjunction with each of its components negated, and express the negation of a disjunction as a conjunction with each of its components negated.

DeMorgan's law: Expressed symbolically as $(A \cup B)' = (A' \cap B')$ or expressed symbolically as $(A \cap B)' = (A' \cup B')$. It allows us to express the complement of a union as an intersection with each of its components complemented, and express the complement of an intersection as a union with each of its components complemented. It is analogous to DeMorgan's equivalence in logic.

Disjoint: Two (or more) sets whose intersection is the empty set, i.e. the sets have no elements in common.

Disjunction: Two *TF* statements joined by the word "or"; the symbol for a disjunction is $\vee$; the statement $p \vee q$ is read "*p* or *q*" or "*p* disjunction *q*".

Disjunctive addition: A valid argument form which states that if a true statement is joined with any other statement using the disjunction connective, the resulting disjunction must be true.

Disjunctive syllogism: A valid argument form which states that if one premise is a disjunction and the other premise is the negation of one of its components, then the remaining component of the disjunction must be true.

Distributive equivalence: Expressed symbolically by one of two compound statements, either $[p \vee (q \wedge r)] \Leftrightarrow [(p \vee q) \wedge (p \vee r)]$, which states that a disjunction distributes over a conjunction, or $[p \wedge (q \vee r)] \Leftrightarrow [(p \wedge q) \vee (p \wedge r)]$, which states that a conjunction distributes over a disjunction.

Distributive law: Expressed symbolically as $A \cup (B \cap C) = (A \cup B) \cap (A \cup C)$ or $A \cap (B \cup C) = (A \cap B) \cup (A \cap C)$, which states that intersection distributes over union and union distributes over intersection. It is analogous to the Distributive equivalence in logic.

Domain: The set of all first coordinates of a relation.

Double complement law: The complement of the complement of a given set A, written $(A')'$, is the set, A.

Double negation equivalence: Expressed symbolically as $\sim (\sim p) \Leftrightarrow p$. It tells us that the negation of the negation of a given statement is equivalent to the given statement.

Electronic circuit: Consists of a wire, a source of electricity S, and a switch or switches that can be turned on and off.

Element: Any member of a set; the symbol $\in$ is read "is an element of", and $\notin$ is read "is not an element of".

Empty set: The set that contains no elements; denoted by either { } or $\varnothing$.

Equal sets: If every element of set A is an element of set B and every element of set B is an element of set A, then $A = B$.

Equivalence: A biconditional statement that is a tautology. The significance of an equivalence is that its LHS and RHS always have the same truth value, regardless of the truth values of the variables involved.

Equivalent sets: Sets that have the same cardinality.

Formal proof: A sequence of statements, each justified by a reason, arranged in a table, which serves to provide rigorous proof of the validity of an argument.

Function from A to B: A relation in which each member of set A is paired with a unique member of set B.

Idempotent law: The union or intersection of any set with itself is equal to the set itself.

Identity law: The four identity laws state the intersection of set A with the universal set is set A; the intersection of set A with the null set is the null set; the union of set A with the universal set is the universal set; and the union of set A with the null set is set A.

Implication: See "Conditional statement."

Independent events: Events, such that the occurrence of one event does not affect the probability of the other event occurring.

Indirect proof: A formal proof that assumes the conclusion of a valid argument is false. If the argument is valid, this, together with the premises, will result in a contradiction, indicating that our assumption was incorrect. This proves that the argument is valid.

Intersection: The intersection of set A with set B, written $A \cap B$, is the set of all elements that are found both in set A and in set B.

Invalid argument: An argument in which the truth of the premises lead to a false conclusion.

Inverse: The inverse of a given conditional statement $p \to q$ is another conditional statement of the form $\sim p \to \sim q$. Note that the LHS and RHS of the given conditional statement have been negated.

Inverse law: The intersection of a set with its complement is the empty set while the union of a set with its complement is the universal set.

Inverse of a Relation: A new relation that is formed by interchanging the first and second coordinates in each ordered pair of the original relation. The inverse of a relation R is denoted by R^{-1}.

Laws of equal sets: Analogous to the equivalences in logic.

LHS: The left hand side of an implication.

Membership table: Displays an element's membership or non-membership in a set; analogous to a truth table in logic.

Mutually exclusive events: Events that cannot both happen at the same time.

Modus ponens: A valid argument form which states that if one premise of an argument is an implication and the other premise is the LHS of the implication, then the RHS of the implication must be true.

Modus tollens: A valid argument form which states that if one premise of an argument is an implication and the other premise is the negation of the RHS of the implication, then the negation of the LHS of the implication must be true.

Null set: See "Empty set."

One-to-one function: A function in which no two elements of the domain are ever paired with the same member of the range.

Parallel circuit: A circuit in which at least one switch has to be on for current to flow.

Power set: The set of all subsets of a given set, denoted as $\mathscr{P}(A)$. Its elements are the subsets of set A.

Premise: A statement in an argument that is always assumed to be true.

Probability: A ratio which divides the number of ways an event may occur by the total number of possible outcomes.

Product set ($A \times B$): The set whose elements are all the ordered pairs that are possible given that x is any member of set A, and y is any member of set B.

Proper subset: Set A is a proper subset of set B, denoted $A \subset B$, if and only if every element of set A is an element of set B, but $A \neq B$ (i.e. there is at least one element in B that is not in set A).

Range: The set of all second coordinates of a relation.

Relation from A to B: Any non-empty subset of the product set $A \times B$.

Relation on set A: A relation from set A to itself; any non-empty subset of $A \times A$.

RHS: The right hand side of an implication.

Series circuit: A circuit in which both switches must be on for current to flow.

Set: Any collection of objects.

Set builder notation: Used to define the elements of a given set. Its general form is $\{x \mid x \text{ has a given property}\}$.

Set difference: Denoted $A - B$, is the set of elements that are in set A but are not in set B.

Set difference law: States that set difference between two sets, $A - B$, is equal to the intersection of set A with the complement of set B, written as $A \cap B'$.

Simple statement: A *TF* statement that does not does contain a connective.

Subset: Set *A* is a subset of set *B*, denoted $A \subseteq B$, if and only if every element of set *A* is an element of set *B*. If a given set *S* contains *n* elements, then 2^n subsets of set *S* exist.

Tautology: A compound statement that is always true regardless of the truth values of the simple statements involved.

***TF* statement:** A statement that can have only one of two truth values, either true or false.

***TF* method–conditional conclusion:** A method employed to test an argument's validity when the argument contains a conditional conclusion. The strategy used is to assume the LHS of the conclusion is true and show that under this assumption, the RHS of the conditional statement must also be true, thus making the argument valid.

***TF* method–direct approach:** A method employed to test an argument's validity in which all the premises are assumed to be true and one is able to arrive at the truth value of the conclusion.

***TF* method–equivalences:** A method employed to test an argument's validity in which one statement is replaced by another logically equivalent statement.

***TF* method–indirect approach:** A method employed to test an argument's validity in cases in which it cannot be deduced directly. The strategy is to assume the conclusion to be false while every premise is true. If there are no contradictions then the assumption was correct and the argument is invalid. If a contraction is found, the original assumption was incorrect and the argument must be valid.

Truth table: A table that indicates all the instances when a *TF* statement is true and when it is false.

Union: The union of set *A* with set *B*, written $A \cup B$, is the set of all elements that are found in either set *A* or in set *B*.

Universal set: Also referred to as "the universe" and expressed symbolically as *U*, is the set of elements from which we construct all other sets in a particular discussion.

Valid argument: An argument in which the true premises always lead to a true conclusion.

Valid argument form: A conditional tautology.

Venn diagram: A pictorial representation of sets and the relationships that exist among them. The interior of a rectangle represents the region in which all the elements of the universe are found. Elements of the sets in a discussion are contained within circles placed in this rectangle.

ANSWERS

Answers for Chapter 1

In-Class Excercises and Problems for Section 1.1

In-Class Exercises

1. $s \wedge b$
2. $v \rightarrow t$
3. $r \vee l$
4. $c \leftrightarrow v$
5. $\sim o$

6. $n \rightarrow s$
7. $\sim (d \wedge l)$
8. $o \rightarrow (c \wedge w)$
9. $(s \wedge c) \rightarrow w$
10. $s \wedge e$

11. $(b \rightarrow \sim w) \wedge (s \rightarrow d)$
12. $(s \wedge h) \rightarrow (l \wedge e)$
13. $s \rightarrow i$
14. $(b \wedge \sim a) \rightarrow s$
15. c

16. She owns a car but doesn't go to college.
17. It is not true that if she lives alone then she owns a car.
18. She doesn't work part-time or she goes to college.
19. If she owns a car then she works part-time and lives alone.
20. She works part-time if and only if she owns a car and goes to college.
21. She lives alone or if she works part-time then she owns a car.

Problems

23. TF; c
25. biconditional; $h \leftrightarrow \sim s$

27. conditional; $s \rightarrow m$
29. conjunction; $e \wedge b$
31. TF; $\sim c$

33. disjunction; $(k \rightarrow c) \vee d$
35. conditional; $n \rightarrow e$
37. biconditional; $(s \wedge r) \leftrightarrow (k \wedge c)$

39. $s \rightarrow (p \vee g)$
41. $\sim (s \rightarrow r)$
43. $o \leftrightarrow (j \wedge d)$

45. $(c \rightarrow w) \wedge [b \rightarrow (l \vee \sim j)]$
47. $[(h \wedge \sim t) \vee o] \rightarrow a$

In-Class Excercises and Problems for Section 1.2

In-Class Exercises

1. T, F
2. T
3. F
4. F
5. F
6. T
7. F
8. F
9. T
10. T
11. T
12. T
13. T
14. F
15. F
16. T
17. F
18. T
19. T
20. F
21. CBD

Problems

23. F
25. T
27. CBD
29. F
31. T
33. CBD
35. T
37. T

Check Your Understanding

1. F
2. T
3. F
4. T
5. $(r \wedge w) \rightarrow e$
6. $b \leftrightarrow (w \wedge \sim s)$

7. $c \vee (m \rightarrow v)$
8. $(m \rightarrow a) \vee (s \rightarrow b)$

9. I will vote in the next presidential election if and only if I am registered to vote and take a political science class.

10. Either I don't take a political science class, or if I am registered to vote I will vote in the next presidential election.

In-Class Excercises and Problems for Section 1.3

In-Class Exercises

1. T	2. F	3. T	4. T	5. T	6. F	7. CBD	8. T
9. T	10. T	11. CBD	12. T	13. F	14. T	15. T	16. F
17. T	18. CBD	19. T	20. F	21. CBD	22. T	23. T	24. F 25. T

Problems

27. F	29. F	31. T	33. T	35. T	37. F
39. CBD	41. F	43. T	45. CBD	47. T	49. CBD
51. F	53. F	55. F			

Check Your Understanding

1. T	2. F	3. CBD	4. F	5. CBD
6. T	7. F	8. CBD	9. F	10. F

In-Class Excercises and Problems for Section 1.4

In-Class Exercises

1. Not a tautology - FTFF
2. Not a tautology - TFTT
3. Not a tautology - FTTF
4. Tautology
5. Tautology
6. Tautology
7. Tautology
8. Tautology
9. Not a tautology - FTFF
10. Tautology

Problems

11. Tautology
13. Tautology
15. Tautology
17. Not a tautology - TTFT
19. Not a tautology - TTFT
21. Not a tautology - TFFT
23. Tautology
25. Not a tautology - TTFT
27. Not a tautology - TTFF
29. Tautology

In-Class Excercises and Problems for Section 1.5

In-Class Exercises

1. Tautology
2. Not a tautology - TTTTFTFT
3. Tautology
4. Tautology
5. Tautology
6. Tautology
7. Tautology
8. Not a tautology - TTTTTFTF
9. Not a tautology - FTFFTTTT
10. Tautology

Problems

11. Tautology
13. Tautology
15. Tautology
17. Not a tautology - TTTTTFTF
19. Not a tautology - TFTTTFTT
21. 8
23. 2, F
25. 8 entries of F,2,1
27. $(a \wedge b) \vee (a \wedge c)$
29. $(m \vee l) \wedge (m \vee \sim w)$

Check your Understanding

1. Not a tautology   2. Tautology

In-Class Excercises and Problems for Section 1.6

In-Class Exercises

1. p

2. $\sim r \wedge w$

3. $(p \wedge q) \vee (p \wedge r)$

4. $\sim p \rightarrow s$

5. $(g \vee s) \vee \sim p$

6. $\sim c \vee (d \wedge \sim b)$

7. $p \wedge (s \rightarrow r)$

8. $\sim a \wedge (b \wedge \sim c)$

9. $(\sim b \vee \sim p) \wedge (\sim b \vee q)$

10. $\sim [\sim (\sim w)]$

11. I pull into a lot or parallel park.

12. We can paint the town red or blue and we can paint it red or green.

13. It is not true that the swan's long neck didn't glisten in the noon day sun.

Problems

15. Distributive Equivalence

17. Associative Equivalence

19. Commutative Equivalence

21. Double Negation Equivalence

22. d   23. e   24. f   25. b   26. c   27. a

29. It is not true that the euro is not the currency of the European Union.

31. Before you enter the restaurant you must wear either shoes and a jacket, or shoes and a tie.

33. The banana is ripe and the raspberry is plump.

Check Your Understanding

1. Distributive Equivalence

2. Associative Equivalence

3. Commutative Equivalence

4. $[r \vee (p \rightarrow w)]$

5. $[\sim a \wedge (b \wedge \sim c)]$

6. $[(\sim b \wedge \sim p) \vee (\sim b \wedge q)]$

7. $(\sim b \rightarrow c)$

8. It is not true that in 2008, an African-American did not win the US presidential election.

9. I will pay my credit card bill on time or I will incur a late fee and I will pay my credit card bill on time or I will incur a finance fee.

10. Roger will either jog five miles or he will exercise at the gym.

In-Class Excercises and Problems for Section 1.7

In-Class Exercises

1. $\sim c \vee k$

2. $\sim p \rightarrow \sim r$

3. $p \vee \sim s$

4. $w \rightarrow \sim z$

5. $b \vee a$

6. $m \vee (a \wedge b)$

7. $\sim p \rightarrow (c \wedge d)$

8. $(p \wedge r) \vee k$

9. $a \vee (r \wedge s)$

10. $(w \wedge h) \rightarrow (a \wedge \sim c)$

11. $h \rightarrow \sim a$

12. $\sim s \rightarrow c$

13. $q \rightarrow p$

14. $\sim (w \wedge a) \rightarrow \sim p$

15. $\sim b \rightarrow (m \vee n)$

16. e

17. a

18. d

19. b

20. c

21. $\sim s \rightarrow r$

22. $\sim p \rightarrow \sim q$

23. $p \rightarrow \sim q$

24. $\sim m \rightarrow r$

25. $\sim s \rightarrow (m \vee p)$

26. $\sim d \rightarrow (b \wedge p)$

27. $(c \vee d) \rightarrow \sim (p \leftrightarrow q)$

28. $\sim (p \wedge a) \rightarrow (p \vee \sim q)$

29. $(p \leftrightarrow q) \rightarrow \sim (c \vee d)$

30. $\sim (r \wedge \sim s) \rightarrow (w \vee c)$

31. If I don't do homework I do not pass.

32. If I didn't quit my job I didn't win the lottery.

33. You won't go to the Chinese restaurant or you'll get a fortune cookie.

34. If the bird doesn't get up early it doesn't catch the worm.

35. If I don't fit into my jeans, I didn't go on a diet.

36. If you don't live in a glass house then you should throw stones.
    If you shouldn't throw stones then you live in a glass house.
37. If you don't go to the birthday party you shouldn't bring a present.
    If you bring a present you should go to the birthday party.
38. If you can stand the heat, then you don't stay out of the kitchen.
    If you stay out of the kitchen then you can't stand the heat.
39. If it's raining you take your umbrella.
    If you don't take your umbrella it's not raining.
40. If the sun is not shining the stars can be seen.
    If the stars can't be seen the sun is shining.

Problems

41. Conditional   43. Contrapositive   45. Conditional   47. Contrapositive
48. f   49. b   50. e   51. g   52. a   53. d   54. c
55. a. If I don't make a profit then revenue does not exceed cost.
    b. Revenue does not exceed cost or I make a profit.
    c. If revenue does not exceed cost then I don't make a profit.
    d. If I make a profit then revenue exceeds cost.
57. a. If a cord of wood doesn't come in handy it doesn't keep snowing.
    b. It doesn't keep snowing or a cord of wood came in handy.
    c. If It doesn't keep snowing then a cord of wood won't come in handy.
    d. If a cord of wood comes in handy then it keeps snowing.
59. a. If polar ice caps don't melt then global warming is controlled.
    b. Global warming is controlled or polar ice caps will melt.
    c. If global warming is controlled then polar ice caps will not melt.
    d. If polar ice caps melt then global warming is not controlled.

Check Your Understanding

1. Conditional Equivalence   2. Contrapositive Equivalence   3. Conditional Equivalence
4. Contrapositive Equivalence   5. $[(p \vee q) \rightarrow\, \sim k]$ 6. $[(p \vee q) \vee k]$ 7. $[k \rightarrow (p \vee q)]$
8. $[k \rightarrow\, \sim (p \vee q)]$ 9. Converse   10. Inverse   11. Contrapositive
12. $q \rightarrow\, \sim r$ 13. $l \rightarrow d$
14. If children grow confident then they are praised.
15. If children don't grow confident then they are not praised.

In-Class Excercises and Problems for Section 1.8

In-Class Exercises

1. $\sim p \wedge q$ 2. $\sim w \vee s$ 3. $p \wedge s$
4. $\sim (\sim b \vee r)$ 5. $\sim (a \vee b)$ 6. $\sim (\sim q \wedge s)$
7. $\sim m \wedge \sim (a \rightarrow b)$ 8. $\sim (p \leftrightarrow q) \vee r$ 9. $\sim [\sim p \wedge (a \rightarrow c)]$
10. $(w \wedge r) \vee \sim (p \rightarrow q)$ 11. I will not take piano lessons or I will practice the scales.
12. It is not true that traveling abroad doesn't require a passport and traveling abroad doesn't require a birth certificate.
13. It is not true that we will go out for dinner or we will not go to the concert.
14. Doctors advise us to eat healthy food and exercise.
15. $m \wedge \sim b$ 16. $\sim s \wedge \sim w$ 17. $\sim (a \rightarrow d)$ 18. $\sim (\sim k \rightarrow\, \sim c)$ 19. $(r \vee s) \wedge \sim w$

20. $\sim[p \rightarrow \sim(w \vee q)]$

21. It is not true that the ozone layer is not protected and life on Earth is not doomed.

22. It is not true that the disposable income rises but the economy doesn't improve.

23. $(p \wedge r) \vee (\sim r \wedge \sim p)$ 24. $(\sim w \wedge \sim s) \vee (s \wedge w)$ 25. $(\sim m \wedge n) \vee (\sim n \wedge m)$

26. You buy a new computer and your current computer is not more than two years or your current computer is more than two years and you do not buy a new one.

27. You dial 911 and there is an alternative or there is no other alternative and you don't dial 911.

Problems

29. $k \wedge \sim w$ 31. $\sim c \vee p$ 33. $\sim p \wedge a$ 35. $(\sim d \wedge \sim g) \vee (g \wedge d)$

37. $q \wedge a \wedge \sim p$ 39. $\sim a \wedge (p \vee m)$

40. e   41. b   42. f   43. d   44. c   45. a

47. The sun wasn't out or it was raining.

49. I don't pass math and I don't have to drop out of school.

51. The best music isn't jazz, or you don't like it nor do you like country.

53. Either the temperature outside isn't 110 degrees, or you didn't hydrate and you didn't become dizzy.

55. A peace treaty is to be signed and either there is not a cease fire or the President doesn't attend the conference, or there is a cease fire and the President attends the conference but the peace treaty is not signed.

57. I will be elected and the economy won't suffer, or the economy will suffer and you will not be out of a job.

59. The patient has signs of vertigo and anxiety but doesn't have a neurological condition.

Chapter 1 Review

1. f   2. d   3. c   4. a   5. i   6. b   7. j   8. g   9. h   10. e

11. Commutative   12. Associative   13. Distributive   14. Biconditional

15. Conditional Negation   16. DeMorgan's   17. Conditional   18. DeMorgan's

19. Conditional   20. Distributive   21. Double Negation

22. Contrapositve   23. Commutative

24. $(c \wedge r) \rightarrow s$

25. $n \vee (l \rightarrow r)$

26. $(s \wedge m) \vee (c \wedge h)$

27. $(s \wedge m) \rightarrow \sim d$

28. $l \leftrightarrow (s \wedge w)$

29. $(f \wedge a) \wedge \sim r$

30. $(w \wedge p) \rightarrow (b \vee f)$

31. $\sim w \wedge m$

32. Avi is happy and he won't smile.

33. I will play poker and I have a test or I don't have a test and I will not play poker.

34. The computer isn't cheap or the monitor is expensive.

35. The answer to part three is true but the answer to part four is false.

36. It is a sentence and it doesn't have a subject or it doesn't have a verb, or it has a subject and a verb but it is not a sentence..

37. In her psychology class, she will neither study Skinner nor Freud.

38. I'll stay in school and either I won't study chemistry or I won't work in the lab.

39. If she doesn't want to be a doctor then she wants to be a lawyer.
40. They will not visit or they will not stay two weeks.
41. If she doesn't have a pedicure then she won't have a manicure.
42. You don't feed the dog or she will not bite.
43. Money is both an evil and a necessity.
44. It is not true that calculators have not made statistics easier or you do not have to do the homework.
45. If the number is not two then the number is either not even or not prime.
46. The enemy is not supplied with missiles and they are not defeated.
47. Either you use the Internet and your research skills will be enhanced, or you use the Internet and you will learn a great deal.
48. If I go to the play then the tickets are not expensive and if the tickets are not expensive then I will go to the play.
49. $(\sim h \vee \sim g) \rightarrow \sim c$ If I don't wear my hat or I don't wear my gloves it isn't cold outside.
50. $\sim c \vee (h \wedge g)$ It isn't cold outside or I wear my hat and gloves.
51. $c \wedge \sim (h \wedge g)$ It is cold outside but I did not wear my hat and gloves.
52. $\sim c \rightarrow (\sim h \vee \sim g)$ If it isn't cold outside I will not wear my hat or I will not wear my gloves .
53. $(h \wedge g) \rightarrow c$ If I wear my hat and gloves it is cold outside .

Sample Exam: Chapter 1

1. d	2. b	3. c	4. b	5. a	6. d	7. b	8. c	9. a	10. b
11. a	12. d	13. c	14. d	15. a	16. b	17. b	18. a	19. a	20. c

Answers for Chapter 2

In-Class Excercises and Problems for Section 2.1

In-Class Excercises

1. Valid

2. Invalid

p	q	r
T	T	T
F	T	T

3. Invalid

r	s	w
F	T	F

4. Invalid

p	r	s
T	T	F

5. Valid

6. Invalid

p	q	r
F	F	T

7. Invalid

a	b	c
F	T	F

8. Valid

9. Invalid

a	b	c	d
F	T	T	F

10. Valid

Problems

11. Invalid

r	s
F	T

13. Invalid

c	k	p
F	F	F

15. Valid

17. Valid

19. Valid

21. Valid

23. Invalid

p	q	s
T	T	F

25. Valid

Check Your Understanding

1. *b*   2. *a*   3. *c*   4. Since the argument is valid, it is impossible to produce any counterexample chart.
5. Valid

In-Class Excercises and Problems for Section 2.2

In-Class Exercises

1. T	2. CBD	3. T	4. F	5. T	6. T
7. CBD	8. T	9. CBD	10. T	11. T	

12. Valid   13. Valid   14. Invalid

a	b	c
F	F	T

15. Invalid

c	m	n	r
F	T	F	F

16. Invalid

p	q	r	s
F	T	T	F

17. Valid

18. Invalid

e	p	s	w
F	T	T	T

19. Valid

20. Valid

21. Invalid

c	p	r	s	w
F	T	F	T	T

22. $\sim w$ Valid

$$\frac{r \rightarrow w}{\sim r}$$

23. $r \rightarrow p$ Invalid

$$\frac{\begin{array}{c} r \vee i \\ \sim i \end{array}}{\sim p}$$

i	p	r
F	T	T

24. $q \rightarrow \sim i$ Valid

$$\frac{\begin{array}{c} q \vee s \\ \sim s \end{array}}{\sim i}$$

25. $b \leftrightarrow p$ Invalid

$$\frac{\sim p \wedge s}{s \rightarrow b}$$

b	p	s
F	F	T

26.   Invalid

$$\frac{\begin{array}{c} i \wedge s \\ (\sim i \wedge \sim s) \rightarrow \sim j \end{array}}{\sim j \rightarrow r}$$

i	j	r	s
T	F	F	T

Problems

27. T   29. F   31. CBD   33. T   35. T   37. No   39. Yes   41. No   43. No   45. No

47. Valid   49. Valid   51. Invalid

a	b	p	w
T	F	T	F

53. Valid

55. $s \rightarrow m$ Valid

$$\frac{\begin{array}{c} \sim s \rightarrow e \\ w \wedge \sim m \end{array}}{e}$$

57. $\sim (h \wedge s) \rightarrow o$ Valid

$$\frac{\sim h}{o}$$

Check Your Understanding

1. Yes   2. No   3. Yes   4. No   5. True   6. CBD

7. False   8. True   9.

a	b	c	g	k
F	T	T	F	F
F	F	T	F	F

10.   Valid

In-Class Excercises and Problems for Section 2.3

In-Class Exercises

1. Valid   2. Invalid

p	q	r	s
T	T	T	T
T	T	F	T
T	F	T	T

3. Invalid

q	r	s	w
F	T	T	T
T	T	T	T

4. Valid   5. Invalid

m	n	p	q	r
F	T	F	F	T

6. Valid   7. Valid   8. Valid   9. Valid   10. Valid

Problems

11. Valid   13. Valid   15. Invalid

b	c	d	e
T	F	T	T
T	F	T	F

17. Valid   19. Invalid

m	q	r	s	w
T	F	F	T	T

Check Your Understanding

1. True   2. True   3. True   4. False   5. True   6. True   7. False   8. CBD   9. Valid

10.

a	b	p	r	w
T	F	F	F	T

In-Class Excercises and Problems for Section 2.4

In-Class Exercises

1. Valid   2. Valid   3. Invalid

b	s	w
F	T	F

4. Valid   5. Valid

6. Valid   7. Invalid

p	q	r	s
T	T	F	T

8. Valid   9. Valid   10. Valid

11. Invalid

d	m	n	p	r	s
F	T	F	T	T	F

12. Invalid

a	b	c	d	p	x	z
T	F	F	T	F	T	T

13. $m \rightarrow h$ Valid

$(a \vee b) \rightarrow j$

$\sim w \rightarrow \sim j$

$h \rightarrow a$

$m \rightarrow w$

14. $\sim (s \vee c)$ Invalid

$(c \vee b) \rightarrow (r \rightarrow s)$

$b \rightarrow r$

b	c	r	s
T	F	F	F

15. $w \rightarrow (h \wedge m)$ Valid

$\sim e \rightarrow \sim h$

$w \vee \sim p$

$p \rightarrow (e \vee o)$

16. $\sim c \vee (s \vee w)$ Invalid

$s \rightarrow p$

$\sim w$

$c \rightarrow \sim p$

c	p	s	w
T	T	T	F

Problems

17. Valid   19. Valid   21. Invalid

a	d	j	m
F	T	F	F

23. Valid   25. Invalid

a	c	d	m	p
T	T	T	T	T

27. Valid

29. $c \rightarrow (j \vee d)$ Invalid

$d \rightarrow \sim x$

$m \vee x$

$\sim m$

$c \rightarrow \sim j$

c	d	j	m	x
T	F	T	F	T

31. $\sim s \rightarrow \sim n$ Valid

$(s \wedge p) \rightarrow r$

$(p \vee g) \rightarrow n$

$p \rightarrow r$

In-Class Excercises and Problems for Section 2.5

In-Class Exercises

1. Invalid

m	p	q	r
F	F	F	F

2. Valid   3. Valid   4. Invalid

a	b	c	p
F	T	F	T

5. Invalid

p	q	r	s	w
T	T	T	T	F
T	F	T	T	F

6. Invalid

a	b	c	d	g
T	T	F	F	F

7. Valid   8. Valid   9. Valid   10. Valid

11. $b \to \sim r$ Valid

$o \to (p \wedge s)$

$b \vee o$

$\overline{}$

$r \to s$

12. $e \to q$ Invalid

c	e	m	p	q
T	T	T	T	T

$\sim q \vee (p \wedge c)$

$\sim m \to p$

$\overline{}$

$e \to \sim m$

13. $a \to (b \wedge \sim c)$ No

a	b	c
F	T	F

$c \to a$

b

$\overline{}$

$a \vee c$

Problems

15. Valid   17. Valid   19. Invalid

a	b	c	d	p
T	T	F	T	T

21. Invalid

n	p	q	r	s
T	T	F	F	T

23. $c \to (l \to a)$ Valid

$c \vee p$

$\sim (l \wedge p)$

$\overline{}$

$a \vee \sim l$

25. $b \to c$ Invalid

b	c	e	p	r	w
F	F	F	F	T	F

$c \to (p \wedge e)$

$r \to (\sim e \vee w)$

$(b \wedge w) \vee (\sim b \wedge r)$

$\overline{}$

$p \vee w$

Check Your Understanding

1.
k	p	w	z
T	T	T	T

2. valid   3. valid

4.
a	d	m	n	r
T	T	T	T	F

5. $h \to (c \vee w)$

$w \to (y \vee n)$

$\sim n$

$\overline{}$

$c \vee y$

c	h	n	w	y
F	F	F	F	F

Chapter 2 Review

1. T   2. T   3. F   4. F   5. CBD

6. T   7. T   8. F   9. CBD   10. F

11. Valid

12. Invalid

a	b	p	q
F	F	F	T

13. Invalid

p	s	w
F	F	T

14. Valid

15. Valid

16. Invalid

a	m	n	p
F	F	F	F

17. Invalid

a	b	p	q	s	w
F	F	T	T	F	F

18. Valid

19. Valid

20. Invalid

a	b	m	r	s	w
T	T	F	T	T	T
T	F	F	T	T	T

21. Valid

22. Valid

23. Invalid

b	p	q	s	w
T	T	F	F	F
T	T	F	F	T
T	F	F	F	T

24. Invalid

a	b	c	p
T	F	T	F
F	F	T	F
F	F	F	F

25. Invalid

a	b	c	p	w
T	F	T	F	F
T	F	T	T	F

26. Valid

27. Valid

28. Valid

29. Invalid

p	q	s	w
F	F	T	T

30. Valid

31. Invalid

p	q	r	s
T	T	T	T

32. Valid

33. Valid

34. Valid

35. Valid

36. Invalid

a	b	c	p	q	r	w
T	T	F	F	T	T	F

37. $d \lor v$ Valid

$d \to s$

$v \to l$

$\dfrac{\sim l}{s}$

38. $g \land w$ Invalid

$w \to [(d \land p) \lor r]$

$\dfrac{p \to \sim l}{r \to \sim l}$

d	g	l	p	r	w
F	T	T	F	T	T
T	T	T	F	T	T

39. $r \to \sim w$ Valid

$(\sim s \lor e) \to w$

$\dfrac{\sim (s \land \sim e)}{\sim r}$

40. $\sim (p \to a)$ Valid

$p \to b$

$b \to (s \lor c)$

$\dfrac{\sim s}{c \leftrightarrow \sim a}$

41. $s \lor r$ Invalid

$(g \lor b) \to s$

$\dfrac{\sim g \to b}{\sim r}$

b	g	r	s
T	F	T	T
F	T	T	T
T	T	T	T

42. $n \to (a \lor b)$ Invalid

$o \to \sim a$

$\dfrac{o \lor (n \land \sim l)}{b \to l}$

a	b	l	n	o
F	T	F	T	T
F	T	F	F	T

43. $\sim w \lor \sim (e \to j)$ Valid

$\dfrac{d \to w}{\sim e \to \sim d}$

44. $c \to (d \land s)$ Valid

$d \to (g \land \sim b)$

$c \lor e$

$\dfrac{\sim e}{g}$

45. $w \to b$ Invalid

$(w \land r) \to p$

$p \to b$

$\dfrac{r \lor \sim b}{\sim w}$

b	p	r	w
T	T	T	T

46. $p \rightarrow (g \wedge n)$ Invalid

    $n \rightarrow t$

    $t \rightarrow (h \vee d)$

    ———————

    $p \rightarrow h$

d	g	h	n	p	t
T	T	F	T	T	T

47. $p \vee i$ Invalid

    $p \rightarrow (f \wedge s)$

    $i \rightarrow \sim c$

    $\sim c \wedge \sim f$

    ———————

    s

c	f	i	p	s
F	F	T	F	F

48. $d \vee r$ Valid

    $r \rightarrow (l \wedge t)$

    $\sim l$

    ———————

    d

Sample Exam: Chapter 2

1. T   2. F   3. CBD   4. F   5. T

6.

a	b	c	d	e	g
T	T	F	T	T	F

7.

a	b	p	q	r	s
T	F	F	F	F	T

8.

a	p	q	r	s
F	T	F	T	F

9.

p	q	r	s	w
T	T	F	T	F
F	T	F	T	F

10. Valid   11. Valid   12. Invalid

m	n	p	q	y	z
F	T	T	F	T	F

13. Invalid

p	q	r	s	w
T	T	T	F	F
T	T	F	F	F

14. Valid   15. Invalid

p	r	s	w
F	F	T	F

16. Valid   17. Invalid

a	b	c	d
F	T	T	T
T	T	T	T

18. Valid   19. Invalid

c	d	p	q	r	s	w
T	F	F	F	T	T	T
T	F	T	F	T	F	T
T	F	F	F	T	F	T

20. Valid   21. Valid

22. $w \rightarrow s$ Valid

$\sim w \rightarrow \sim p$

$\dfrac{g \wedge \sim s}{\sim p}$

23. $y \rightarrow (p \vee d)$ Invalid

$\sim d \rightarrow \sim c$

$\dfrac{c}{y}$

c	d	p	y
T	T	T	F
T	T	F	F

24. $c \rightarrow j$ Valid

$j \rightarrow m$

$m \rightarrow \sim w$

$\dfrac{w}{\sim c}$

25. $\sim (i \rightarrow p)$ Valid

$p \vee \sim f$

$\dfrac{h \rightarrow f}{\sim h \vee m}$

26. $g \rightarrow \sim s$ Invalid

$s \rightarrow (w \vee n)$

$\dfrac{n \rightarrow c}{g \rightarrow c}$

c	g	n	s	w
F	T	F	F	T
F	T	F	F	F

Answers for Chapter 3

In-Class Excercises and Problems for Section 3.1
In-Class Exercises

1. MT   2. MP   3. MT   4. MT   5. MP   6. MP   7. MT   8. MT

9.

S	R
1. $\sim a$	P
2. $\sim a \rightarrow b$	P
3. b	MP(1)(2)
4. $c \rightarrow \sim b$	P
5. $\sim c$	MT(3)(4)

10.

S	R
1. $\sim w$	P
2. $s \rightarrow w$	P
3. $\sim s$	MT(1)(2)
4. $\sim r \rightarrow s$	P
5. r	MT(3)(4)
6. $r \rightarrow p$	P
7. p	MP(5)(6)

11.

a	b	r	w
T	T	F	F
T	F	F	F
F	T	F	F

12.

S	R
1. $\sim (p \vee r)$	P
2. $s \rightarrow (p \vee r)$	P
3. $\sim s$	MT(1)(2)
4. $\sim q \rightarrow s$	P
5. q	MT(3)(4)

13.

S	R
1. $p \rightarrow q$	P
2. $(p \rightarrow q) \rightarrow (r \rightarrow s)$	P
3. $r \rightarrow s$	MP(1)(2)
4. $(r \rightarrow s) \rightarrow (m \rightarrow b)$	P
5. $m \rightarrow b$	MP(3)(4)

14.

k	p	s	w
F	T	T	F
F	T	F	T
F	T	F	F

15.

c	p	r	s	w
F	F	F	F	T

16.

S	R
1. $\sim p$	P
2. $(r \vee \sim s) \rightarrow p$	P
3. $\sim (r \vee \sim s)$	MT(1)(2)
4. $c \rightarrow (r \vee \sim s)$	P
5. $\sim c$	MT(3)(4)
6. $\sim c \rightarrow d$	P
7. d	MP(5)(6)

17.

S	R
1. $\sim (a \rightarrow c)$	P
2. $(p \vee r) \rightarrow (a \rightarrow c)$	P
3. $\sim (p \vee r)$	MT(1)(2)
4. $\sim (p \vee r) \rightarrow s$	P
5. s	MP(3)(4)
6. $s \rightarrow (k \wedge g)$	P
7. $k \wedge g$	MP(5)(6)

18.

k	l	p	q	r	w
F	F	F	T	T	F

Problems

19. Reason
    1. P
    2. P
    3. MP(1)(2)
    4. P
    5. MP(3)(4)

21. Reason
    1. P
    2. P
    3. MT(1)(2)
    4. P
    5. MP(3)(4)

23. Reason
    1. P
    2. P
    3. MT(1)(2)
    4. P
    5. MP(3)(4)
    6. P
    7. MP(5)(6)

25.

S	R
1. $\sim s$	P
2. $\sim r \rightarrow s$	P
3. r	MT(1)(2)

27.

S	R
1. $\sim (p \rightarrow r)$	P
2. $s \rightarrow (p \rightarrow r)$	P
3. $\sim s$	MT(1)(2)

29.

S	R
1. $\sim q$	P
2. $r \rightarrow q$	P
3. $\sim r$	MT(1)(2)
4. $p \rightarrow r$	P
5. $\sim p$	MT(3)(4)

31.

p	q	r
F	F	F

33.

S	R
1. $\sim (p \wedge a)$	P
2. $w \rightarrow (p \wedge a)$	P
3. $\sim w$	MT(1)(2)
4. $\sim w \rightarrow k$	P
5. k	MP(3)(4)

35.

a	d	g	m	r
T	T	T	F	F
T	T	F	F	F
T	F	T	F	F

37.

S	R
1. $c \vee d$	P
2. $(c \vee d) \rightarrow q$	P
3. q	MP(1)(2)
4. $q \rightarrow \sim r$	P
5. $\sim r$	MP(3)(4)
6. $a \rightarrow r$	P
7. $\sim a$	MT(5)(6)
8. $b \rightarrow a$	P
9. $\sim b$	MT(7)(8)

39.

a	b	m	n	p	q	w
T	F	F	F	T	T	F

41. b, c   43. a, c

45. $(b \vee c) \rightarrow k$
    $k \rightarrow \sim d$
    $p \rightarrow d$
    $\sim p \rightarrow (m \wedge s)$
    $\underline{b \vee c}$
    $m \wedge s$

S	R
1. $b \vee c$	P
2. $(b \vee c) \rightarrow k$	P
3. k	MP(1)(2)
4. $k \rightarrow \sim d$	P
5. $\sim d$	MP(3)(4)
6. $p \rightarrow d$	P
7. $\sim p$	MT(5)(6)
8. $\sim p \rightarrow (m \wedge s)$	P
9. $m \wedge s$	MP(7)(8)

47. $h \rightarrow r$
    $h \rightarrow w$
    $r \rightarrow i$
    $\underline{\sim i}$
    $\sim w$

h	i	r	w
F	F	F	T

Check Your Understanding

1.

Reason
1. P
2. P
3. MP (1)(2)
4. P
5. MP (3)(4)
6. P
7. MT (5)(6)

2.

S	R
1. $a \rightarrow b$	P
2. $w \rightarrow \sim (a \rightarrow b)$	P
3. $\sim w$	MT (1)(2)
4. $\sim w \rightarrow p$	P
5. p	MP (3)(4)
6. $p \rightarrow \sim n$	P
7. $\sim n$	MP (5)(6)

In-Class Excercises and Problems for Section 3.2

In-Class Excercises

1. DS   2. DS   3. DA   4. DA   5. DS   6. DA   7. MT   8. DS   9. MP   10. DA

11.
S	R
1. a	P
2. $\sim a \vee b$	P
3. b	DS(1)(2)
4. $b \to c$	P
5. c	MP(3)(4)

12.
S	R
1. r	P
2. $r \to (s \vee q)$	P
3. $s \vee q$	MP(1)(2)
4. $\sim s$	P
5. q	DS(3)(4)

13.
p	q	r
F	T	F

14.
S	R
1. w	P
2. $w \to c$	P
3. c	MP(1)(2)
4. $c \vee p$	DA(3)
5. $(c \vee p) \to b$	P
6. b	MP(4)(5)

15.
S	R
1. $\sim (q \wedge r)$	P
2. $p \vee (q \wedge r)$	P
3. p	DS(1)(2)
4. $p \to s$	P
5. s	MP(3)(4)

16.
S	R
1. $\sim w$	P
2. $p \to w$	P
3. $\sim p$	MT(1)(2)
4. $\sim p \vee s$	DA(3)
5. $(\sim p \vee s) \to k$	P
6. k	MP(4)(5)

17.
S	R
1. $\sim c$	P
2. $(a \to b) \vee c$	P
3. $a \to b$	DS(1)(2)
4. $\sim b$	P
5. $\sim a$	MT(3)(4)
6. $\sim a \vee p$	DA(5)

18.
l	m	p	w
T	T	F	T
F	T	F	T

19.
S	R
1. b	P
2. $a \to \sim b$	P
3. $\sim a$	MT(1)(2)
4. $c \to a$	P
5. $\sim c$	MT(3)(4)
6. $c \vee d$	P
7. d	DS(5)(6)

20.

S	R
1. p	P
2. $\sim p \vee r$	P
3. r	DS(1)(2)
4. $s \rightarrow \sim r$	P
5. $\sim s$	MT(3)(4)
6. $\sim s \rightarrow w$	P
7. w	MP(5)(6)

21.

a	d	p	s
F	F	F	T

22.

S	R
1. $\sim q$	P
2. $p \rightarrow q$	P
3. $\sim p$	MT(1)(2)
4. $\sim p \rightarrow w$	P
5. w	MP(3)(4)
6. $w \vee z$	DA(5)
7. $(w \vee z) \rightarrow k$	P
8. k	MP(6)(7)

23.

S	R
1. $\sim q$	P
2. $d \vee q$	P
3. d	DS(1)(2)
4. $\sim c \rightarrow \sim d$	P
5. c	MT(3)(4)
6. $c \vee (p \rightarrow s)$	DA(5)

24.

S	R
1. $r \rightarrow w$	P
2. $(r \rightarrow w) \rightarrow (s \vee g)$	P
3. $s \vee g$	MP(1)(2)
4. $\sim s$	P
5. g	DS(3)(4)
6. $g \vee p$	DA(5)

25.

a	b	c	d
T	T	F	F

26.

S	R
1. s	P
2. $m \vee \sim s$	P
3. m	DS(1)(2)
4. $m \vee q$	DA(3)
5. $n \rightarrow \sim (m \vee q)$	P
6. $\sim n$	MT(4)(5)

27.

S	R
1. s	P
2. $r \vee \sim s$	P
3. r	DS(1)(2)
4. $r \vee w$	DA(3)
5. $(r \vee w) \rightarrow p$	P
6. p	MP(4)(5)
7. $q \rightarrow \sim p$	P
8. $\sim q$	MT(6)(7)

28.

S	R
1. $\sim p \rightarrow q$	P
2. $\sim(\sim p \rightarrow q) \vee [r \rightarrow (s \vee c)]$	P
3. $r \rightarrow (s \vee c)$	DS(1)(2)
4. $\sim(s \vee c)$	P
5. $\sim r$	MT(3)(4)
6. $\sim r \vee w$	DA(5)
7. $(\sim r \vee w) \rightarrow d$	P
8. d	MP(6)(7)

29. $w \rightarrow i$
$(i \vee p) \rightarrow \sim s$
$s \vee c$
$\underline{w}$
c

S	R
1. w	P
2. $w \rightarrow i$	P
3. i	MP(1)(2)
4. $i \vee p$	DA(3)
5. $(i \vee p) \rightarrow \sim s$	P
6. $\sim s$	MP(4)(5)
7. $s \vee c$	P
8. c	DS(6)(7)

30. $s \vee c$
$c \rightarrow (p \vee m)$
$\sim s$
$\underline{m}$
$\sim p$

c	m	p	s
T	T	T	F

31. $e \rightarrow c$
$c \rightarrow \sim w$
$w \vee l$
$\underline{\sim l}$
$\sim e$

S	R
1. $\sim l$	P
2. $w \vee l$	P
3. w	DS(1)(2)
4. $c \rightarrow \sim w$	P
5. $\sim c$	MT(3)(4)
6. $e \rightarrow c$	P
7. $\sim e$	MT(5)(6)

Problems

33. Reason
    1. P
    2. P
    3. MP(1)(2)
    4. DA(3)
    5. P
    6. MP(4)(5)

35. Reason
    1. P
    2. P
    3. MP(1)(2)
    4. P
    5. MT(3)(4)
    6. P
    7. DS (5)(6)
    8. DA (7)

37.

S	R
1. $\sim q$	P
2. $q \lor \sim p$	P
3. $\sim p$	DS(1)(2)
4. $\sim r \to p$	P
5. r	MT(3)(4)

39.

S	R
1. c	P
2. $c \to m$	P
3. m	MP(1)(2)
4. $m \lor p$	DA(3)
5. $(m \lor p) \to s$	P
6. s	MP(4)(5)

41.

S	R
1. s	P
2. $s \to a$	P
3. a	MP(1)(2)
4. $a \lor b$	DA(3)
5. $(a \lor b) \to c$	P
6. c	MP(4)(5)

43.

S	R
1. w	P
2. $w \to (p \to q)$	P
3. $p \to q$	MP(1)(2)
4. p	P
5. q	MP(3)(4)
6. $q \lor r$	DA(5)

45.

S	R
1. n	P
2. $m \lor \sim n$	P
3. m	DS(1)(2)
4. $m \to (p \lor r)$	P
5. $p \lor r$	MP(3)(4)
6. $\sim r$	P
7. p	DS(5)(6)

47.

S	R
1. a	P
2. $a \lor p$	DA(1)
3. $(a \lor p) \to k$	P
4. k	MP(2)(3)
5. $k \to s$	P
6. s	MP(4)(5)
7. $s \lor q$	DA(6)
8. $(s \lor q) \to w$	P
9. w	MP(7)(8)

49.

S	R
1. $\sim a$	P
2. $(b \to c) \lor a$	P
3. $b \to c$	DS(1)(2)
4. b	P
5. c	MP(3)(4)
6. $c \lor w$	DA(5)
7. $(c \lor w) \to p$	P
8. p	MP(6)(7)

51.

S	R
1. $\sim s$	P
2. $s \lor w$	P
3. w	DS(1)(2)
4. $w \to a$	P
5. a	MP(3)(4)
6. $a \lor b$	DA(5)
7. $(a \lor b) \to p$	P
8. p	MP(6)(7)
9. $p \lor q$	DA(8)

53.

S	R
1. ~ s	P
2. b → s	P
3. ~ b	MT(1)(2)
4. ~ p	P
5. ~ p → (a ∨ b)	P
6. a ∨ b	MP(4)(5)
7. a	DS(3)(6)
8. a → q	P
9. q	MP(7)(8)

55.

S	R
1. ~ (m ↔ n)	P
2. (m ↔ n) ∨ r	P
3. r	DS(1)(2)
4. r → p	P
5. p	MP(3)(4)
6. ~ (a → m) → ~ p	P
7. a → m	MT(5)(6)
8. a	P
9. m	MP(7)(8)

57. b, d   59. a, b, d   61. d   63. b, c

65. b → w
s → ~ w
s ∨ i
b
―――
i

S	R
1. b	P
2. b → w	P
3. w	MP(1)(2)
4. s → ~ w	P
5. ~ s	MT(3)(4)
6. s ∨ i	P
7. i	DS(5)(6)

67. m → (b ∨ c)
b → r
~ r
m
―――
c ∨ n

S	R
1. m	P
2. m → (b ∨ c)	P
3. b ∨ c	MP(1)(2)
4. ~ r	P
5. b → r	P
6. ~ b	MT(4)(5)
7. c	DS(3)(6)
8. c ∨ n	DA(7)

Check Your Understanding

1. MT   2. DS   3. MP   4. DA   5. p ∨ w   6. p ∨ w   7. r ∧ ~ s

8. <u>Reason</u>

1. P

2. P

3. DS (1)(2)

4. P

5. MP (3)(4)

6. P

7. MT (5)(6)

8. DA (7)

In-Class Excercises and Problems for Section 3.3
In-Class Exercises
1. CS 2. CA 3. CA 4. CS 5. DA 6. CA 7. DS 8. MT 9. CS 10. DA 11. MP 12. MT

13.

S	R
1. p	P
2. $\sim p \vee s$	P
3. s	DS(1)(2)
4. $p \wedge s$	CA(1)(3)
5. $(p \wedge s) \rightarrow q$	P
6. q	MP(4)(5)

14.

b	p	q	r
T	T	F	T

15.

S	R
1. $m \wedge n$	P
2. m	CS(1)
3. $\sim m \vee p$	P
4. p	DS(2)(3)
5. $p \rightarrow w$	P
6. w	MP(4)(5)

16.

S	R
1. $s \wedge g$	P
2. s	CS(1)
3. $s \rightarrow (a \vee w)$	P
4. $a \vee w$	MP(2)(3)
5. $d \rightarrow \sim (a \vee w)$	P
6. $\sim d$	MT(4)(5)

17.

S	R
1. p	P
2. $\sim p \vee (\sim s \rightarrow \sim w)$	P
3. $\sim s \rightarrow \sim w$	DS(1)(2)
4. $s \rightarrow \sim p$	P
5. $\sim s$	MT(1)(4)
6. $\sim w$	MP(3)(5)
7. $\sim s \wedge \sim w$	CA(5)(6)

18.

S	R
1. $a \wedge b$	P
2. a	CS(1)
3. $a \rightarrow p$	P
4. p	MP(2)(3)
5. b	CS(1)
6. $b \wedge p$	CA(4)(5)
7. $(b \wedge p) \rightarrow s$	P
8. s	MP(6)(7)

19.

p	r	s	w
T	T	T	T

20.

S	R
1. $a \wedge k$	P
2. k	CS(1)
3. $k \rightarrow \sim g$	P
4. $\sim g$	MP(2)(3)
5. $g \vee \sim h$	P
6. $\sim h$	DS(4)(5)
7. a	CS(1)
8. $a \wedge \sim h$	CA(6)(7)

21.

S	R
1. $\sim a \wedge b$	P
2. $\sim a$	CS(1)
3. $a \vee c$	P
4. c	DS(2)(3)
5. $d \rightarrow \sim c$	P
6. $\sim d$	MT(4)(5)
7. $\sim d \rightarrow e$	P
8. e	MP(6)(7)

22.

S	R
1. ~ b	P
2. s → b	P
3. ~ s	MT(1)(2)
4. s ∨ c	P
5. c	DS(3)(4)
6. c ∧ ~ s	CA(3)(5)
7. (c ∧ ~ s) → k	P
8. k	MP(6)(7)

23.

S	R
1. s	P
2. r ∨ ~ s	P
3. r	DS(1)(2)
4. r → (p ∧ q)	P
5. p ∧ q	MP(3)(4)
6. p	CS(5)
7. ~ w → ~ p	P
8. w	MT(6)(7)
9. w ∨ c	DA(8)

24.

S	R
1. c ∧ g	P
2. c	CS(1)
3. c → ~ j	P
4. ~ j	MP(2)(3)
5. j ∨ p	P
6. p	DS(4)(5)
7. p → s	P
8. s	MP(6)(7)
9. g	CS(1)
10. s ∧ g	CA(8)(9)

25.

S	R
1. k	P
2. c ∨ ~ k	P
3. c	DS(1)(2)
4. c → (a ∧ b)	P
5. a ∧ b	MP(3)(4)
6. b	CS(5)
7. b ∨ z	DA(6)
8. (b ∨ z) → m	P
9. m	MP(7)(8)

26.

S	R
1. p ∧ b	P
2. p	CS(1)
3. p → q	P
4. q	MP(2)(3)
5. r → ~ q	P
6. ~ r	MT(4)(5)
7. s ∨ r	P
8. s	DS(6)(7)
9. s ∨ w	DA(8)
10. (s ∨ w) → m	P
11. m	MP(9)(10)

27.

S	R
1. $\sim p$	P
2. $(\sim a \wedge s) \vee p$	P
3. $\sim a \wedge s$	DS(1)(2)
4. $\sim a$	CS(3)
5. $\sim a \to j$	P
6. j	MP(4)(5)
7. $j \vee c$	DA(6)
8. $(j \vee c) \to d$	P
9. d	MP(7)(8)
10. $d \to k$	P
11. k	MP(9)(10)
12. s	CS(3)
13. $k \wedge s$	CA(11)(12)

28.

a	b	c	d	e	p
T	F	T	T	T	T

29. $c \vee s$
$s \to b$
$(c \wedge \sim s) \to m$
$\sim b$
———
m

S	R
1. $\sim b$	P
2. $s \to b$	P
3. $\sim s$	MT(1)(2)
4. $c \vee s$	P
5. c	DS(3)(4)
6. $c \wedge \sim s$	CA(3)(5)
7. $(c \wedge \sim s) \to m$	P
8. m	MP(6)(7)

30. $t \to p$
$p \to (c \wedge d)$
$d \to (r \vee u)$
$t \wedge \sim u$
———
$c \wedge r$

S	R
1. $t \wedge \sim u$	P
2. t	CS(1)
3. $t \to p$	P
4. p	MP(2)(3)
5. $p \to (c \wedge d)$	P
6. $c \wedge d$	MP(4)(5)
7. d	CS(6)
8. $d \to (r \vee u)$	P
9. $r \vee u$	MP(7)(8)
10. $\sim u$	CS(1)
11. r	DS(9)(10)
12. c	CS(6)
13. $c \wedge r$	CA(11)(12)

31. $(g \vee m) \wedge \sim e$

	e	g	h	m	s
	F	F	T	T	F

$\sim h \rightarrow \sim m$

$s \rightarrow h$

$\underline{\sim g}$

s

Problems

33. Reason

1. P
2. P
3. MP(1)(2)
4. CS(3)
5. CS(3)
6. P
7. DS(4)(6)
8. CA(5)(7)

35. Reason

1. P
2. P
3. DS(1)(2)
4. CS(3)
5. CS(3)
6. P
7. MT(5)(6)
8. CA(4)(7)
9. P
10. MP(8)(9)
11. CA(4)(10)

37.

S	R
1. m	P
2. n	P
3. $m \wedge n$	CA(1)(2)
4. $\sim (m \wedge n) \vee q$	P
5. q	DS(3)(4)

39.

S	R
1. $p \wedge q$	P
2. q	CS(1)
3. $q \rightarrow s$	P
4. s	MP(2)(3)
5. $w \rightarrow \sim s$	P
6. $\sim w$	MT(4)(5)

41.

S	R
1. w	P
2. $w \rightarrow r$	P
3. r	MP(1)(2)
4. $\sim (a \wedge c) \rightarrow \sim r$	P
5. $a \wedge c$	MT(3)(4)
6. a	CS(5)

43.

S	R
1. s	P
2. $s \vee q$	DA(1)
3. $(s \vee q) \rightarrow p$	P
4. p	MP(2)(3)
5. $p \wedge s$	CA(1)(4)
6. $\sim (p \wedge s) \vee a$	P
7. a	DS(5)(6)

45.

S	R
1. $(\sim p \rightarrow \sim q) \wedge (\sim r \vee s)$	P
2. $\sim p \rightarrow \sim q$	CS(1)
3. q	P
4. p	MT(2)(3)
5. $\sim r \vee s$	CS(1)
6. r	P
7. s	DS(5)(6)
8. $p \wedge s$	CA(4)(7)

47.

a	p	q	s
T	T	T	T

49.

S	R
1. $(p \lor q) \land w$	P
2. $p \lor q$	CS(1)
3. $\sim p$	P
4. q	DS(2)(3)
5. $r \to \sim q$	P
6. $\sim r$	MT(4)(5)
7. $\sim r \to z$	P
8. z	MP(6)(7)

51.

S	R
1. $q \land r$	P
2. r	CS(1)
3. $\sim p \to \sim r$	P
4. p	MT(2)(3)
5. $p \lor s$	DA(4)
6. $(p \lor s) \to w$	P
7. w	MP(5)(6)

53.

S	R
1. $w \land s$	P
2. w	CS(1)
3. $\sim w \lor (a \land b)$	P
4. $a \land b$	DS(2)(3)
5. b	CS(4)
6. s	CS(1)
7. $b \land s$	CA(5)(6)
8. $(b \land s) \to q$	P
9. q	MP(7)(8)
10. $q \lor z$	DA(9)
11. $(q \lor z) \to c$	P
12. c	MP(10)(11)

55.

S	R
1. $\sim p \land r$	P
2. $\sim p$	CS(1)
3. $\sim p \to q$	P
4. q	MP(2)(3)
5. $\sim q \lor s$	P
6. s	DS(4)(5)
7. r	CS(1)
8. $s \land r$	CA(6)(7)
9. $(s \land r) \to w$	P
10. w	MP(8)(9)
11. $w \lor b$	DA(10)
12. $(w \lor b) \to c$	P
13. c	MP(11)(12)

57. a, c   59. a, d   61. a, b, c

63. $j \rightarrow (c \wedge \sim t)$

	S	R
$j \vee a$	1. $\sim a$	P
$c \rightarrow l$	2. $j \vee a$	P
$\sim a$	3. j	DS(1)(2)
$l \vee s$	4. $j \rightarrow (c \wedge \sim t)$	P
	5. $c \wedge \sim t$	MP(3)(4)
	6. c	CS(5)
	7. $c \rightarrow l$	P
	8. l	MP(6)(7)
	9. $l \vee s$	DA(8)

65. $d \rightarrow c$

	S	R
$d \vee h$	1. $s \wedge i$	P
$s \rightarrow \sim h$	2. s	CS(1)
$s \wedge i$	3. $s \rightarrow \sim h$	P
$i \wedge c$	4. $\sim h$	MP(2)(3)
	5. $d \vee h$	P
	6. d	DS(4)(5)
	7. $d \rightarrow c$	P
	8. c	MP(6)(7)
	9. i	CS(1)
	10. $i \wedge c$	CA(8)(9)

Check Your Understanding

1. $\sim (a \wedge \sim b)$ 2. b 3. $\sim g$ or $a \rightarrow \sim b$ 4. a 5. $g \rightarrow \sim k$ or a 6. $\sim k$

7. <u>Reason</u>

   1. P

   2. CS (1)

   3. P

   4. MP (2)(3)

   5. DA (4)

   6. P

   7. MT (5)(6)

   8. CS (1)

   9. CA (7)(8)

   10. P

   11. MP (9)(10)

   12. P

   13. DS (11)(12)

In-Class Exercises and Problems for Section 3.4
In-Class Exercises
1. CN   2. CE   3. DM   4. DA   5. CP   6. MT   7. DE
8. DS   9. DM   10. MP   11. CS   12. DM   13. CE   14. CN

15.

S		R
1. $\sim (p \vee q)$		P
2. $\sim p \wedge \sim q$		DM(1)
3. $\sim p$		CS(2)
4. $\sim p \rightarrow s$		P
5. s		MP(3)(4)

16.

S		R
1. b		P
2. $b \rightarrow (a \rightarrow c)$		P
3. $a \rightarrow c$		MP(1)(2)
4. $\sim a \vee c$		CE(3)
5. $\sim (\sim a \vee c) \vee n$		P
6. n		DS(4)(5)

17.

S		R
1. $\sim (s \rightarrow w)$		P
2. $s \wedge \sim w$		CN(1)
3. $\sim w$		CS(2)
4. $w \vee p$		P
5. p		DS(3)(4)
6. $a \rightarrow \sim p$		P
7. $\sim a$		MT(5)(6)

18.

S		R
1. $a \vee \sim b$		P
2. $\sim (\sim a \wedge b)$		DM(1)
3. $s \rightarrow (\sim a \wedge b)$		P
4. $\sim s$		MT(2)(3)
5. $\sim s \rightarrow c$		P
6. c		MP(4)(5)

19.

S		R
1. $\sim p \rightarrow m$		P
2. $\sim m \rightarrow p$		CP(1)
3. $(\sim m \rightarrow p) \rightarrow q$		P
4. q		MP(2)(3)
5. $a \vee \sim q$		P
6. a		DS(4)(5)
7. $a \vee b$		DA(6)

20.

S	R
1. $d \lor c$	P
2. $\sim d \rightarrow c$	CE(1)
3. $s \rightarrow \sim (\sim d \rightarrow c)$	P
4. $\sim s$	MT(2)(3)
5. $\sim s \lor p$	DA(4)
6. $(\sim s \lor p) \rightarrow k$	P
7. k	MP(5)(6)

21.

S	R
1. $\sim (w \rightarrow s)$	P
2. $w \land \sim s$	CN(1)
3. $\sim s$	CS(2)
4. $s \lor p$	P
5. p	DS(3)(4)
6. $p \rightarrow (a \rightarrow \sim b)$	p
7. $a \rightarrow \sim b$	MP(5)(6)
8. $b \rightarrow \sim a$	CP(7)
9. $r \rightarrow \sim (b \rightarrow \sim a)$	P
10. $\sim r$	MT(8)(9)

22.

S	R
1. $\sim (p \lor \sim w)$	P
2. $\sim p \land w$	DM(1)
3. w	CS(2)
4. $w \rightarrow s$	P
5. s	MP(3)(4)
6. $s \lor b$	DA(5)
7. $m \rightarrow \sim (s \lor b)$	P
8. $\sim m$	MT(6)(7)
9. $m \lor (q \rightarrow d)$	P
10. $q \rightarrow d$	DS(8)(9)
11. $\sim q \lor d$	CE(10)

23.

S	R
1. $(\sim p \lor q) \land s$	P
2. $\sim p \lor q$	CS(1)
3. $p \rightarrow q$	CE(2)
4. $(p \rightarrow q) \rightarrow \sim r$	P
5. $\sim r$	MP(3)(4)
6. $r \lor c$	P
7. c	DS(5)(6)

24.

a	b	p	w
T	T	T	F

25.

S	R
1. $\sim p \lor a$	P
2. $\sim (p \land \sim a)$	DM(1)
3. $r \rightarrow (p \land \sim a)$	P
4. $\sim r$	MT(2)(3)
5. $\sim r \rightarrow q$	P
6. q	MP(4)(5)

26.

S	R
1. $r \wedge (\sim q \to k)$	P
2. $\sim q \to k$	CS(1)
3. $\sim k \to q$	CP(2)
4. $\sim (\sim k \to q) \vee p$	P
5. p	DS(3)(4)
6. $p \vee w$	DA(5)
7. $(p \vee w) \to n$	P
8. n	MP(6)(7)
9. r	CS(1)
10. $r \wedge n$	CA(8)(9)

27.

S	R
1. $\sim (\sim s \to r)$	P
2. $\sim s \wedge \sim r$	CN(1)
3. $\sim r$	CS(2)
4. $r \vee (m \to n)$	P
5. $m \to n$	DS(3)(4)
6. $\sim (d \vee n)$	P
7. $\sim d \wedge \sim n$	DM(6)
8. $\sim n$	CS(7)
9. $\sim m$	MT(5)(8)

28.

S	R
1. $c \wedge (\sim p \vee w)$	P
2. $\sim p \vee w$	CS(1)
3. $p \to w$	CE(2)
4. $(p \to w) \to s$	P
5. s	MP(3)(4)
6. c	CS(1)
7. $s \wedge c$	CA(5)(6)
8. $\sim r \to \sim (s \wedge c)$	P
9. r	MT(7)(8)

29.

S	R
1. $g \to \sim k$	P
2. $\sim g \vee \sim k$	CE(1)
3. $(\sim k \vee \sim g) \to p$	P
4. p	MP(2)(3)
5. $(w \vee \sim s) \to \sim p$	P
6. $\sim (w \vee \sim s)$	MT(4)(5)
7. $\sim w \wedge s$	DM(6)
8. s	CS(7)

30.

a	b	p	r	w
F	F	T	F	T
F	F	F	F	T

31.

S	R
1. $p \to q$	P
2. $\sim p \vee q$	CE(1)
3. $(\sim p \vee q) \to w$	P
4. w	MP(2)(3)
5. $(s \vee \sim r) \to \sim w$	P
6. $\sim (s \vee \sim r)$	MT(4)(5)
7. $\sim s \wedge r$	DM(6)
8. r	CS(7)
9. $\sim r \vee c$	P
10. c	DS(8)(9)

32.

S	R
1. $\sim (w \rightarrow \sim s)$	P
2. $w \wedge s$	CN(1)
3. w	CS(2)
4. $\sim w \vee p$	P
5. p	DS(3)(4)
6. s	CS(2)
7. $p \wedge s$	CA(5)(6)
8. $(p \wedge s) \rightarrow q$	P
9. q	MP(7)(8)
10. $q \rightarrow (\sim a \vee b)$	P
11. $\sim a \vee b$	MP(9)(10)
12. $a \rightarrow b$	CE(11)

33.

a	b	c	h	p	s
T	T	T	F	T	T
T	T	F	F	T	T

34.

S	R
1. $(a \wedge \sim b) \vee (a \wedge c)$	P
2. $a \wedge (\sim b \vee c)$	DE(1)
3. a	CS(2)
4. $\sim b \vee c$	CS(2)
5. $b \rightarrow c$	CE(4)
6. $(b \rightarrow c) \rightarrow q$	P
7. q	MP(5)(6)
8. $(w \vee \sim s) \rightarrow \sim q$	P
9. $\sim (w \vee \sim s)$	MT(7)(8)
10. $\sim w \wedge s$	DM(9)
11. s	CS(10)
12. $s \wedge a$	CA(3)(11)
13. $(s \wedge a) \rightarrow p$	P
14. p	MP(12)(13)

35.

$\sim (h \rightarrow \sim s)$
$(h \wedge s) \rightarrow c$
$\sim c \vee i$
___
i

S	R
1. $\sim (h \rightarrow \sim s)$	P
2. $h \wedge s$	CN(1)
3. $(h \wedge s) \rightarrow c$	P
4. c	MP(2)(3)
5. $\sim c \vee i$	P
6. i	DS(4)(5)

36. $r \rightarrow (n \wedge \sim b)$

	S	R
$\sim r \rightarrow (o \vee m)$	1. $\sim (\sim m \rightarrow o)$	P
$\sim (\sim m \rightarrow o)$	2. $\sim m \wedge \sim o$	CN(1)
$n \vee d$	3. $\sim (m \vee o)$	DM(2)
	4. $\sim r \rightarrow (o \vee m)$	P
	5. r	MT(3)(4)
	6. $r \rightarrow (n \wedge \sim b)$	P
	7. $n \wedge \sim b$	MP(5)(6)
	8. n	CS(7)
	9. $n \vee d$	DA(8)

37. $o \rightarrow (n \wedge s)$

	b	n	o	r	s
$\sim (r \rightarrow b)$	F	T	F	T	T

$r \rightarrow n$

$b \vee s$

o

Problems

39. Reason
1. P
2. CN(1)
3. CS(2)
4. CS(2)
5. P
6. DS(4)(5)
7. CA(3)(6)
8. P
9. MP(7)(8)

41. Reason
1. P
2. CP(1)
3. P
4. MT(2)(3)
5. P
6. DS(4)(5)
7. CE(6)

43. Reason
1. P
2. DA(1)
3. CE(2)
4. CP(3)
5. P
6. MP(4)(5)
7. CN(6)
8. CS(7)
9. CS(7)
10. DA(8)
11. P
12. DS(10)(11)
13. CA(9)(12)

45.

S	R
1. $m \vee q$	P
2. $\sim m \rightarrow q$	CE(1)
3. $(\sim m \rightarrow q) \rightarrow p$	P
4. p	MP(2)(3)
5. $n \rightarrow \sim p$	P
6. $\sim n$	MT(4)(5)

47.

S	R
1. $q \rightarrow\, \sim p$	P
2. $p \rightarrow\, \sim q$	CP(1)
3. $\sim (p \rightarrow\, \sim q) \vee w$	P
4. w	DS(2)(3)
5. $\sim r \rightarrow\, \sim w$	P
6. r	MT(4)(5)

49.

S	R
1. $\sim p \vee \sim w$	P
2. $\sim (p \wedge w)$	DM(1)
3. $r \rightarrow (p \wedge w)$	P
4. $\sim r$	MT(2)(3)
5. $r \vee s$	P
6. s	DS(4)(5)

51.

S	R
1. $g \rightarrow\, \sim a$	P
2. $a \rightarrow\, \sim g$	CP(1)
3. $(p \vee \sim r) \rightarrow\, \sim (a \rightarrow\, \sim g)$	P
4. $\sim (p \vee \sim r)$	MT(2)(3)
5. $\sim p \wedge r$	DM(4)
6. r	CS(5)
7. $r \rightarrow s$	P
8. s	MP(6)(7)
9. $s \vee w$	DA(8)
10. $(s \vee w) \rightarrow m$	P
11. m	MP(9)(10)

53.

S	R
1. $\sim (a \rightarrow b)$	P
2. $a \wedge \sim b$	CN(1)
3. $\sim (\sim a \vee b)$	DM(2)
4. $w \vee (\sim a \vee b)$	P
5. w	DS(3)(4)
6. $w \vee p$	DA(5)
7. $(w \vee p) \rightarrow c$	P
8. c	MP(6)(7)

55.

S	R
1. $\sim p \vee s$	P
2. $\sim (p \wedge \sim s)$	DM(1)
3. $m \rightarrow (p \wedge \sim s)$	P
4. $\sim m$	MT(2)(3)
5. $\sim m \vee a$	DA(4)
6. $(\sim m \vee a) \rightarrow w$	P
7. w	MP(5)(6)

57.

S	R
1. $d \rightarrow \sim n$	P
2. $n \rightarrow \sim d$	CP(1)
3. $\sim (n \rightarrow \sim d) \vee w$	P
4. w	DS(2)(3)
5. $w \rightarrow (a \wedge b)$	P
6. $a \wedge b$	MP(4)(5)
7. b	CS(6)
8. $b \vee c$	DA(7)

59.

a	q	r	s	w
F	T	T	T	F
T	T	T	T	F
T	F	T	T	F

61.

S	R
1. $s \wedge \sim w$	P
2. $\sim (s \rightarrow w)$	CN(1)
3. $\sim (c \wedge p) \rightarrow (s \rightarrow w)$	P
4. $c \wedge p$	MT(2)(3)
5. c	CS(4)
6. s	CS(1)
7. $c \wedge s$	CA(5)(6)
8. $(c \wedge s) \rightarrow n$	P
9. n	MP(7)(8)

63.

S	R
1. $\sim r \to s$	P
2. $r \lor s$	CE(1)
3. $(r \lor s) \to (w \to \sim a)$	P
4. $w \to \sim a$	MP(2)(3)
5. $a \to \sim w$	CP(4)
6. $(a \to \sim w) \to \sim b$	P
7. $\sim b$	MP(5)(6)
8. $p \land q$	P
9. p	CS(8)
10. $p \land \sim b$	CA(7)(9)

65.

S	R
1. $\sim (a \to b)$	P
2. $a \land \sim b$	CN(1)
3. $(a \land \sim b) \to c$	P
4. c	MP(2)(3)
5. $c \lor q$	DA(4)
6. $(c \lor q) \to (w \to \sim s)$	P
7. $w \to \sim s$	MP(5)(6)
8. s	P
9. $\sim w$	MT(7)(8)
10. a	CS(2)
11. $a \land \sim w$	CA(9)(10)

67. c   69. a, d   71. a

73. $r \lor h$

h	m	p	r	v
F	T	T	T	T
F	T	F	T	T

$\sim (m \to \sim v)$

$(m \lor p) \to r$

h

75. $(b \land g) \lor (b \land h)$

$b \to (i \to e)$

$i \land \sim h$

$e \land g$

	S	R
1.	$(b \land g) \lor (b \land h)$	P
2.	$b \land (g \lor h)$	DE(1)
3.	b	CS(2)
4.	$b \to (i \to e)$	P
5.	$i \to e$	MP(3)(4)
6.	$i \land \sim h$	P
7.	i	CS(6)
8.	e	MP(5)(7)
9.	$\sim h$	CS(6)
10.	$g \lor h$	CS(2)
11.	g	DS(9)(10)
12.	$e \land g$	CA(8)(11)

Check Your Understanding

1. DM (1)   2. CE (1)   3. $(d \land c) \lor (d \land \sim g)$ 4. CN (1)   5. $b \to \sim k$ 6. $\sim (l \land \sim n)$

7. $m \to \sim p$ 8. $\sim (r \to \sim s)$ 9. Step 2. $b \land \sim a$ CN(1)   Step 3. $\sim (\sim b \lor a)$

10. Step 2. $s \lor w$ DM (1)   Step 3. $\sim s \to w$ CE(2)

In-Class Exercises and Problems for Section 3.5
In-Class Exercises

1.

	S	R
1.	c	ACP
2.	$c \to (d \land g)$	P
3.	$d \land g$	MP(1)(2)
4.	g	CS(3)
5.	$c \to g$	CP(1)(4)

2.

	S	R
1.	r	ACP
2.	$\sim p \to \sim r$	P
3.	p	MT(1)(2)
4.	$\sim p \lor q$	P
5.	q	DS(3)(4)
6.	$r \to q$	CP(1)(5)

3.

	S	R
1.	$\sim r$	ACP
2.	$r \lor s$	P
3.	s	DS(1)(2)
4.	$\sim a \lor \sim s$	P
5.	$\sim a$	DS(3)(4)
6.	$\sim r \to \sim a$	CP(1)(5)

4.

h	s	w
F	F	T

5.

S	R
1. ~ k	ACP
2. ~ k → (p ∧ b)	P
3. p ∧ b	MP(1)(2)
4. b	CS(3)
5. ~ b ∨ a	P
6. a	DS(4)(5)
7. ~ k → a	CP(1)(6)

6.

S	R
1. h	ACP
2. h ∨ p	DA(1)
3. ~ (h ∨ p) ∨ l	P
4. l	DS(2)(3)
5. s →~ l	P
6. ~ s	MT(4)(5)
7. h →~ s	CP(1)(6)

7.

S	R
1. a	ACP
2. p	P
3. p ∧ a	CA(1)(2)
4. (p ∧ a) → n	P
5. n	MP(3)(4)
6. ~ n ∨ (c ∧ d)	P
7. c ∧ d	DS(5)(6)
8. c	CS(7)
9. a → c	CP(1)(8)

8.

p	q	r	s
T	T	F	T

9.

S	R
1. r	ACP
2. r ∨ s	DA(1)
3. (r ∨ s) → q	P
4. q	MP(2)(3)
5. ~ p →~ q	P
6. p	MT(4)(5)
7. p ∧ r	CA(1)(6)
8. (p ∧ r) → w	P
9. w	MP(7)(8)
10. r → w	CP(1)(9)

10.

S	R
1. p	ACP
2. $p \lor q$	DA(1)
3. $(p \lor q) \rightarrow w$	P
4. w	MP(2)(3)
5. a	P
6. $a \rightarrow (w \rightarrow s)$	P
7. $w \rightarrow s$	MP(5)(6)
8. s	MP(4)(7)
9. $\sim s \lor r$	P
10. r	DS(8)(9)
11. $p \rightarrow r$	CP(1)(10)

11.

S	R
1. d	P
2. $\sim d \lor p$	P
3. p	DS(1)(2)
4. $p \rightarrow (\sim m \rightarrow n)$	P
5. $\sim m \rightarrow n$	MP(3)(4)
6. $\sim n$	ACP
7. m	MT(5)(6)
8. $m \land d$	CA(1)(7)
9. $(m \land d) \rightarrow r$	P
10. r	MP(8)(9)
11. $r \lor a$	DA(10)
12. $\sim n \rightarrow (r \lor a)$	CP(6)(11)

12.

a	b	c	d	p	s	w
T	F	F	T	F	T	T
T	F	F	T	F	F	T

Problems

13.

Reason
1. ACP
2. P
3. DS (1) (2)
4. P
5. MT (3) (4)
6. P
7. MP(5)(6)
8. CP (1)(7)

15.

S	R
1. p	ACP
2. $c \lor p$	DA(1)
3. $(c \lor p) \rightarrow s$	P
4. s	MP(2)(3)
5. $s \lor b$	DA(4)
6. $p \rightarrow (s \lor b)$	CP (1)(5)

17.

S	R
1. w	ACP
2. $\sim a \rightarrow \sim w$	P
3. a	MT(1)(2)
4. $a \lor d$	DA(3)
5. $(a \lor d) \rightarrow \sim p$	P
6. $\sim p$	MP(4)(5)
7. $w \rightarrow \sim p$	CP (1)(6)

19.

S	R
1. $\sim p$	P
2. s	ACP
3. $\sim p \wedge s$	CA(1)(2)
4. $(\sim p \wedge s) \rightarrow q$	P
5. q	MP(3)(4)
6. $\sim q \vee \sim r$	P
7. $\sim r$	DS(5)(6)
8. $s \rightarrow \sim r$	CP (2)(7)

21.

S	R
1. r	ACP
2. $r \rightarrow (a \wedge \sim b)$	P
3. $a \wedge \sim b$	MP(1)(2)
4. a	CS(3)
5. $d \rightarrow \sim a$	P
6. $\sim d$	MT(4)(5)
7. $d \vee \sim q$	P
8. $\sim q$	DS(6)(7)
9. $r \rightarrow \sim q$	CP (1)(8)

23.

S	R
1. $\sim d$	P
2. $d \vee \sim s$	P
3. $\sim s$	DS(1)(2)
4. $\sim s \rightarrow (\sim c \vee q)$	P
5. $\sim c \vee q$	MP(3)(4)
6. c	ACP
7. q	DS(5)(6)
8. $q \rightarrow p$	P
9. p	MP(7)(8)
10. $c \rightarrow p$	CP (6)(9)

25.

a	b	c	p	w
T	F	T	T	F

27.

S	R
1. $\sim (a \wedge \sim p)$	P
2. $\sim a \vee p$	DM(1)
3. a	ACP
4. p	DS(2)(3)
5. $\sim w \rightarrow \sim a$	P
6. w	MT(3)(5)
7. $w \wedge p$	CA(4)(6)
8. $\sim (w \wedge p) \vee s$	P
9. s	DS(7)(8)
10. $s \rightarrow b$	P
11. b	MP(9)(10)
12. $a \rightarrow b$	CP (3)(11)

29.

S	R
1. $\sim (a \rightarrow b)$	P
2. $a \wedge \sim b$	CN(1)
3. $(a \wedge \sim b) \rightarrow (c \vee d)$	P
4. $c \vee d$	MP(2)(3)
5. g	ACP
6. $g \rightarrow (w \wedge \sim d)$	P
7. $w \wedge \sim d$	MP(5)(6)
8. $\sim d$	CS(7)
9. c	DS(4)(8)
10. $\sim c \vee z$	P
11. z	DS(9)(10)
12. $g \rightarrow z$	CP(5)(11)

31. b   33. c   35. a, c

37. a

$\sim l \rightarrow \sim a$

$e \rightarrow \sim j$

$l \rightarrow (j \vee k)$

$\sim k \rightarrow (\sim e \vee b)$

S	R
1. a	P
2. $\sim l \rightarrow \sim a$	P
3. l	MT(1)(2)
4. $l \rightarrow (j \vee k)$	P
5. $j \vee k$	MP(3)(4)
6. $\sim k$	ACP
7. j	DS(5)(6)
8. $e \rightarrow \sim j$	P
9. $\sim e$	MT(7)(8)
10. $\sim e \vee b$	DA(9)
11. $\sim k \rightarrow (\sim e \vee b)$	CP (6)(10)

39. $c \vee d$

$c \rightarrow \sim b$

$m \rightarrow (\sim b \vee \sim h)$

$\sim d \rightarrow m$

	b	c	d	h	m
	F	T	F	T	F
	F	T	F	F	F

In-Class Exercises and Problems for Section 3.6
In-Class Exercises

1.

S	R
1. ~ n	AIP
2. m ∨ n	P
3. m	DS(1)(2)
4. m → n	P
5. n	MP(3)(4)
6. n	CD(1)(5)

2.

S	R
1. ~ [(a ∧ c) → r]	AIP
2. (a ∧ c)∧ ~ r	CN(1)
3. a ∧ c	CS(2)
4. a	CS(3)
5. a → (c → r)	P
6. c → r	MP(4)(5)
7. c	CS(3)
8. r	MP(6)(7)
9. ~ r	CS(2)
10. (a ∧ c) → r	CD(8)(9)

3.

S	R
1. ~ (g ∨ m)	AIP
2. ~ g∧ ~ m	DM(1)
3. ~ g	CS(2)
4. c → g	P
5. ~ c	MT(3)(4)
6. c ∨ l	P
7. l	DS(5)(6)
8. l → m	P
9. m	MP(7)(8)
10. ~ m	CS(2)
11. g ∨ m	CD(9)(10)

4.

d	k	p	r	s
F	F	F	F	F

5.

a	b	c	g	n	p
F	F	T	F	F	F
F	F	F	F	F	F

6.

S	R
1. ~ (~ w →~ r)	AIP
2. ~ w ∧ r	CN(1)
3. r	CS(2)
4. r →~ p	P
5. ~ p	MP(3)(4)
6. ~ (s∨ ~ n)	P
7. ~ s ∧ n	DM(6)
8. ~ s	CS(7)
9. ~ s → (~ p → w)	P
10. ~ p → w	MP(8)(9)
11. w	MP(5)(10)
12. ~ w	CS(2)
13. ~ w →~ r	CD(11)(12)

7.

S	R
1. $\sim (p \vee s)$	AIP
2. $\sim p \wedge \sim s$	DM(1)
3. $\sim s$	CS(2)
4. $d \rightarrow s$	P
5. $\sim d$	MT(3)(4)
6. $\sim a \vee w$	P
7. $a \rightarrow w$	CE(6)
8. $(a \rightarrow w) \rightarrow (p \vee d)$	P
9. $p \vee d$	MP(7)(8)
10. p	DS(5)(9)
11. $\sim p$	CS(2)
12. $p \vee s$	CD(10)(11)

8.

S	R
1. $r \wedge s$	AIP
2. r	CS(1)
3. $r \vee p$	DA(2)
4. $(r \vee p) \rightarrow m$	P
5. m	MP(3)(4)
6. $\sim m \vee (w \rightarrow \sim s)$	P
7. $w \rightarrow \sim s$	DS(5)(6)
8. s	CS(1)
9. $\sim w$	MT(7)(8)
10. $(p \vee s) \rightarrow w$	P
11. $\sim (p \vee s)$	MT(9)(10)
12. $\sim p \wedge \sim s$	DM(11)
13. $\sim s$	CS(12)
14. $\sim (r \wedge s)$	CD(8)(13)

9.

S	R
1. $\sim (c \rightarrow \sim d)$	AIP
2. $c \wedge d$	CN(1)
3. d	CS(2)
4. $d \rightarrow a$	P
5. a	MP(3)(4)
6. $\sim h$	P
7. $\sim s \vee h$	P
8. $\sim s$	DS(6)(7)
9. $\sim s \rightarrow (a \rightarrow \sim c)$	P
10. $a \rightarrow \sim c$	MP(8)(9)
11. $\sim c$	MP(5)(10)
12. c	CS(2)
13. $c \rightarrow \sim d$	CD(11)(12)

10.

a	b	c	p	q
F	F	F	T	F

11.

S	R
1. k	P
2. $\sim z \rightarrow \sim k$	P
3. z	MT(1)(2)
4. $\sim z \vee (\sim s \rightarrow w)$	P
5. $\sim s \rightarrow w$	DS(3)(4)
6. $\sim (w \vee \sim a)$	AIP
7. $\sim w \wedge a$	DM(6)
8. $\sim w$	CS(7)
9. s	MT(5)(8)
10. $a \rightarrow \sim s$	P
11. $\sim a$	MT(9)(10)
12. a	CS(7)
13. $w \vee \sim a$	CD(11)(12)

12.

S	R
1. r	P
2. $\sim (c \rightarrow \sim p)$	AIP
3. $c \wedge p$	CN(2)
4. p	CS(3)
5. $p \wedge r$	CA(1)(4)
6. $\sim (p \wedge r) \vee s$	P
7. s	DS(5)(6)
8. $s \rightarrow (c \rightarrow w)$	P
9. $c \rightarrow w$	MP(7)(8)
10. c	CS(3)
11. w	MP(9)(10)
12. $(k \vee c) \rightarrow \sim w$	P
13. $\sim (k \vee c)$	MT(11)(12)
14. $\sim k \wedge \sim c$	DM(13)
15. $\sim c$	CS(14)
16. $c \rightarrow \sim p$	CD(10)(15)

Problems

13. Reason

1. AIP
2. P
3. DS(1)(2)
4. CS(3)
5. CS(3)
6. P
7. MP(5)(6)
8. P
9. MT(7)(8)
10. DM(9)
11. CS(10)
12. CD(4)(11)

15. Reason

1. AIP
2. DM(1)
3. CS(2)
4. CS(2)
5. P
6. DS(3)(5)
7. CS(6)
8. CS(6)
9. P
10. MP(8)(9)
11. DA(4)
12. P
13. MP(11)(12)
14. MP(10)(13)
15. CD(7)(14)

17.

S	R
1. $\sim k$	AIP
2. $\sim k \rightarrow h$	P
3. h	MP(1)(2)
4. $\sim h \vee k$	P
5. k	DS(3)(4)
6. k	CD(1)(5)

19.

S	R
1. $\sim (\sim p \rightarrow r)$	AIP
2. $\sim p \wedge \sim r$	CN(1)
3. $\sim r$	CS(2)
4. $\sim k \vee r$	P
5. $\sim k$	DS(3)(4)
6. $\sim s \vee k$	P
7. $\sim s$	DS(5)(6)
8. $(p \rightarrow q) \rightarrow s$	P
9. $\sim (p \rightarrow q)$	MT(7)(8)
10. $p \wedge \sim q$	CN(9)
11. p	CS(10)
12. $\sim p$	CS(2)
13. $\sim p \rightarrow r$	CD(11)(12)

21.

S	R
1. $\sim (\sim q \vee s)$	AIP
2. $q \wedge \sim s$	DM(1)
3. q	CS(2)
4. $\sim w \vee \sim q$	P
5. $\sim w$	DS(3)(4)
6. $(n \vee b) \rightarrow w$	P
7. $\sim (n \vee b)$	MT(5)(6)
8. $\sim n \wedge \sim b$	DM(7)
9. $\sim n$	CS(8)
10. $\sim n \rightarrow s$	P
11. s	MP(9)(10)
12. $\sim s$	CS(2)
13. $\sim q \vee s$	CD(11)(12)

23.

S	R
1. $\sim (\sim p \vee a)$	AIP
2. $p \wedge \sim a$	DM(1)
3. $\sim a$	CS(2)
4. $\sim a \rightarrow (\sim p \vee c)$	P
5. $\sim p \vee c$	MP(3)(4)
6. $a \vee (c \rightarrow \sim p)$	P
7. $c \rightarrow \sim p$	DS(3)(6)
8. p	CS(2)
9. $\sim c$	MT(7)(8)
10. $\sim p$	DS(5)(9)
11. $\sim p \vee a$	CD(8)(10)

25.

c	d	e	z
T	F	F	F

27.

a	c	d	p	q	s
T	F	T	T	T	T
F	F	T	T	T	T

29.

S	R
1. $\sim (d \to l)$	P
2. $d \wedge \sim l$	CN(1)
3. d	CS(2)
4. $d \to (b \vee s)$	P
5. $b \vee s$	MP(3)(4)
6. $\sim (b \vee \sim c)$	AIP
7. $\sim b \wedge c$	DM(6)
8. $\sim b$	CS(7)
9. s	DS(5)(8)
10. $s \to (m \wedge \sim c)$	P
11. $m \wedge \sim c$	MP (9)(10)
12. $\sim c$	CS(11)
13. c	CS(7)
14. $b \vee \sim c$	CD(12)(13)

31.

S	R
1. $\sim r$	P
2. $s \to r$	P
3. $\sim s$	MT(1)(2)
4. $\sim s \to (d \to l)$	P
5. $d \to l$	MP(3)(4)
6. $\sim (\sim l \to c)$	AIP
7. $\sim l \wedge \sim c$	CN(6)
8. $\sim l$	CS(7)
9. $\sim d$	MT(5)(8)
10. $\sim c$	CS(7)
11. $c \vee d$	P
12. d	DS(10)(11)
13. $\sim l \to c$	CD(9)(12)

33. $c \lor \sim v$	S	R
$(f \lor d) \to \sim c$	1. $\sim (v \to b)$	AIP
$\sim f \to b$	2. $v \land \sim b$	CN(1)
$v \to b$	3. $\sim b$	CS(2)
	4. $\sim f \to b$	P
	5. f	MT(3)(4)
	6. $f \lor d$	DA(5)
	7. $(f \lor d) \to \sim c$	P
	8. $\sim c$	MP(6)(7)
	9. $c \lor \sim v$	P
	10. $\sim v$	DS(8)(9)
	11. v	CS(2)
	12. $v \to b$	CD(10)(11)

35. $p \lor c$	S	R
$g \to (\sim j \lor p)$	1. $\sim p$	AIP
$(g \land j) \lor \sim c$	2. $p \lor c$	P
p	3. c	DS(1)(2)
	4. $(g \land j) \lor \sim c$	P
	5. $g \land j$	DS(3)(4)
	6. g	CS(5)
	7. $g \to (\sim j \lor p)$	P
	8. $\sim j \lor p$	MP(6)(7)
	9. j	CS(5)
	10. p	DS(8)(9)
	11. p	CD(1)(10)

Check Your Understanding

Proof A	Proof B
1. Reason	Reason
1. ACP	1. AIP
2. DA (1)	2. CN (1)
3. P	3. CS (2)
4. MP (2)(3)	4. CS (2)
5. P	5. P
6. DS (4)(5)	6. MT (4)(5)
7. P	7. P
8. MP (6)(7)	8. DS (6)(7)
9. CP (1)(8)	9. P
	10. MT (8)(9)
	11. DM (10)
	12. CS(11)
	13. CD (3)(12)

Chapter 3 Review

1. d   2. g   3. b   4. c   5. i   6. h   7. f   8. a   9. j   10. e

11. Reason	12. Reason	13. Reason
1. P	1. ACP	1. AIP
2. CS(1)	2. P	2. DM(1)
3. CS(1)	3. MP(1)(2)	3. CS(2)
4. DA(2)	4. P	4. CS(2)
5. P	5. DS(3)(4)	5. P
6. MP(4)(5)	6. P	6. MP(3)(5)
7. P	7. MT(5)(6)	7. P
8. MT(6)(7)	8. P	8. MT(4)(7)
9. DM(8)	9. MP(7)(8)	9. CN(8)
10. CS(9)	10. CS(9)	10. CS(9)
11. P	11. CP(1)(10)	11. CS(9)
12. MP(10)(11)		12. DA(10)
13. DS(3)(12)		13. P
		14. MP(12)(13)
		15. MT(6)(11)
		16. CD(14)(15)

14.	S	R
1.	$\sim p$	P
2.	$\sim p \vee q$	DA(1)
3.	$(\sim p \vee q) \to r$	P
4.	r	MP(2)(3)
5.	$a \to \sim r$	P
6.	$\sim a$	MT(4)(5)

15.	S	R
1.	$(c \vee d) \wedge a$	P
2.	$c \vee d$	CS(1)
3.	$b \to \sim (c \vee d)$	P
4.	$\sim b$	MT(2)(3)
5.	$b \vee \sim s$	P
6.	$\sim s$	DS(4)(5)

16.	S	R
1.	$\sim p$	AIP
2.	$\sim p \to \sim d$	P
3.	$\sim d$	MP(1)(2)
4.	$d \vee \sim s$	P
5.	$\sim s$	DS(3)(4)
6.	$(\sim d \vee a) \to s$	P
7.	$\sim (\sim d \vee a)$	MT(5)(6)
8.	$d \wedge \sim a$	DM(7)
9.	d	CS(8)
10.	p	CD(3)(9)

17.

S	R
1. $a \land \sim q$	P
2. a	CS(1)
3. $a \to (\sim s \lor \sim c)$	P
4. $\sim s \lor \sim c$	MP(2)(3)
5. $\sim q$	CS(1)
6. $\sim (p \land s) \to q$	P
7. $p \land s$	MT(5)(6)
8. p	CS(7)
9. s	CS(7)
10. $\sim c$	DS(4)(9)
11. $p \land \sim c$	CA(8)(10)

18.

S	R
1. $\sim c$	P
2. $b \to c$	P
3. $\sim b$	MT(1)(2)
4. $\sim b \lor \sim p$	DA(3)
5. $(\sim b \lor \sim p) \to s$	P
6. s	MP(4)(5)
7. $s \lor q$	DA(6)

19.

S	R
1. $\sim d \to c$	P
2. $d \lor c$	CE(1)
3. $(d \lor c) \to k$	P
4. k	MP(2)(3)
5. $k \to p$	P
6. p	MP(4)(5)

20.

S	R
1. $m \land p$	P
2. m	CS(1)
3. $m \to \sim n$	P
4. $\sim n$	MP(2)(3)
5. $\sim n \to (a \land b)$	P
6. $a \land b$	MP(4)(5)
7. b	CS(6)
8. $\sim b \lor s$	P
9. s	DS(7)(8)

21.

S	R
1. $\sim l$	ACP
2. $l \lor a$	P
3. a	DS(1)(2)
4. $a \to (p \to \sim s)$	P
5. $p \to \sim s$	MP(3)(4)
6. p	P
7. $\sim s$	MP(5)(6)
8. $\sim s \lor k$	DA(7)
9. $(\sim s \lor k) \to b$	P
10. b	MP(8)(9)
11. $\sim l \to b$	CP(1)(10)

22.

a	g
T	F

23.

S	R
1. $\sim (p \vee a) \wedge \sim s$	P
2. $\sim (p \vee a)$	CS(1)
3. $(\sim p \wedge \sim a)$	DM(2)
4. $\sim p$	CS(3)
5. $p \vee (\sim r \rightarrow s)$	P
6. $\sim r \rightarrow s$	DS(4)(5)
7. $\sim s$	CS(1)
8. r	MT(6)(7)

24.

S	R
1. $\sim k$	ACP
2. $k \vee \sim (a \vee \sim b)$	P
3. $\sim (a \vee \sim b)$	DS(1)(2)
4. $\sim a \wedge b$	DM(3)
5. b	CS(4)
6. $b \rightarrow c$	P
7. c	MP(5)(6)
8. $c \vee n$	DA(7)
9. $\sim k \rightarrow (c \vee n)$	CP(1)8

25.

a	b	p	s	w
T	T	F	T	F

26.

S	R
1. $\sim l$	P
2. $(a \vee g) \rightarrow l$	P
3. $\sim (a \vee g)$	MT(1)(2)
4. $\sim a \wedge \sim g$	DM(3)
5. $\sim g$	CS(4)
6. $g \vee (p \vee q)$	P
7. $p \vee q$	DS(5)(6)
8. $\sim p \rightarrow q$	CE(7)

27.

a	g	l	m	w
T	F	T	T	F

28.

S	R
1. $a \wedge \sim b$	P
2. a	CS(1)
3. $a \vee s$	DA(2)
4. $(a \vee s) \to n$	P
5. n	MP(3)(4)
6. $\sim n \vee \sim p$	P
7. $\sim p$	DS(5)(6)
8. $\sim b$	CS(1)
9. $\sim p \wedge \sim b$	CA(7)(8)
10. $\sim (p \vee b)$	DM(9)

29.

S	R
1. $\sim (s \to \sim q)$	P
2. $s \wedge q$	CN(1)
3. q	CS(2)
4. $\sim q \vee p$	P
5. p	DS(3)(4)
6. $p \vee d$	DA(5)
7. $(p \vee d) \to c$	P
8. c	MP(6)(7)

30.

S	R
1. $\sim a$	ACP
2. $a \vee \sim b$	P
3. $\sim b$	DS(1)(2)
4. $(\sim p \vee q) \to b$	P
5. $\sim (\sim p \vee q)$	MT(3)(4)
6. $p \wedge \sim q$	DM(5)
7. $\sim q$	CS(6)
8. $\sim s \to q$	P
9. s	MT(7)(8)
10. $\sim a \to s$	CP(1)(9)

31.

a	b	d	e	p	w
F	T	F	F	T	F
F	F	F	F	T	F

32.

	S	R
1.	a	P
2.	$a \vee b$	DA(1)
3.	$(a \vee b) \rightarrow r$	P
4.	r	MP(2)(3)
5.	$\sim r \vee w$	P
6.	w	DS(4)(5)
7.	$(m \vee n) \rightarrow \sim w$	P
8.	$\sim (m \vee n)$	MT(6)(7)
9.	$\sim m \wedge \sim n$	DM(8)
10.	$\sim m$	CS(9)

33.

	S	R
1.	$\sim w$	AIP
2.	$\sim w \vee q$	DA(1)
3.	$(\sim w \vee q) \rightarrow \sim b$	P
4.	$\sim b$	MP(2)(3)
5.	$\sim a \vee b$	P
6.	$\sim a$	DS(4)(5)
7.	$(\sim r \rightarrow s) \rightarrow a$	P
8.	$\sim (\sim r \rightarrow s)$	MT(6)(7)
9.	$\sim r \wedge \sim s$	CN(8)
10.	$\sim r$	CS(9)
11.	$r \vee (s \wedge p)$	P
12.	$s \wedge p$	DS(10)(11)
13.	s	CS(12)
14.	$\sim s$	CS(9)
15.	w	CD(13)(14)

34.

	S	R
1.	a	ACP
2.	$w \rightarrow \sim a$	P
3.	$\sim w$	MT(1)(2)
4.	$\sim (a \wedge \sim b)$	P
5.	$\sim a \vee b$	DM(4)
6.	b	DS(1)(5)
7.	$\sim w \wedge b$	CA(3)(6)
8.	$(\sim w \wedge b) \rightarrow c$	P
9.	c	MP(7)(8)
10.	$\sim c \vee p$	P
11.	p	DS(9)(10)
12.	$a \rightarrow p$	CP(1)(11)

35.

a	k	p	q	r	s	w
T	T	F	F	T	T	T

36. a, d   **37.** a, c

Sample Exam: Chapter 3

1. DS   2. MT   3. CS   4. CN   5. CE   6. DA   7. DM   8. MP

9. b   10. d   11. c   12. d   13. c   14. c

15. Reason	16. S	R
1. P	1. d	P
2. CN(1)	2. $(\sim s \vee r) \rightarrow \sim d$	P
3. CS(2)	3. $\sim (\sim s \vee r)$	MT(1)(2)
4. CS(2)	4. $s \wedge \sim r$	DM(3)
5. P	5. $\sim r$	CS(4)
6. MP(3)(5)	6. $r \vee \sim b$	P
7. MT(4)(6)	7. $\sim b$	DS(5)(6)
8. P	8. $\sim b \vee n$	DA(7)
9. DS(7)(8)		
10. AIP		
11. DM(10)		
12. CS(11)		
13. CS(11)		
14. P		
15. MP(13)(14)		
16. DS(9)(15)		
17. CD(12)(16)		

17. S	R
1. d	ACP
2. $d \rightarrow (a \wedge \sim c)$	P
3. $a \wedge \sim c$	MP(1)(2)
4. a	CS(3)
5. $\sim s \rightarrow \sim a$	P
6. s	MT(4)(5)
7. $s \vee q$	DA(6)
8. $(s \vee q) \rightarrow r$	P
9. r	MP(7)(8)
10. $d \rightarrow r$	CP(1)(9)

18.

S	R
1. $\sim a \wedge p$	P
2. $\sim a$	CS(1)
3. $\sim a \rightarrow (k \rightarrow b)$	P
4. $k \rightarrow b$	MP(2)(3)
5. p	CS(1)
6. $(b \vee \sim c) \rightarrow \sim p$	P
7. $\sim (b \vee \sim c)$	MT(5)(6)
8. $\sim b \wedge c$	DM(7)
9. $\sim b$	CS(8)
10. $\sim k$	MT(4)(9)

19.

S	R
1. $g \rightarrow k$	P
2. $\sim g \vee k$	CE(1)
3. $(\sim g \vee k) \rightarrow p$	P
4. p	MP(2)(3)
5. $p \vee a$	DA(4)
6. $(p \vee a) \rightarrow c$	P
7. c	MP(5)(6)
8. $c \wedge p$	CA(4)(7)

20.

a	b	n	r	s	w
T	T	F	T	T	T

21.

a	b	k	p	s
F	F	F	T	F

22.

$c \rightarrow p$
$\sim c \rightarrow g$
$e \rightarrow \sim p$
$\underline{e\hphantom{xxxxx}}$
g

S	R
1. e	P
2. $e \rightarrow \sim p$	P
3. $\sim p$	MP(1)(2)
4. $c \rightarrow p$	P
5. $\sim c$	MT(3)(4)
6. $\sim c \rightarrow g$	P
7. g	MP(5)(6)

23.

$c \rightarrow g$
$g \rightarrow (f \vee s)$
$\underline{f \rightarrow (p \wedge l)}$
$c \rightarrow l$

c	f	g	l	p	s
T	F	T	F	T	T
T	F	T	F	F	T

Answers for Chapter 4

In-Class Excercises and Problems for Section 4.1

In-Class Exercises

1. false   2. true   3. false   4. true   5. false   6. true

7. true   8. false   9. false   10. true   11. true   12. false   13. false

14. false ($x = 1$)   15. All calculus texts are not hard to read.   16. Some subway stations are clean.

17. Some people at the DMV are not extremely helpful.   18. Some chefs do not have a signature dish.

19. Everyone at the gym does not use the upright bicycles.   20. All bank customers do not pay for checks.

Problems

21. true   23. true   25. true   27. true   29. false

31. true   33. All roads in the city do not have potholes.

35. All gas stations do not offer full service.

37. Some ATM machines do not charge a transaction fee.

In-Class Excercises and Problems for Section 4.2

In-Class Exercises

1. $P(x)$: "x is a country song." $Q(x)$: "x is sad." $\forall x \; P(x) \rightarrow Q(x)$

2. $P(x)$: "x is a car. $Q(x)$: "x is a hybrid." $\exists x \; P(x) \wedge \sim Q(x)$

3. $P(x)$: "x is a berry found." $Q(x)$: "x is edible." $\exists x \; P(x) \rightarrow \sim Q(x)$

4. $P(x)$: "x is a carnival ride." $Q(x)$: "x has a height requirment." $\forall x \; P(x) \rightarrow Q(x)$

5. $P(x)$: "x is a restaurant." $Q(x)$: "x offers a take-out menu." $\exists x \; P(x) \wedge Q(x)$

6. $P(x)$: "x is a vigorous exercise." $Q(x)$: "x requires stretching first." $\forall x \; P(x) \rightarrow Q(x)$

7. $P(x)$: "x is a movie theater." $Q(x)$: "x has surround sound", $R(x)$: "x has previews." $\forall x \; P(x) \rightarrow (Q(x) \wedge R(x))$

8. $P(x)$: "x is a diner", $Q(x)$: "x offers soup as a first course", $R(x)$: "x offers salad as a first course." $\exists x \; P(x) \wedge (Q(x) \vee R(x))$

9. $P(x)$: "x is a road." $Q(x)$: "x is closed during a snow storm." $\exists x \; P(x) \wedge \sim Q(x)$

10. $P(x)$: "x is a good citizen.", $Q(x)$: "x breaks the law." $\forall x \; P(x) \rightarrow \sim Q(x)$

Problems

11. $P(x)$: "x is a commodity." $Q(x)$: "x is imported." $\exists x \; P(x) \wedge Q(x)$

13. $P(x)$: "x is a prime number." $Q(x)$: "x has only two factors." $\forall x \; P(x) \rightarrow Q(x)$

15. $P(x)$: "x is a quadratic equation." $Q(x)$: "x has at most two solutions." $\forall x \; P(x) \rightarrow Q(x)$

17. $P(x)$: "x is country." $Q(x)$: "x has suffered civil wars." $\exists x \; P(x) \wedge Q(x)$

19. $P(x)$: "x is a mammal." $Q(x)$: "x can survive without water." $R(x)$: "x can survive without food."
   $\forall x \; P(x) \rightarrow \sim (Q(x) \vee R(x))$

In-Class Excercises and Problems for Section 4.3

In-Class Exercises

1.

Statement	Reason
1. $\exists x \; S(x) \wedge L(x)$	Premise
2. $\forall x \; S(x) \rightarrow C(x)$	Premise
3. $S(c) \wedge L(c)$	EI (1)
4. $S(c) \rightarrow C(c)$	UI (2)
5. $S(c)$	CS (3)
6. $C(c)$	MP (4)(5)
7. $L(c)$	CS (3)
8. $C(c) \wedge L(c)$	CA (6)(7)
9. $\exists x \; C(x) \wedge L(x)$	EG (8)

2.

Statement	Reason
1. $\forall x \; S(x) \rightarrow O(x)$	Premise
2. $\exists x \; A(x) \wedge \sim O(x)$	Premise
3. $A(c) \wedge \sim O(c)$	EI (2)
4. $S(c) \rightarrow O(c)$	UI (1)
5. $\sim O(c)$	CS (4)
6. $\sim S(c)$	MT(3)(5)
7. $A(c)$	CS(4)
8. $A(c) \wedge \sim S(c)$	CA(6)(7)
9. $\exists x \; A(x) \wedge \sim S(x)$	EG(8)

3.

Statement	Reason
1. $\forall x \; L(x) \rightarrow \sim D(x)$	Premise
2. $\exists x \; E(x) \wedge L(x)$	Premise
3. $E(c) \wedge L(c)$	EI (2)
4. $L(c) \rightarrow \sim D(c)$	UI (1)
5. $L(c)$	CS (3)
6. $\sim D(c)$	MP (4)(5)
7. $E(c)$	CS (3)
8. $E(c) \wedge \sim D(c)$	CA (6)(7)
9. $\exists x \; E(x) \wedge \sim D(x)$	EG (8)

4.

Statement	Reason
1. $\forall x \; P(x) \rightarrow B(x)$	Premise
2. $Ax \; T(x) \rightarrow I(x)$	Premise
3. $\exists x \; T(x) \wedge \sim B(x)$	Premise
4. $T(c) \wedge \sim B(c)$	EI (3)
5. $P(c) \rightarrow B(c)$	UI (1)
6. $T(c) \rightarrow I(c)$	UI (2)
7. $\sim B(c)$	CS (4)
8. $\sim P(x)$	MT (5)(7)
9. $T(c)$	CS (4)
10. $I(c)$	MP (6)(9)
11. $I(c) \wedge \sim P(c)$	CA (8)(10)
12. $\exists x \; I(x) \wedge \sim P(x)$	EG (11)

5.

Statement	Reason
1. $\forall x\ W(x) \rightarrow D(x)$	Prem
2. $\forall x\ J(x) \rightarrow L(x)$	Prem
3. $\exists x\ J(x) \wedge W(x)$	Prem
4. $J(c) \wedge W(c)$	EI (3)
5. $W(c) \rightarrow D(c)$	UI (1)
6. $W(c)$	CS (4)
7. $D(c)$	MP (5)(6)
8. $J(c)$	CS (4)
9. $J(c) \rightarrow L(c)$	UI (2)
10. $L(c)$	MP (8)(9)
11. $L(c) \wedge D(c)$	CA (7)(10)
12. $\exists x\ L(x) \wedge D(x)$	EG (11)

6.

Statement	Reason
1. $\exists x\ C(x) \wedge I(x)$	Prem
2. $\forall x\ I(x) \rightarrow\sim E(x)$	Prem
3. $C(c) \wedge I(c)$	EI (1)
4. $I(c) \rightarrow\sim E(c)$	UI (2)
5. $I(c)$	CS (3)
6. $\sim E(c)$	MP (4)(5)
7. $C(c)$	CS (3)
8. $C(c) \wedge \sim E(c)$	CA (6)(7)
9. $\exists x\ C(x) \wedge \sim E(x)$	EG (8)

7.

Statement	Reason
1. $\exists x\ V(x) \wedge \sim F(x)$	Prem
2. $\forall x\ G(x) \rightarrow F(x)$	Prem
3. $V(c) \wedge \sim F(c)$	EI (1)
4. $G(c) \rightarrow F(c)$	UI (2)
5. $\sim F(c)$	CS (3)
6. $\sim G(c)$	MT (4)(5)
7. $V(c)$	CS (3)
8. $V(c) \wedge \sim G(c)$	CA (6)(7)
9. $\exists x\ V(x) \wedge \sim G(x)$	EG (8)

8.

Statement	Reason
1. $\forall x\ I(x) \wedge \sim A(x)$	Prem
2. $\exists x\ A(x) \wedge \sim R(x)$	Prem
3. $A(c) \wedge \sim R(c)$	EI (2)
4. $I(c) \rightarrow\sim A(c)$	UI (1)
5. $A(c)$	CS (3)
6. $\sim I(c)$	MT (4)(5)
7. $\sim R(c)$	CS (3)
8. $\sim I(c) \wedge \sim R(c)$	CA (6)(7)
9. $\exists x\ \sim I(x) \wedge \sim R(x)$	EG (8)

Problems

9.

Statement	Reason
1. $\forall x\ D(x) \rightarrow F(x)$	Prem
2. $\exists x\ D(x) \wedge S(x)$	Prem
3. $D(c) \wedge S(c)$	EI (2)
4. $D(c) \rightarrow F(c)$	UI (1)
5. $D(c)$	CS (3)
6. $F(c) \wedge \sim C(c)$	MP (4)(5)
7. $S(c)$	CS (3)
8. $F(c) \wedge S(c)$	CA (6)(7)
9. $\exists x\ F(x) \wedge S(x)$	EG (8)

11.

Statement	Reason
1. $\forall x\ B(x) \rightarrow S(x)$	Prem
2. $\forall x\ S(x) \rightarrow M(x)$	Prem
3. $\forall x\ M(x) \rightarrow C(x)$	Prem
4. $\exists x\ \sim C(x) \wedge \sim T(x)$	Prem
5. $\sim C(c) \wedge \sim T(c)$	EI (4)
6. $\sim C(c)$	CS (5)
7. $M(c) \rightarrow C(c)$	UI (3)
8. $\sim M(c)$	MT (6)(7)
9. $S(c) \rightarrow M(c)$	UI (2)
10. $\sim S(c)$	MT (8)(9)
11. $B(c) \rightarrow S(c)$	UI (1)
12. $\sim B(c)$	MT (10)(11)
13. $\sim T(c)$	CS (5)
14. $\sim T(c) \wedge \sim B(c)$	CA (12)(13)
15. $\exists x\ \sim T(x) \wedge \sim B(x)$	EG (14)

13.

Statement	Reason
1. $\forall x\ C(x) \rightarrow H(x)$	Prem
2. $\exists x\ B(x) \wedge \sim H(x)$	Prem
3. $\forall x\ T(x) \rightarrow C(x)$	Prem
4. $B(c) \wedge \sim H(c)$	EI (2)
5. $\sim H(c)$	CS (4)
6. $C(c) \rightarrow H(c)$	UI (1)
7. $\sim C(c)$	MT (5)(6)
8. $T(c) \rightarrow C(c)$	UI (3)
9. $\sim T(c)$	MT (7)(8)
10. $B(c)$	CS (4)
11. $B(c) \wedge \sim T(c)$	CA (9)(10)
12. $\exists x\ B(x) \wedge \sim T(x)$	EG (11)

15.

Statement	Reason
1. $\forall x\ P(x) \rightarrow \sim D(x)$	Prem
2. $\exists x\ P(x) \wedge S(x)$	Prem
3. $P(c) \wedge S(c)$	EI (2)
4. $P(c) \rightarrow \sim D(c)$	UI (1)
5. $P(c)$	CS (3)
6. $\sim D(c)$	MP (4)(5)
7. $S(c)$	CS (3)
8. $S(c) \wedge \sim D(c)$	CA (6)(7)
9. $P(c) \wedge S(c) \wedge \sim D(c)$	CA (5)(8)
10. $\exists x\ P(x) \wedge S(x) \wedge \sim D(x)$	EG (9)

In-Class Excercises and Problems for Section 4.4

In-Class Exercises

1-8. Current flows in all cases except cases 7 and 8.   9. Yes   10. No   11. Yes   12. Yes   13. Yes

14.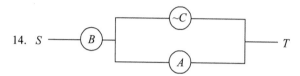

Problems

15. No   17. Yes   19. No   21. Yes

23. C 25. $F \vee (E \wedge G)$ 27. $A \wedge B$ 29. $A \vee \sim B$

31. No circuit   33. a. $A \wedge B$, $A \wedge C$, $B \wedge C$, $A \wedge B \wedge C$ b. Union

c. One solution is $[A \wedge (B \vee C)] \vee (B \wedge C)$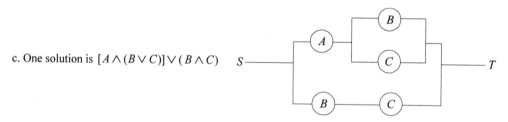

d. $A \vee (B \wedge C)$

Chapter 4 Review

1. False

2. All baseball players do not make the Hall of Fame.

3. Some math classes are not boring.

4. $D(x)$: "x is a driver.", $R(x)$: "x is rude.", $C(x)$: "x is courteous." $\exists x \; D(x) \wedge R(x) \wedge \sim C(x)$

5.

Statement	Reason
1. $\forall x \; P(x) \to M(x)$	Prem
2. $\exists x \; P(x) \wedge R(x)$	Prem
3. $P(c) \wedge R(c)$	EI(2)
4. $P(c) \to M(c)$	UI(1)
5. $P(c)$	CS(3)
6. $M(c)$	MP(4)(5)
7. $R(c)$	CS(3)
8. $R(c) \wedge M(c)$	CA(6)(7)
9. $\exists x \; R(x) \wedge M(x)$	EG(8)

6-13. Current flows only in cases 6,7, and 8

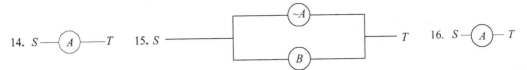

14. S—A—T 15. S ... T 16. S—A—T

Sample Exam: Chapter 4

1. false   2. $P(x)$: x is rap music, $Q(x)$: x is loud, $\forall x\ P(x) \to Q(x)$

3. $P(x)$: x is a car, $Q(x)$: x flies, $\forall x\ P(x) \to \sim Q(x)$

4. $P(x)$: x is a mushroom, $Q(x)$: x is edible, $\forall x\ P(x) \to \sim Q(x)$

5. $P(x)$: x is an NBA player, $Q(x)$: x is of Asian descent, $\exists x\ P(x) \land Q(x)$

6.

Statement	Reason
1. $\forall x\ G(x) \to S(x)$	Prem
2. $\exists x\ E(x) \land \sim S(x)$	Prem
3. $E(c) \land \sim S(c)$	EI(2)
4. $G(c) \to S(c)$	UI(1)
5. $\sim S(c)$	CS(3)
6. $\sim G(c)$	MT(4)(5)
7. $E(c)$	CS(3)
8. $E(c) \land \sim G(c)$	CA(6)(7)
9. $\exists x\ E(x) \land \sim G(x)$	EG(8)

7-14. Current flows only in cases 7 and 8

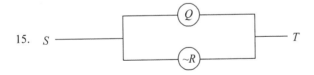

15. S ... T

Answers for Chapter 5

In-Class Excercises and Problems for Section 5.1

In-Class Exercises

1. {0,4,8,12,16}   2. {20,25,30,35,40}   3. {on,no,one,eon}   4. ∅   5. {Monday, Friday}

6. The even numbers from zero to ten   7.The odd numbers between 30 and 40

8. The first four letters of the alphabet

9. a. F   b. T   c. F   d. F   e. T   f. T

   g. F   h. T   i. T   j. T   k. F   l. T

10. 5   11. 32   12. 31   13. 32   14. 64   15. 63   16. 16   17. 15   18. 32

19. $\{\{2\},\{8\},\{10\},\{2,8\},\{2,10\},\{8,10\},\{2,8,10\}, \varnothing \}$

Problems

21. $\{4, 5, 6, 7, 8, 9\}$

23. {dog, god, cod, cog}   25. $\varnothing$ 27. The set of all odd numbers less than 22

29. The set of the first six integers squared

31. The set of all days of the week beginning with the letter z

33. True   35. False   37. True   39. True   41. False

43. 8   45. 63   47. Yes   49. False   51. False

Check Your Understanding

1. $\{17,19,23\}$ 2. The last four letters of the alphabet   3. True   4. False

5. False   6. True   7. False   8. False   9. 127

10. $\{\{NY\}, \{NJ\}, \{CT\}, \{NY, NJ\}, \{NY, CT\}, \{NJ, CT\}, \{NY, NJ, CT\}, \varnothing \}$

In-Class Excercises and Problems for Section 5.2

In-Class Exercises

1. $\{1,2,3,4,5,6\}$ 2. $\{2\}$ 3. $\{1,5,6\}$ 4. $\{3,4\}$ 5. $\{1,2,4,5,6,7\}$

6. $\{3\}$ 7. $\varnothing$ 8. $\{1,3,5,7\}$ 9. $\{1,5,7\}$ 10. $\{6\}$

11. $\{1,3,4,5,7\}$ 12. $\{1,5\}$ 13. $\{2,4,6\}$ 14. U 15. $\{1,2,3,5,7\}$

16. $\{1,3,4,5,6,7\}$ 17. $\{1,3,5,6,7\}$ 18. $\{4,7\}$ 19. $\{6\}$ 20. $\{2,3,6,7\}$

21. {Monday, Friday}   22. {Thursday}   23. {Saturday}   24. {Tuesday}

Problems

25. $\{3, 7\}$ 27. $\{1, 4, 5, 6\}$ 29. $\{2, 5, 6, 9\}$ 31. $\{1, 2, 3, 4, 8, 9\}$

33. $\{1, 2, 3, 4, 5, 6, 8, 9\}$ 35. {April, September, November}

37. {May, June, July, August}   39. {May, July, August}

41. {May, June}   43. {May, June, July, August, September, October}

Check Your Understanding

1. U 2. $\{6, 10\}$ 3. $\{4, 8\}$ 4. $\{1, 2, 3, 5\}$ 5. $\{5, 7, 9\}$

6. T 7. F 8. T 9. T 10. F 11. T

In-Class Excercises and Problems for Section 5.3

In-Class Exercises

1. 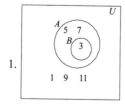   2. 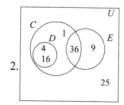   3.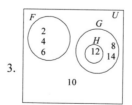

4. 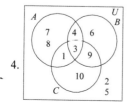   5. 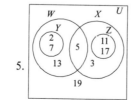   6.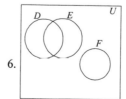

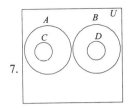
7.

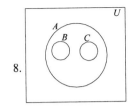
8.

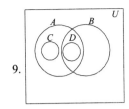
9.

10. III, IV   11. I, III   12. I, II   13. II, IV   14. III   15. IV   16. I, II, IV   17. I, II, III

18. I, IV   19. II, IV   20. I, II, V, VI   21. II, IV, VI, VIII   22. I, V   23. I, III, V, VI, VII, VIII

24. II, VI   25. III   26. II, V, VI   27. III, IV, VI, VIII   28. III, V, VI, VII   29. I, II, IV, V, VI

30. III, IV, VI, VIII   31. IV   32. V, VI   33. V   34. II, IV, V, VII

35. $A \cap B \cap C$ 36. $A - B$ 37. $(B \cup C) - A$

38. $[A - (B \cup C)] \cup [B - (A \cup C)]$ or $[(A \cup B) - (A \cap B)] - C$

Problems

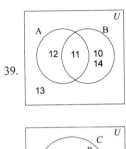
39.

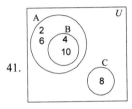
41.

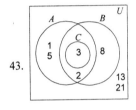
43.

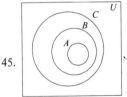
45.

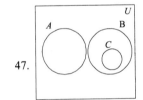
47.

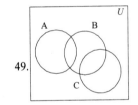
49.

51. II   53. IV   55. II   57. I, III   59. I, II, III, IV

61. I, II   63. VII, VIII   65. I, II   67. I, III, V   69. III, IV, VII   71. IV

73. II, III, IV, V, VI, VII, VIII   75. ∅   77. VII   79. VI   81. VIII   83. I, III, IV, V

85. $(A \cap B \cap C)'$ 87. $(B \cap C) - A$ 89. $(A \cup B \cup C)'$ 91. $A \cup (B \cap C)$

Check Your Understanding

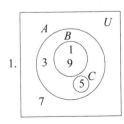
1.

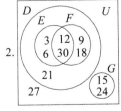
2.

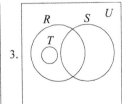
3.

4.   VI   5. VII   6. III, IV   7. V

In-Class Excercises and Problems for Section 5.4

In-Class Exercises

1. 4. 2. a 5 b. 10 3. a. 22 b. 24 c. 45 d. 35 4. a 960 b. 180 c. 720 d. 40 e. 580

5. a. 690 b. 10 c. 100 d. 350 e. 50 f. 360 g. 430 h.190 6. a. 23 b. 3 c. 2 d. 6 e. 5

7. a. 5 b. 15 c. 20

Problems

9. a. 4 b. 5 c. 3 11. a. 0 b. 3 c. 6 d. 25 13. a. 0 b. 200 c. 400 d.100

In-Class Excercises and Problems for Section 5.5

In-Class Exercises

1. a. Set Difference b. DeMorgan's Law c. Distributive Law d. Inverse Law e. Identity Law

2. a. Absorption Law b. Set Difference c. Distributive Law d. Idempotent Law
   e. Commutative Law f. Absorption Law

6. $B \cap C$ 7. A' 8. B' 9. A 10. U

Problems

11.

A	B	B'	$A \cap B$	$A - B'$
$\in$	$\in$	$\notin$	$\in$	$\in$
$\in$	$\notin$	$\in$	$\notin$	$\notin$
$\notin$	$\in$	$\notin$	$\notin$	$\notin$
$\notin$	$\notin$	$\in$	$\notin$	$\notin$

13.

A	B	C	$B \cup C$	$A \cap B$	$A \cap C$	$A \cap (B \cup C)$	$(A \cap B) \cup (A \cap C)$
$\in$	$\in$	$\in$	$\in$	$\in$	$\in$	$\in$	$\in$
$\in$	$\in$	$\notin$	$\in$	$\in$	$\notin$	$\in$	$\in$
$\in$	$\notin$	$\in$	$\in$	$\notin$	$\in$	$\in$	$\in$
$\in$	$\notin$	$\notin$	$\notin$	$\notin$	$\notin$	$\notin$	$\notin$
$\notin$	$\in$	$\in$	$\in$	$\notin$	$\notin$	$\notin$	$\notin$
$\notin$	$\in$	$\notin$	$\in$	$\notin$	$\notin$	$\notin$	$\notin$
$\notin$	$\notin$	$\in$	$\in$	$\notin$	$\notin$	$\notin$	$\notin$
$\notin$	$\notin$	$\notin$	$\notin$	$\notin$	$\notin$	$\notin$	$\notin$

15. III, IV 17. I, II, III, IV, V, VI, VII 19. I, II, III, IV

21. a. DeMorgan's Law b. Distributive Law c. Inverse Law d. Identity Law
    e. Set Difference

23. A' 25. $A \cap C$ 27. B' 29. C' 31. $P \cap R$

Chapter 5 Review

1. {y}   2. {Africa, Antarctica, Asia, Australia, Europe, North America, South America 3. ∅

4. {April, June, September, November}

5. {Mercury, Venus, Earth, Mars, Jupiter, Saturn, Uranus, Neptune}

6. {5,9}   7. {5}   8. ∅   9. ∅   10. {1,2,3,4,5,7,9}

11. true 12. true   13. false   14. true 15. true 16. true 17. true   18. false

19.   20.

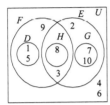

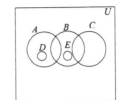

21. 4   22. 16   23. 128   24. 127   25. {{h}, ∅}

26. Yes   27. Yes   28. No   29. No

30.   31.

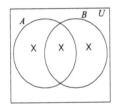

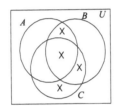

32. a. II   b. II, IV   c. VII, VIII   d. V, VII

33. a. $[(A \cap B) - C] \cup [(B \cap C) - A]$ b. $B - (A \cup C)$

   c. $(A \cap B \cap C) \cup (A - (B \cup C)) \cup (C - (A \cup B))$ d. $(A \cap C) - B$

   e. $(C - (A \cup B)) \cup (A \cup B \cup C)'$

34. Yes, the compound statement is true.

35. *i.* d   *ii.* c   *iii.* a   *iv.* e   *v.* b

36. Sets A, B, C and D are proper subsets of the universal set. $C \subset A, D \subset A, C \cap D \neq \varnothing$ and neither C nor

   D is a subset of the other set, $A \cap B \neq \varnothing$. Neither set A nor set B is a subset of the other set.

   $B \cap C = \varnothing$, $B \cap D \neq \varnothing$ and neither B nor D is a subset of the other.

37.a. 2. Set Difference	3. DeMorgan's Law	4. Set Difference	5. Commutative Law
6. Distributive Law	7. Inverse Law	8. Identity Law	9. Set Difference
b. 2. Set Difference	3. Absorption Law	4. Set Difference	5. DeMorgan's Law
6. Commutative Law	7. Associative Law	8. Idempotent Law	9. Absorption Law

38. a. $(A - B) \cup (A - C') = (A \cap B') \cup (A \cap C) = A \cap (B' \cup C) = A - (B' \cup C)' = A - (B \cap C')$

   b. $[P' \cap (P' \cup S)] - (P' \cap Q) = P' - (P' \cap Q) = P' \cap (P' \cap Q)' = P' \cap (P \cup Q')$

      $= (P' \cap P) \cup (P' \cap Q') = \varnothing \cup (P' \cap Q') = P' \cap Q' = (P \cup Q)'$

   c. $(A \cup C)' - [A' \cup (A' \cap B)] = (A \cup C)' - A' = (A \cup C)' \cap A = (A' \cap C') \cap A$

      $= (C' \cap A') \cap A = C' \cap (A' \cap A) = C' \cap \varnothing = \varnothing$

39. a. 1265   b. 100   c. 1245   d. 910   e. 435
40. a. 28   b. 12   c. 5
41. a. 1   b. 23   c. 16   d. 8

Sample Exam: Chapter 5

1. {3,6,9}   2. U 3. {1,3,10}   4. {5,7,8}   5. {1,7,10}   6. 8   7. yes
8. no   9. b   10. d   11. true   12. true   13. true   14. false
15. false   16. false   17. false   18. true   19. true   20. false   21. c

22.

23. not true   24. d   25. Set Difference   26. Commutative Law

27. DeMorgan's Law   28. Commutative Law   29. Associative Law   30. Idempotent Law
31. Distributive Law   32. Inverse Law   33. Identity Law

34. $W \cap (S \cap W)' = W \cap (S' \cup W') = (W \cap S') \cup (W \cap W') = (W \cap S') \cup \varnothing = W \cap S' = W - S$
35. a. 26   b. 8   c. 11   d. 21   e. 23
36. a. 64   b. 50   c. 25
37. b

Answers to Chapter 6

In-Class Excercises and Problems for Section 6.1
In-Class Exercises

1. Invalid   2. Valid   3. Valid   4. Valid   5. Invalid   6. Invalid   7. Valid   8. Valid
Problems

9. Invalid   11. Valid   13. Valid   15. Invalid
Check Your Understanding

1. $S' \cup B$ 2. $(W \cap R')'$ 3. $(A \cup C)'$ or $A' \cap C'$ 4. yes   5. no   6. yes   7. valid

In-Class Excercises and Problems for Section 6.2
In-Class Exercises

1. a. $\dfrac{1}{26}$ b. $\dfrac{1}{26}$ c. $\dfrac{25}{26}$ d. $\dfrac{21}{26}$ e. $\dfrac{3}{26}$ f. $\dfrac{22}{26}$ g. $\dfrac{24}{26}$

2. a. $\dfrac{1}{6}$ b. 1   c. $\dfrac{2}{6}$ d. $\dfrac{3}{6}$ e. $\dfrac{5}{6}$ f. $\dfrac{5}{6}$ g. $\dfrac{4}{6}$ h. 0   i. $\dfrac{3}{6}$ j. $\dfrac{4}{6}$ k. $\dfrac{5}{6}$ l. 0

3. a. $\dfrac{5}{100}$ b. $\dfrac{35}{100}$ c. $\dfrac{50}{100}$ d. $\dfrac{50}{100}$

4. a. $\dfrac{13}{52}$ b. $\dfrac{39}{52}$ c. $\dfrac{26}{52}$ d. $\dfrac{48}{52}$ e. $\dfrac{1}{52}$ f. $\dfrac{8}{52}$ g. $\dfrac{44}{52}$ h. $\dfrac{16}{52}$

Problems

5. a. $\dfrac{2}{7}$ b. $\dfrac{4}{7}$ c. $\dfrac{1}{7}$ d. $\dfrac{5}{7}$ e. $\dfrac{6}{7}$ f. $\dfrac{2}{7}$ g. 0 h. $\dfrac{6}{7}$

7. a. $\dfrac{40}{100}$ b. $\dfrac{30}{100}$ c. $\dfrac{90}{100}$ d. $\dfrac{70}{100}$ e. $\dfrac{30}{100}$

9. a. $\dfrac{1}{12}$ b. $\dfrac{11}{12}$ c. $\dfrac{6}{12}$ d. 0 e. 1 f. $\dfrac{2}{12}$ g. $\dfrac{10}{12}$ h. $\dfrac{9}{12}$ i. $\dfrac{5}{12}$

Check Your Understanding
1. 7/24   2. 9/11   3. a. 9/21 b. 17/21   4. a. 27/100 = 0.27   b. 81/100 = 0.81   c. 67/100 = 0.67

In-Class Excercises and Problems for Section 6.3

In-Class Exercises

1. yes   2. no   3. yes   4. no   5. yes

6.

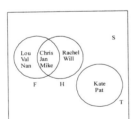

7. a. $\dfrac{6}{10}$ b. $\dfrac{5}{10}$ c. $\dfrac{3}{10}$ d. $\dfrac{8}{10}$ e. $\dfrac{2}{10}$

f. 0 g. $\dfrac{8}{10}$ h. $\dfrac{5}{10}$ i. $\dfrac{2}{10}$ j. $\dfrac{3}{10}$

8. $\dfrac{12}{17}$ 9. $\dfrac{13}{17}$ 10. $\dfrac{10}{17}$ 11. $\dfrac{14}{17}$ 12. $\dfrac{7}{17}$

13. $\dfrac{10}{17}$ 14. $\dfrac{16}{17}$ 15. $\dfrac{16}{17}$ 16. $\dfrac{5}{17}$ 17. $\dfrac{7}{17}$

18. $\dfrac{1}{75}$ 19. $\dfrac{29}{75}$ 20. $\dfrac{45}{75}$ 21. $\dfrac{74}{75}$ 22. $\dfrac{37}{75}$ 23. $\dfrac{38}{75}$

24. $\dfrac{9}{75}$ 25. $\dfrac{5}{75}$ 26. $\dfrac{42}{75}$ 27. $\dfrac{33}{75}$ 28. $\dfrac{30}{75}$ 29. $\dfrac{30}{75}$

30. $\dfrac{4}{26}$ 31. $\dfrac{6}{26}$ 32. $\dfrac{2}{26}$ 33. 1   34. 0

35. $\dfrac{8}{26}$ 36. $\dfrac{18}{26}$ 37. $\dfrac{3}{26}$ 38. $\dfrac{23}{26}$ 39. $\dfrac{10}{26}$

Problems

41. a. no   b. no   d. yes   d. yes

43.

a. $\dfrac{6}{10}$ b. $\dfrac{5}{10}$ c. $\dfrac{3}{10}$ d. $\dfrac{8}{10}$

e. $\dfrac{2}{10}$ f. $\dfrac{7}{10}$ g. $\dfrac{7}{10}$ h. $\dfrac{2}{10}$

45. a. $\dfrac{26}{31}$ b. $\dfrac{24}{31}$ c. $\dfrac{9}{31}$ d. $\dfrac{22}{31}$ e. $\dfrac{28}{31}$ f. $\dfrac{28}{31}$ g. $\dfrac{22}{31}$

47. a. $\dfrac{39}{52}$ b. $\dfrac{32}{52}$ c. $\dfrac{16}{52}$ d. $\dfrac{24}{52}$ e. $\dfrac{40}{52}$ f. $\dfrac{51}{52}$

g. $\dfrac{36}{52}$ h. $\dfrac{27}{52}$ i. $\dfrac{24}{52}$ j. $\dfrac{48}{52}$ k. 1   l. $\dfrac{8}{52}$

Check Your Understanding

1. $1 - P(A \cup B)$ 2. Yes   3. 12/52   4. 16/52   5. 28/52   6. 24/52

In-Class Excercises and Problems for Section 6.4

In-Class Exercises

1. a. yes   b. no   c. no   2. a. yes   b. yes   c. no   d. no

3. a. $\dfrac{1}{6}$ b. $\dfrac{3}{12}$ c. $\dfrac{3}{24}$ d. $\dfrac{5}{8}$ e. $\dfrac{5}{8}$ f. $\dfrac{1}{8}$ g. $\dfrac{2}{24}$ h. $\dfrac{21}{24}$

4. a. $\dfrac{1}{25}$ b. $\dfrac{1}{25}$ c. $\dfrac{1}{25}$ d. $\dfrac{2}{25}$ e. $\dfrac{2}{25}$

   f. $\dfrac{1}{5}$ g. $\dfrac{4}{25}$ h. $\dfrac{16}{25}$ i. $\dfrac{8}{25}$

5. a. $\dfrac{1}{36}$ b. $\dfrac{1}{36}$ c. $\dfrac{2}{36}$ d. $\dfrac{5}{36}$ e. $\dfrac{6}{36}$

Problems

7. $\dfrac{2}{12}$ 9. a. $\dfrac{16}{2704}$ b. $\dfrac{16}{2704}$ c. $\dfrac{156}{2704}$ d. $\dfrac{2}{2704}$ e. $\dfrac{32}{2704}$ f. $\dfrac{1}{52}$

Check Your Understanding

1. $m \times n$ ways.   2. $P(A) \times P(B)$ 3. a. 64.   4. Yes   5. 4/8 or ½
6. 16/64 or ¼   7. No   8. 16/64 or ¼.

In-Class Excercises and Problems for Section 6.5

In-Class Exercises

1. $\{(a,1),(a,3),(a,5),(b,1),(b,3),(b,5)\}$ 2. $\{(a,a),(a,b),(b,a),(b,b)\}$

3. a. $\{a,b\}$ b. $\{1,3\}$ c. yes   4. a. $\{a,b\}$ b. $\{1,3,5\}$ c. no

5. a. $\{a,b\}$ b. $\{3,5\}$ c. no   6. a. $\{a,b\}$ b. $\{5\}$ c. yes

7. a. $\{a\}$ b. $\{1,3,5\}$ c. no   8. a. $\{b\}$ b. $\{3\}$ c. no

9. no   10. yes   11. no   12. yes

Problems

13. $\{(5,1),(5,2),(5,3),(5,4),(5,5),(7,1),(7,2),(7,3),(7,4),(7,5)\}$

15. a. $\{1,2,3,4\}$ b. $\{7,8,10\}$ c. no

17. a. $\{1,2,3,4,5\}$ b. $\{6,7,8,9,10,11\}$ c. no

19. a. $\{1,2,3,4,5\}$ b. $\{6,8,9,11\}$ c. yes

21. no   23. yes   25. yes   27. no

Chapter 6 Review

1. Invalid   2. Valid

3. $\dfrac{5,472}{1,000,000}$ 4. $\dfrac{5,427}{1,000,000}$ 5. $\dfrac{15,884}{1,000,000}$ 6. $\dfrac{3,096}{1,000,000}$ 7. $\dfrac{753,571}{1,000,000}$

8. 0.09   9. 0.46   10. 0.60   11. 0.81   12. yes   13. no   14. yes   15. no

16. yes   17. yes

Sample Exam: Chapter 6

1. Valid   2. Invalid   3. $\dfrac{260}{1,352}$ 4. $\dfrac{1,764}{6,084}$ 5. $\dfrac{676}{1,352}$ 6. $\dfrac{448}{1,352}$

7. $R \times R = \{(\{a,b\},\{a,b\}),(\{a,b\},\{1,2,3\}),(\{1,2,3\},\{a,b\}),(\{1,2,3\},\{1,2,3\})\}$

8. yes   9. yes   10. a. no   b. yes

Index